Springer Series in Optical Sciences   Volume 53
Edited by Koichi Shimoda

Springer Series in Optical Sciences

Volume 42 **Principles of Phase Conjugation**
By B. Ya. Zel'dovich, N. F. Pilipetsky, and V. V. Shkunov

Volume 43 **X-Ray Microscopy**
Editors: G. Schmahl and D. Rudolph

Volume 44 **Introduction to Laser Physics**
By K. Shimoda 2nd Edition

Volume 45 **Scanning Electron Microscopy**
Physics of Image Formation and Microanalysis
By L. Reimer

Volume 46 **Holography and Deformation Analysis**
By W. Schumann, J.-P. Zürcher, and D. Cuche

Volume 47 **Tunable Solid State Lasers**
Editors: P. Hammerling, A. B. Budgor, and A. Pinto

Volume 48 **Integrated Optics**
Editors: H. P. Nolting and R. Ulrich

Volume 49 **Laser Spectroscopy VII**
Editors: T. W. Hänsch and Y. R. Shen

Volume 50 **Laser-Induced Dynamic Gratings**
By H. J. Eichler, P. Günter, and D. W. Pohl

Volume 51 **Tunable Solid State Lasers for Remote Sensing**
Editors: R. L. Byer, E. K. Gustafson, and R. Trebino

Volume 52 **Tunable Solid-State Lasers II**
Editors: A. B. Budgor, L. Esterowitz, and L. G. DeShazer

Volume 53 **The CO_2 Laser**
By W. J. Witteman

Volume 54 **Lasers, Spectroscopy and New Ideas**
Editors: M. D. Levenson and W. M. Yen

Volumes 1–41 are listed on the back inside cover

W. J. Witteman

The CO$_2$ Laser

With 135 Figures

Springer-Verlag Berlin Heidelberg New York
London Paris Tokyo

Professor Dr. W.J. Witteman

Department of Applied Physics, Twente University of Technology,
P.O. Box 217, Enschede, Netherlands

ISBN 3-540-17657-8 Springer-Verlag Berlin Heidelberg New York Tokyo
ISBN 0-387-17657-8 Springer-Verlag New York Heidelberg Berlin Tokyo

Library of Congress Cataloging-in-Publication Data. Witteman, W.J. The CO₂ laser. (Springer series in optical sciences ; v. 53). 1. Carbon dioxide lasers. I. Title. II. Series. TA1695.W57 1987 621.36'63 87-9847

Offset printing: Druckhaus Beltz, 6944 Hemsbach/Bergstr.
Bookbinding: J. Schäffer GmbH & Co. KG., 6718 Grünstadt.
2153/3150-543210

To *Ellen*
    Willem
    Agnes
    Marc
    Gemma

Preface

The field of CO_2 lasers has grown enormously in the last two decades. It has not only provided us with much insight into the features typical of molecular lasers and demonstrated different device concepts, but has also found a wide range of applications. This monograph is primarily devoted to those developments of the CO_2 laser that have become well established. It gives an extensive treatment of the relevant molecular physics, gas kinetics, excitation and relaxation processes, and laser physics pertinent to carbon dioxide. Furthermore, it provides a thorough theoretical background to specific technologies used in various devices. Many numerical values of physical constants and accurate spectroscopic data for CO_2 isotopes are also included. In fact, the book is based on the history of the CO_2 laser starting in 1964 and it reflects the subsequent developments in which I was actively engaged from the beginning.

Most of the new device concepts were initiated as the result of pure scientific research in many laboratories all over the world, irrespective of applications. On the other hand, many important potential applications have had a very stimulating effect on the further development of dedicated devices. This is particularly true for the spectacular progress made in high-energy pulsed systems in the range of 100 kJ to be used in experiments connected with laser fusion. Although the physics of short-pulse amplification and the technical problems related to these so-called e-beam sustained CO_2 laser systems are treated, a full description of the huge systems for laser fusion is beyond the scope of this book.

There are several books on laser physics, but none is devoted exclusively to CO_2-laser physics and devices, nor are their treatments of this subject very detailed. The aim of the present monograph is to bridge this gap. It will contribute to the understanding of different laser performances and will provide insight into the great potential of the CO_2 laser. The theories presented here are developed to show their relevance both to basic physics and to technology, and in most cases experimental results are compared with theory. Topics that are more general in laser physics, such as the optical properties of resonators, which can easily be found elsewhere, are not described extensively but only as far as is relevant to a particular system's performance. This book should be a useful handbook for experimental physicists and engineers in the field of CO_2 lasers. The work will also be useful to the graduate student or applied scientist with a scant background in laser physics.

I wish to thank many of my co-workers who have contributed with their own research to much of the material presented in this book. I would especially like to mention Dr. R.J.M. Bonnie, Dr. G.J. Ernst, Dr. F.A. van Goor, Dr. A.H.M. Olbertz, and Dr. R.A. Rooth. I am also indebted to Simone Sloot for the accurate typing of the manuscript and to H.T.M. Prins for the illustrations.

Enschede, December 1986   *W.J. Witteman*

Contents

1. Introduction 1
   1.1 Applications 2
   1.2 Efficiency and Output Considerations 2
   1.3 Stable or Unstable Resonators 4

2. Rotational-Vibrational Structure of CO_2 8
   2.1 The Normal Modes of Vibrations 8
   2.2 Wave Equation 10
   2.3 Rotational Line Distribution 16
   2.4 Fermi Resonance 18
   2.5 Transitions of the Regular Bands 19
   2.6 The Gain Ratios of Regular Bands 20
   2.7 Absolute Frequencies of CO_2 Isotopes 22

3. Laser Processes in CO_2 53
   3.1 Spontaneous Emission 53
   3.2 Stimulated Emission 56
   3.3 Laser Gain 58
   3.4 Line Shape 58
   3.5 Gain Saturation 62
   3.6 The Temperature Model of the Laser Process 64
   3.7 Vibrational Excitation of the Upper Laser Level 64
   3.8 Relaxation Phenomena and Vibrational Temperatures 68
   3.9 Gain Measurements and Vibrational Temperatures 71
   3.10 The Role of He, H_2O, and Xe in the Laser Mixture 74
   3.11 Power Extraction 76

4. Continuous Discharge Lasers 81
   4.1 The Behavior of the Discharge 81
   4.2 Elementary Theory of the Positive Column 83
   4.3 The Similarity Rules 85
   4.4 Thermal Effects and Similarity 87
   4.5 Optical Aspects of Single Mode Operation 92
       4.5.1 Gain of a Gaussian Beam 94

		4.5.2	Width of a Gaussian Beam in an Oscillator	97
		4.5.3	Saturation Parameter Measurements	101
	4.6	Sealed-off CO_2 Lasers	104
	4.7	Single Mode CO_2 Lasers	108
	4.8	The Sequence- and Hot-Band Lasers	111
	4.9	Transition Selection with Adjustable Outcoupling	112
		4.9.1	Analysis of Three-Mirror Configuration	114
		4.9.2	Experiments with a Tunable Outcoupling	117
		4.9.3	Performance at High Stability	118
	4.10	Frequency and Output Stabilization by the Opto-Voltaic Effect	119
		4.10.1	Opto-Voltaic Signal in the Water-Cooling Jacket .	125
5.	Fast Flow Systems	127
	5.1	Convection-Cooled Laser	127
	5.2	Principles of Laser Design	129
	5.3	Gain and Saturation Parameter	134
6.	Pulsed Systems	138
	6.1	Basic Principles of Laser Operation	139
	6.2	Electron-Energy Transfer and $I-V$ Characteristics of Laser-Gas Mixtures	140
	6.3	Derivation of Boltzmann's Equation	141
		6.3.1	Near-Isotropic Expansion of Boltzmann's Equation	142
		6.3.2	Energy Transport by Elastic Collisions	144
		6.3.3	Energy Transfer by the Applied Field	147
		6.3.4	Energy Transfer by Inelastic Collisions	148
		6.3.5	Stationary Boltzmann's Equation for the Electron Energy Distribution	149
	6.4	Solving Boltzmann's Equation for the CO_2-Laser Discharge	150
		6.4.1	Transport Coefficients	155
		6.4.2	Predictions of Optimum Laser Efficiency	157
		6.4.3	Operating E/N Values for Self-Sustained Glow Discharges	161
	6.5	Analysis of the Pulse-Forming Process	162
	6.6	Double-Discharge UV Preionized Systems	168
	6.7	Uniform-Field Electrode Profile (Chang Profile)	170
	6.8	Minimum-Width Electrode (Ernst Profile)	174
	6.9	Dielectric Corona Preionization	178
	6.10	Single-Discharge Corona Preionized Systems	182
	6.11	Electron-Beam Controlled Systems	185
		6.11.1	Recombination-Limited Plasma	187
		6.11.2	Cold Cathode	190
		6.11.3	Simulation of e-Beam Sustained Discharge	191

6.11.4   Optimized Output from the e-Beam Sustained
System 192

7.   **AM Mode Locking of TEA Lasers** 195
7.1   Acousto-Optic Modulation 196
7.2   Time-Domain Analysis of AM Mode Locking 204
    7.2.1   Self-Consistency of the Ideal, Circulating Pulse 205
    7.2.2   Self-Consistency of the Circulating Pulse with
Detuning 208
7.3   Frequency-Domain Analysis of AM Mode Locking 210
7.4   Experimental Investigations of AM Mode Locked Systems 217
    7.4.1   Stabilization of an AM Mode Locked TEA
Laser with an Intracavity Low-Pressure CO_2-Laser
Amplifier 219
    7.4.2   Stabilization of an AM Mode-Locked TEA Laser
with Injection of Continuous Radiation 222
    7.4.3   The Transient Evolution of an AM Mode Locked
TEA Laser 224
    7.4.4   Numerical Results of the AM Mode Locking 229

8.   **FM Mode Locking of TEA Lasers** 232
8.1   Electro-Optic Phase Modulation 232
8.2   Time-Domain Analysis of FM Mode Locking 236
    8.2.1   Self-Consistency of the Circulating Pulse with
Detuning 238
8.3   Frequency-Domain Analysis of FM Mode Locking 241
8.4   Experiments with FM Mode-Locked TEA Lasers 249

9.   **Passive Mode Locking** 251
9.1   Basic Principles 251
9.2   Mode Locking with Fast Saturable Absorption 253
    9.2.1   The Fundamental Equation 253
    9.2.2   Saturable Absorber and Laser Gain 256
    9.2.3   Pulse Form 258
    9.2.4   General Mode-Locking Criteria 261
    9.2.5   Mode Locking the CO_2 System 262
9.3   Mode Locking by P-Type Germanium 264
    9.3.1   Active and Passive Mode Locking 265
    9.3.2   Colliding Pulse and Locking 266

10. **Short-Pulse Amplification** 267
10.1   Pulse Propagation in a Two-Level System 268
10.2   Pulse Propagation with Rotational Relaxation 270
    10.2.1   Rotational Relaxation 270

	10.2.2	Short Pulse Multiline Energy Extraction with Rotational Relaxation	272
	10.2.3	The Effect of Rotational Relaxation on the Inversion and Pulse Form	276
	10.2.4	Multiband Energy Extraction	279
10.3		Experimental Technique for Generating Multi-Line Short Pulses	281
	10.3.1	Experimental Set-up	282
	10.3.2	Multi-Line Experiments	283
10.4		Single Pulse Selection	285
10.5		Prepulse, Retropulse and Parasitic Radiation Protection	288
10.6		Experimental Studies of Short Pulse Multi-Line Energy Extraction	290
10.7		Multiple-Pass Pulse Amplification	293
	10.7.1	Intramode Vibrational Relaxation	295
	10.7.2	Fermi Relaxation	296

References 299

Subject Index 307

1. Introduction

The development of various types of CO_2 lasers has been spectacular. The first reported laser had a continuous output power of only a few milliwatts [1.1,2]. Nowadays continuous lasers are built with output powers of more than 20 kW [1.3]. For pulsed systems it started in 1968 with a few joules [1.4]. These developments, especially stimulated by the large laser-fusion projects, have led to huge systems in the range of 100 kJ. Not only the pulse energy but also its time profile has been intensively studied. The search for pulse-compression technology has resulted to pulse durations of less than one picosecond [1.5].

When the potentialities of the CO_2 laser were recognized many industrial laboratories, institutes, and university groups were highly interested in this type of laser research. Not only new versions with very different output formats appeared during the course of time, but most of the research was devoted to the optimization of the performance, i.e. studies on extractable power, stability, beam divergence, mode structure, pulse shape, spectral purity, etc. The limits of the output power and pulse energy are, apart from costs, probably only restricted by the damage threshold values of transmitting and reflecting optical components.

The large interest in CO_2 lasers can also be understood from the fact that the efficiency of conversion of electrical energy into laser radiation combined with the maximum available power or pulse energy is by far superior to that of other laser systems.

The creation of the various categories of CO_2 lasers and the continuous improvements of them during the last decades were made possible by the broad interest of physicists and engineers with different backgrounds. In particular, the breakthrough in pulsed systems was possible because of many fundamental studies on glow discharges of molecular systems at atmospheric pressure and the introduction of new techniques for preionization.

With respect to applications, the CO_2 lasers have generally proven to be very versatile, simple to operate, and relatively cheap on investment and maintenance.

1.1 Applications

A large-scale application of the CO_2 laser is its use for material processing. In many cases the laser turns out to be superior to conventional tools. The advantages are the excellent control of the beam power, the absence of mechanical contact with the work piece, and the sharp focusing of the laser beam in the region of interest so that the applied energy is restricted and undesired deformation or decomposition of material can be avoided.

The technology of cutting and welding of metals with CO_2 lasers has been very successful. Steel plates of several centimeters thickness can be easily cut with a 10 kW system; and thin foils with low-power systems. Various metals that cannot be welded by conventional techniques can now be treated with CO_2 lasers.

Low-power lasers below 100 W have found their way onto the production floor where they perform a variety of micromachining and microsoldering tasks. Ablation, melting of thin metal layers, and vaporization without thermal side effects are possible. The $10\,\mu$m wavelength of the CO_2 laser is better absorbed and hence more efficient than the near-infrared radiation of solid-state lasers for processing many synthetic materials. For that reason the CO_2 is also very efficient for marking.

Very short pulses with high energy are of great interest for laser fusion. When a small pellet with a diameter of a few tenths of a millimeter filled with a mixture of deuterium and tritium, is irradiated with a high-energy laser pulse of the order of 100 kJ it is observed that the expanding plasma created by the strong absorption compresses and heats the center part of the pellet at values that initiate fusion. Although it is not our intention to describe the huge systems that have been developed for this technology, the large programs on laser fusion had a tremendous impact on the technology of amplifying and generating short pulses.

Low-power systems have also proven to be successful in many specialized surgical applications. Using a micromanipulator and an endoscopic coupler, the accuracy, reduced bleeding, and reduced post operative pain have made the CO_2 laser a valuable addition to the otolaryngology logic armamentarium. Neurosurgery too is using these lasers with great benefits to patients.

Finally we mention that military systems are often equipped with lasers. CO_2 lasers are used for tactical systems like target sensing, precision guidance, and coded communication.

1.2 Efficiency and Output Considerations

The CO_2 laser like any other gas laser consumes electrical power and delivers less power in its radiation at about $10\,\mu$m wavelength. The difference

2

between input and output power is waste heat that must be consumed. To increase the output power one must increase the input power, but technical and physical constraints limit the amount of input energy. In fact, the choice of heat sink characterizes the type of laser, examples being continuous sealed-off systems, fast-flow high-power systems, and pulsed systems.

The CO_2 laser operates by exchanging energy between low-lying vibrational-rotational energy levels of the CO_2 molecule. Molecules being in the higher energy state are transferred by the radiation field into a vibrational energy mode with a lower-energy state. The difference of energy of these upper and lower states is converted into infrared radiation. The upper level is excited by an electrical discharge either from the ground state or from a resonance transfer by vibrationally excited nitrogen. In both cases the input energy is at least the energy of the upper state. The maximum efficiency that is achievable is then 38 %, as can be directly inferred from an energy level diagram. However, this efficiency is the theoretical limit and no actual CO_2 laser will ever achieve this quantum efficiency. Well-designed systems reach at most 20 % of the electrical input energy.

The lasing process of all CO_2 lasers is sensitive to the temperature of the gas molecules in the discharge, i.e. active medium, which imposes another limit. As more power goes into the gas, its temperature rises and with this the thermal population of the lower laser level. Consequently, more radiation will be lost by the absorption of the lower-level density. In practice, therefore, one observes for increasing input energy, at first an increasing output but at a certain stage the output levels off and begins to decrease when the gas temperature rises above about 150°C.

Knowing these principal limitations one may estimate the output capability of a system. It depends entirely on the method by which the attendant waste heat is removed from the system. There are three basic routes to dispose of this waste energy. The simplest one is to make use of the thermal conductivity of the gas. The waste heat will be conducted to the cooled walls of the vessel containing the discharge. This method is always found in continuous systems with output powers in the order of 100 W or less. A second method to get rid of the waste heat is the removal of the hot gas itself. This is the case for the so-called fast-flow systems which operate in the range of many kilowatts of output power. The third method takes advantage of the heat capacity of the active medium. The maximum input energy, again limited by the gas temperature, is roughly 300 J per liter laser gas at atmospheric pressure. The output power is then about 40 J per liter active medium at one atmosphere. Both input and output energies are proportional to the gas density.

1.3 Stable or Unstable Resonators

The structure of the optical resonator of the laser system determines, to a large extent, the radiation distribution within the active medium and the optical quality of the outcoming beam. For obtaining nearly complete energy extraction from the laser medium the beam must not only fill the entire medium but its intensity must be above the saturation value in order to convert the inverted population into radiation. The design of the resonator, therefore, has a large influence on the laser performance.

The resonator can be either stable or unstable. For a stable resonator the mirror configuration corresponds to a stable periodic focusing system [1.6, 7]. The outcoming beam is transmitted through one or both mirrors. The mirror surfaces intersect the beam along phase fronts so that the reflected waves coincide with the incident wave. The stability relations can be obtained from the laws of geometrical optics and are given by

$$0 \leq (1 - L/R_1)(1 - L/R_2) \leq 1 \ , \tag{1.1}$$

where L is the distance between the mirrors, and R_1 and R_2 are the radii of curvature of the mirrors (a concave mirror has a positive R value and a convex mirror has a negative R value).

For an unstable resonator the radiation is not confined to narrow beams but experiences defocusing by bouncing back and forth between the mirrors. The divergent resonator system expands the radiation field on repeated bounces to fill the entire cross section of at least one mirror. The radiation leaving the cavity propagates around rather than passing through the outcoupling mirror. Unstable configurations are found when conditions (1.1) are not fulfilled. Typical examples of a stable and an unstable resonator are shown in Fig. 1.1.

Physical optics teaches that the oscillating beam in a stable or unstable cavity consists of one or more modes, each having its own spatial distribution. For a not too small Fresnel number (defined below) the stable

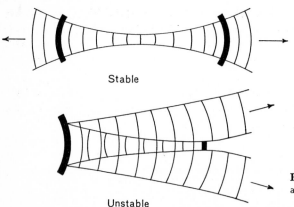

Stable

Unstable

Fig. 1.1. Examples of stable and unstable resonators

resonator produces, for the lowest-order mode near the axis, a Gaussian field distribution perpendicular to the optical axis [1.8]. The diameter of this lowest order or Gaussian mode is roughly given by a few times $(L\lambda)^{1/2}$, λ being the wavelength. It is generally less than the diameters of the mirrors. The Fresnel number, defined as $N_F = a^2/L\lambda$ with $2a$ being the diameter of the cavity or laser medium, indicates the number of different modes that are more or less free to oscillate. For a Fresnel number much larger than unity the Gaussian modes fill about a fraction $1/N_F$ of the laser medium. The whole medium can then only be filled with higher-order transversal modes. A great number of modes are needed to extract the available energy. The beam quality or beam divergence of the multi-mode system is inferior compared to a single-mode beam. Since each mode is characterized by a frequency, the frequency band of the multi-mode output beam can be a considerable fraction of the line width of the laser transition.

For the unstable resonator the situation is very different. Theoretical [1.9] and experimental studies [1.10] have clearly shown that unstable resonators can have large mode volumes even in short resonators and that the lowest-order mode may fill the total volume enclosed between the mirror geometry. This mode having the lowest loss per bounce will suppress the oscillation of other modes so that this type of resonator has a large discrimination against higher-order modes. The active medium is then occupied by a single-mode radiation field.

A theoretical analysis for numerical computations of the diffraction losses, i.e. outcoupling, and the field distribution is based on the Huygens integral equation [1.7,11]. This method is complicated and laborious. A much easier approach is reached by a geometrical description that relies on the focusing or defocusing effects of the mirrors. It produces an outcoupling factor which turns out to be in good agreement with the more detailed physical analysis based on Huygens' integral. In fact, it provides a more or less accurate zero-order approximation for the lowest-order mode, but not for any of the higher-order modes [1.12–14].

We shall describe the geometrical approach of the unstable resonator [1.12]. In Fig. 1.2 the two mirrors with radii of curvature of respectively R_1 and R_2 are shown. The separation distance is L. Both mirrors reflect the

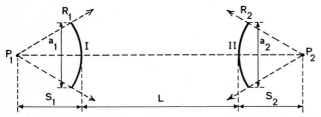

Fig. 1.2. Schematic drawing of the mirror configuration with the virtual images P_1 and P_2

diverging beam, as indicated by the virtual light points P_1 and P_2. The beam from P_1 must be imaged by the mirror configuration at point P_2, and vice versa. The well-known lens formula of geometrical optics will give the following two relations

$$\frac{1}{S_1 + L} - \frac{1}{S_2} = \frac{2}{R_2} \quad \text{and} \tag{1.2a}$$

$$\frac{1}{S_2 + L} - \frac{1}{S_1} = \frac{2}{R_1} \; . \tag{1.2b}$$

Solving these equations for S_1 and S_2 we obtain

$$S_1 = \frac{-L(R_2 - l) \pm [L^2(R_2 - L)^2 - LR_1(R_2 - L)(R_1 + R_2 - 2L)]^{1/2}}{R_1 + R_2 - 2L} \; , \tag{1.3a}$$

$$S_2 = \frac{-L(R_1 - L) \pm [L^2(R_1 - L)^2 - LR_2(R_1 - L)(R_1 + R_2 - 2L)]^{1/2}}{R_1 + R_2 - 2L} \; . \tag{1.3b}$$

Using the mirror diameters, a_1 and a_2 respectively, we can calculate the laser output as a diffraction-coupled beam propagating around the mirrors. Assuming a uniform beam reflected by Mirror I, only the fraction $a_2^2 S_1^2 / a_1^2 (S_1 + L)^2$ will reach the surface of Mirror II. Next we follow this fraction back to Mirror I. The part of it that reaches Mirror I is given by $a_1^2 S_2^2 / a_2^2 (S_2 + L)^2$. The geometrically calculated loss of the unstable resonator arises from the defocused beam that passes the mirrors. Considering a round trip of the beam the part of the radiation, Γ, that remains within the resonator is then simply given by the product of these reflected fractions, i.e.

$$\Gamma = \frac{S_1^2 S_2^2}{(S_1 + L)^2 (S_2 + L)^2} \; . \tag{1.4}$$

The fractional outcoupling per round trip now becomes

$$T = 1 - \Gamma \; . \tag{1.5}$$

It is interesting to note that this fractional power loss is independent of the mirror sizes or the Fresnel number. Especially for a large Fresnel number, say about 10, this geometrical outcoupling factor agrees quite well with the more exact predictions.

The most useful form of unstable resonator appears to be the positive-branch confocal or telescopic unstable resonator [1.10, 15] where the extrac-

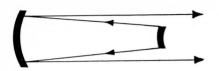

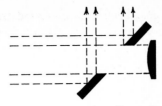

Fig. 1.3. Postive-branch confocal or telescopic unstable resonator

Fig. 1.4. An intracavity diagonal coupling mirror of an unstable resonator

tion of radiation is a parallel beam in one direction, as shown in Fig. 1.3. This configuration produces a collimated output beam. The geometry of this type of resonator is derived from the condition that the reflected outcoming beam is parallel. We obtain from (1.2), for example with $S_1 \to \infty$,

$$S_2 = -\tfrac{1}{2}R_2 \quad \text{and} \quad R_1 + R_2 = 2L \ . \tag{1.6}$$

Substituting the last expressions into (1.4) the outcoupling T_c of a confocal unstable resonator is simply given by

$$T_c = 1 - \left(\frac{R_2}{R_1}\right)^2 \ . \tag{1.7}$$

The outcoupling can be accomplished with a floating output mirror mounted on a non-reflecting window or on a radial spider. However, the best technique appears to be an intracavity coupling mirror under 45 degrees with the optical axis having a carefully cut hole for the cavity radiation, as shown in Fig. 1.4. The annular near-field output is then coupled out of the laser system.

The unstable resonator can, in prinicple, be used for all different types of CO_2 laser systems. However, for low-power continuous-wave (CW) systems, say below 100 W, the unstable configuration is not attractive because of the small Fresnel number. A stable resonator with a transmitting mirror is most practical. For kilowatt systems with large diameters one may apply both kinds of resonators. The power handling capacity of the transmitting mirror of a stable resonator can be a problem at power levels above $1 \, \text{kW/cm}^2$. In that case cooled mirrors in an unstable configuration may solve the problem. The multi-mode structure of the stable resonator will, in general, also fill the whole active medium so that both resonator configurations will extract the same power level. Pulsed systems are usually shorter and may have much larger apertures. They are very suitable for applying the unstable configurations. As a general rule one may prefer an unstable resonator for gains above 50 % per pass and Fresnel numbers above, say, 5. For such a laser system, at any power level, the unstable configuration is most suitable for extracting all available energy from the medium, combined with high beam quality.

2. Rotational-Vibrational Structure of CO_2

Laser action in any CO_2 system occurs between low-lying vibrational-rotational levels of the ground electronic state. Knowledge of the rotational-vibrational structure, the corresponding energy levels, and their transition probabilities is essential for the understanding of the laser process. For a determination of these molecular properties it is necessary to calculate the wave functions. The exact solution of the wave equation describing the motion of the individual atoms of a molecule (relative to the center of mass) is a difficult problem, because molecules have, as a rule, a rather complex structure. However, the empirical results of molecular spectroscopy on CO_2 as, for instance, obtained by *Dennison* [2.1], show that the energy values bear a simple relationship to one another, so that the energy of the molecule can be conveniently considered to be made up of two parts, called, respectively, the vibrational energy and the rotational energy. This permits a simpler solution, because the spectroscopic data suggest that it is possible to treat the vibration and rotation of the molecule quite separately and then to combine the results of the two calculations to present the behaviour of the three atoms in the CO_2 molecule. Thus we neglect any perturbation of the vibrational state due to the rotation of the molecule, and therefore the wave functions are products of rotational and vibrational wave functions.

For the understanding of the vibrational motions and their energies one can not simply consider harmonic oscillations. For this it is essential to include into the Hamiltonian of the wave equation also the deviations from harmonic force fields. These anharmonic force constants can be deduced from some observed spectra [2.2]. The energy levels and transition probabilities can then be calculated accurately by treating these anharmonic force constants with the usual perturbation theory. This means that the wave function of a particular vibrational energy state will be described as a linear combination of the unperturbed vibrational wave functions.

2.1 The Normal Modes of Vibrations

The CO_2 molecule is a linear symmetric molecule with an axis of symmetry along the nuclei and a plane of symmetry perpendicular to this axis. The

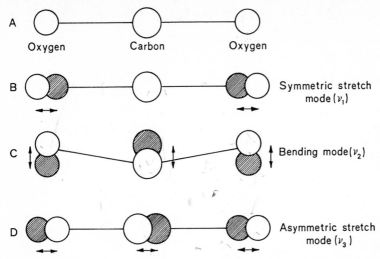

Fig. 2.1. Vibrational motion of CO_2. (A: unexcited CO_2; B–D illustrate three normal modes of vibration)

molecule has thus only two degrees of rotational freedom and hence we have $n = 3 \times 3 - 5$ or 4 vibrational degrees of freedom. The corresponding vibrational motions are shown in Fig. 2.1.

Vibration ν_1 with normal coordinates S_1 is longitudinal and symmetric (valence or symmetric stretch mode). The twofold degenerate vibration ν_2 with normal coordinates S_{21} and S_{22} desribes the motion of the C atom in a plane perpendicular to the axis of symmetry (bending mode). Vibration ν_3 with normal coordinate S_3 is longitudinal and asymmetric (asymmetric valence or stretch mode). The vibrations ν_2 and ν_3 have the property in common that during the motion the distance between the O atoms remains unchanged. In the vibration ν_1 the C atom remains stationary. The four vibrations are associated with different wave numbers. In the case where the vibrational motion is only described by the unperturbed harmonic oscillators the wave numbers are [2.2]

$$\omega_1 = 1351.2 \, \text{cm}^{-1} \, , \quad \omega_2 = 672.2 \, \text{cm}^{-1} \, , \quad \omega_3 = 2396.4 \, \text{cm}^{-1} \, .$$

We may describe the positions of the atoms of a CO_2 molecule with a perpendicular set of Cartesian coordinates, the molecular axis being along the z-axis, as indicated in Fig. 2.2.

Let the O atoms each have mass m and coordinates x_1, y_1, z_1, and x_3, y_3, z_3, while the C atom with mass M has the coordinates x_2, y_2, z_2. The kinetic energy is then given by

$$T = \frac{m}{2}(\dot{x}_1^2 + \dot{y}_1^2 + \dot{z}_1^2 + \dot{x}_3^2 + \dot{y}_3^2 + \dot{z}_3^2) + \frac{M}{2}(\dot{x}_2^2 + \dot{y}_2^2 + \dot{z}_2^2) \, . \tag{2.1}$$

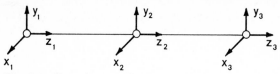

Fig. 2.2. Cartesian coordinates for the positions of the atoms of a CO_2 molecule

We can easily express this kinetic energy in terms of the normal coordinates. We find for the Cartesian coordinates

$$\dot{x}_1 = \dot{x}_3 = \frac{M}{2m+M}\dot{S}_{21} , \qquad \dot{x}_2 = -\frac{2m}{2m+M}\dot{S}_{21} ,$$

$$\dot{y}_1 = \dot{y}_3 = \frac{M}{2m+M}\dot{S}_{22} , \qquad \dot{y}_2 = -\frac{2m}{2m+M}\dot{S}_{22} ,$$

$$\dot{z}_1 = \tfrac{1}{2}\dot{S}_1 + \frac{M}{2m+M}\dot{S}_3 , \qquad \dot{z}_2 = -\frac{2m}{2m+M}\dot{S}_3 .$$

$$\dot{z}_3 = -\tfrac{1}{2}\dot{S}_1 + \frac{M}{2m+M}\dot{S}_3 ,$$

Substituting these values in the expression T we find

$$T = \frac{\mu_1}{2}\dot{S}_1^2 + \frac{\mu_2}{2}(\dot{S}_{21}^2 + \dot{S}_{22}^2 + \dot{S}_3^2) , \quad \text{where}$$

$$\mu_1 = \frac{m}{2} , \quad \text{and} \quad \mu_2 = \frac{2mM}{2m+M} .$$

The potential energy can also be expressed in terms of these normal coordinates. However, mathematically it is advantageous to introduce dimensionless variables in terms of these normal coordinates, as will be done in the next section.

2.2 Wave Equation

It is clear from the foregoing discussion that the approximate wave equation for the rotation and vibration of the CO_2 molecule has the form

$$(H_r + H_v)\psi_{\text{mol}} = E_{rv}\psi_{\text{mol}} \tag{2.2}$$

where E_{rv} is the sum total of the rotational and vibrational energies, and ψ_{mol} is the wave function describing rotation and vibration.

The Hamiltonian of the rotational motion in terms of the usual angles θ and ϕ is given by

$$H_r = -\frac{\hbar^2}{2I}\left[\frac{1}{\sin\theta}\frac{\partial}{\partial\theta}\left(\sin\theta\frac{\partial}{\partial}\right) + \frac{1}{\sin^2\theta}\frac{\partial^2}{\partial\phi^2}\right] \;,$$

I being the moment of inertia of the molecule (equal to $71.1 \times 10^{-40}\,\text{g\,cm}^2$).

The Hamiltonian H_v of the vibrational motion can be described in terms of the following dimensionless variables:

$$\sigma = 2\pi(\omega_1\mu_1 c/h)^{1/2}S_1 \;,$$
$$\xi = 2\pi(\omega_2\mu_2 c/h)^{1/2}S_{21} \;,$$
$$\eta = 2\pi(\omega_2\mu_2 c/h)^{1/2}S_{22} \;,$$
$$\varsigma = 2\pi(\omega_3\mu_2 c/h)^{1/2}S_3 \;, \tag{2.3}$$

and has the following form [2.2, 3]

$$\begin{aligned}
H_v = {}&(2\pi^2 c/h)[\omega_1 p_\sigma^2 + \omega_2(p_\xi^2 + p_\eta^2) + \omega_3 p_\varsigma^2]\\
&+ \tfrac{1}{2}hc(\omega_1\sigma^2 + \omega_2\varrho^2 + \omega_3\varsigma^2) + hc(a\sigma^3 + b\sigma\varrho^2 + c'\sigma\varsigma^2)\\
&+ hc(d\sigma^4 + e\varrho^4 + f\varsigma^4 + g\sigma^2\varrho^2 + h\sigma^2\varsigma^2 + i\varrho^2\varsigma^2) \tag{2.4}
\end{aligned}$$

with $\varrho^2 = \xi^2 + \eta^2$. This Hamiltonian describes the four nonlinear oscillations. The first parenthesis contains the kinetic energy, the second the harmonic part of the potential energy, and the third and fourth describe the anharmonic contributions. The symbols p_σ, p_ξ, p_η, and p_ς are the canonically conjugate momenta of the dimensionless variables σ, ξ, η, and ς. The anharmonic force constants are in units of $[\text{cm}^{-1}]$:

$$a = -30.0 \;, \quad b = 71.3 \;, \quad c' = -250.0 \;, \quad d = 1.5 \;, \quad e = 0.5 \;,$$
$$f = 6.4 \;, \quad g = 1.9 \;, \quad h = 8.9 \;, \quad i = -25.7 \;.$$

Since we neglect any perturbation of a vibrational state due to the rotation of the molecule we can write ψ_{mol} as the product of rotational and vibrational wave functions, i.e.,

$$\psi_{\text{mol}} = \psi_r\psi_v \;. \tag{2.5}$$

By substituting this into (2.2) and dividing by $\psi_v\psi_r$, we find that the left-hand side of the equation consists of the sum of two parts, one depending only on the rotational coordinates and the other only on the vibrational coordinates. Each part must be equal to a constant. These two equations are

$$H_r \psi_r = E_r \psi_r \ ,$$ (2.6a)

$$H_v \psi_v = E_v \psi_v \ ,$$ (2.6b)

where $E_{rv} = E_r + E_v$.

The rotational wave equation (2.6a) can be further separated into the coordinates ϕ and θ and then one finds the solution to be a spherical harmonic [2.4]

$$\psi_r = \frac{1}{\sqrt{2\pi}} e^{im\phi} N_{jm}^{1/2} P_j^{|m|}(\cos \theta) \ ,$$ (2.7a)

where m is a positive or negative integer or zero, j a positive integer or zero and N_{jm} is a normalization constant given by

$$N_{jm} = \frac{(2j+1)(j-|m|)!}{2(j+|m|)!} \ .$$ (2.7b)

The function P_j^m is called the associated Legendre function. For a particular j value ψ_r is $(2j+1)$-fold degenerate. The energy values of rotation, which are determined by the eigenvalues of the equation, form a discrete set and are given by

$$E_r = Bhcj(j+1) \ ,$$ (2.8)

where B is the rotational constant.

The constants j and m are called the rotational and magnetic quantum numbers. The different rotational levels, as described by the m values, correspond to the number of possible orientations of the angular momentum. The number $g(j)$ of degenerate levels is given by

$$g(j) = 2j + 1 \ .$$ (2.9)

Going back to the vibrational part of the wave equation we consider first the Hamiltonian with only the harmonic part of the potential energy. It is then possible to separate the variables in (2.6b) by means of the product of the wave functions of the four vibrations. Each vibration is treated separately as a harmonic oscillator. The valence and the asymmetric valence vibrations have according to standard theory, respectively, the following mathematical form

$$\psi_{n_1} = N_{n_1} e^{-\sigma^2/2} H_{n_1}(\sigma) \ ,$$ (2.10)

$$\psi_{n_3}(\varsigma) = N_{n_3} e^{-\varsigma^2/2} H_{n_3}(\varsigma) \ ,$$ (2.11)

where H_{n_i} is a so-called Hermite polynomial of the n_ith degree. The constant N_{n_i} is used to normalize the wavefunction to unity. Naturally each one of the mutually degenerate pair of the bending vibrations can be described by a corresponding vibrational eigenfunction. The total vibrational energy will then be given by

$$E_v^0 = hc[\omega_1(n_1 + \tfrac{1}{2}) + \omega_2(n_2 + 1) + \omega_3(n_3 + \tfrac{1}{2})] \ . \tag{2.12}$$

It is seen that if the degenerate vibration is excited by one quantum there are two eigenfunctions for the state $n_2 = 1$. If there are two quanta excited $(n_2 = 2)$, there are three possible eigenfunctions for that state: we have two possibilities in which two quanta are in one of the two bending vibrations and the possibility that each vibration has one quantum. In general, the degree of degeneracy if n_2 quanta are excited is equal to the number of different ways in which n_2 can be written as the sum of two integers (equal to $n_2 + 1$). However, the degeneracy exists only as long as strictly harmonic vibrations are assumed. As will be shown later, the anharmonic forces produce a partial splitting of this degeneracy. In order to understand this partial splitting we describe the degenerate vibration by polar coordinates

$$\xi = \varrho \cos \phi \ , \quad \eta = \varrho \sin \phi \ . \tag{2.13}$$

The sum of a possible combination in which the product of harmonic-oscillator wave functions corresponds to n_2 quanta can be written in the form [2.5]

$$\psi_{n_2}^l = N_{n_2} e^{-\varrho^2/2} F_{n_2}^{|l|}(\varrho) e^{il\phi} \tag{2.14}$$

where $F_{n_2}^l$ is a polynomial of degree n_2 in ϱ and where l can take the values

$$|l| = n_2, \ n_2 - 2, \ n_2 - 4, \ldots, 1 \quad \text{or} \quad 0 \ , \tag{2.15}$$

depending on whether n_2 is odd or even.

Some of the lowest-order functions are

$$F_0^0 = 1 \ , \quad F_1^1 = -\varrho \ , \quad F_2^2 = 2\varrho^2 \ , \quad F_2^0 = 1 - \varrho^2 \ .$$

The factor $\exp(il\phi)$ indicates the presence of an angular momentum in the vibration. If the two perpendicular vibrations of the bending mode are both excited, the motion of the C atom can be seen classically as two oscillations with a phase difference of $90°$. For instance, for $n_2 = 2$ we have $l = 0, l = 2$ or $l = -2$. The value $l = 0$ can be considered as a linear motion, whereas $l = 2$ and $l = -2$ are circular motions in opposite directions. During these circular motions the internuclear distances are unchanged and therefore they have no coupling with the ν_1 or ν_3 vibration. The vibration $n_2 = 2$ with $l = 0$ is coupled to the ν_1 and ν_3 vibrations by the anharmonic forces. This explains why there is no optical transition between states terminating with $n_2 = 2, l = 2$ or -2 and $n_3 = 1$. Consequently, the unperturbed vibrational state with n_1 quanta excited in the ν_1 vibration, n_2 quanta with angular momentum l excited in the ν_2 vibration, and n_3 quanta excited in the ν_3 vibration is indicated with the symbol $(n_1 n_2^l n_3)$. The total vibrational wave function corresponding to this state becomes

$$\psi_v = \psi_{n_1} \psi_{n_2}^l \psi_{n_3} \ . \tag{2.16}$$

Going back to find the solution of (2.6b) in which the Hamiltonian is given by (2.4) one can, according to standard perturbation theory, assume a linear combination of functions of the form (2.16). This combination can be inserted into (2.6b). The next step is to multiply the resulting equation by the complex conjugate of the various unperturbed functions as given by (2.16) and integrate in the usual way over all space. This problem has been successfully solved by *Statz* et al. [2.3]. Their results are presented in Table 2.1 in terms of the unperturbed wave functions.

According to the preceding discussion the total energy of vibration and rotation of the CO_2 molecule including nonlinearities is then given by [2.6]

$$E_{rv} = hc \left[\sum_i \omega_i \left(n_i + \frac{d_i}{2} \right) + \sum_i \sum_k x_{ik} \left(n_i + \frac{d_i}{2} \right) \left(n_k + \frac{d}{2} \right) \right.$$
$$\left. + \sum_i g_{ii} l_i^2 + B_v j(j+1) - D_v j^2 (j+1)^2 \right] \ , \tag{2.17}$$

Table 2.1. Wave functions for selected states [2.3]

State	Wave function
$(00^00)'$	$0.9982\ (00^00) + 0.0514\ (10^00) + 0.0206\ (10^02) - 0.0184\ (12^00) + \ldots$
$(02^00)'$	$-\ 0.7726\ (02^00) + 0.6289\ (10^00) + 0.0635\ (20^00) - 0.0294\ (14^00) -$
	$0.0277\ (00^00) + 0.0188\ (12^00) + 0.0188\ (00^02) + 0.0183\ (20^02) -$
	$0.0156\ (22^00) - 0.0147\ (12^02) + 0.0134\ (04^00) + \ldots$
$(10^00)'$	$-\ 0.7666\ (10^00) - 0.6326\ (02^00) - 0.0831\ (20^00) + 0.0441\ (00^00) -$
	$0.0318\ (00^02) - 0.0236\ (14^00) - 0.0223\ (20^02) + 0.0204\ (22^00) -$
	$0.0144\ (12^02) + 0.0133\ (12^00) + 0.0118\ (04^00) + \ldots$
$(00^01)'$	$0.9828\ (00^01) + 0.1772\ (10^01) + 0.0352\ (10^03) + 0.0257\ (20^01) +$
	$0.0184\ (02^01) - 0.0155\ (12^01) + \ldots$
$(01^10)'$	$0.9992\ (01^10) - 0.0263\ (13^10) + 0.0207\ (11^12) + 0.0143\ (11^10) +$
	$0.0106\ (03^10) + \ldots$

where $d_i = 1$ for the ν_1 and ν_3 vibration (nondegenerate) and $d_i = 2$ for the ν_2 vibration, and x_{ik} are anharmonicity constants. The term $\sum_i g_{ii} l^2$ is due to the vibrational angular momentum of the bending mode. The small constant g_{ii} is of the order of x_{ik}. There is a double degeneracy for each $|l|$ value due to the two directions of the angular momentum l. (For large j values this double degeneracy can be removed by Coriolis forces). The last term in (giving a correction to the rotational energy, see (2.8), results from the non-rigidity of the molecule. The centrifugal force in the rotating molecule gives a slight increase in the internuclear distances and therefore changes the angular momentum. The constants B_v and D_v depend slightly on the vibrational state. In Table 2.2 some low-lying vibrational transitions relevant to the various laser processes are listed. A diagram is shown in Fig. 2.3.

The transitions belong to the infrared and Raman bands of CO_2. Strong infrared bands belong to the ν_2 and ν_3 vibrations. One would expect only Raman transitions corresponding to the symmetric vibration ν_1. This, however, is not the case. It is very typical that two strong Raman lines are observed at 1285.8 and 1388.1 cm^{-1}. These lines correspond to the ν_1 vi-

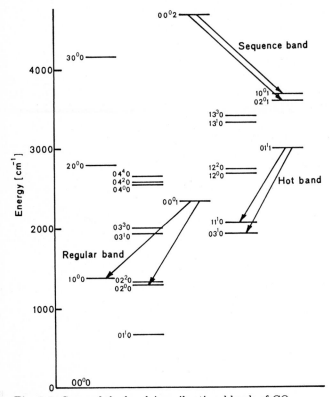

Fig. 2.3. Some of the low-lying vibrational levels of CO_2

15

bration and to twice the ν_2 vibration. This can be explained as follows. Since $2\nu_2$ is very close to ν_1 there is a Fermi resonance (see below) between corresponding states. These states are mixtures of (10^00) and (02^00), as can be seen in Table 2.1. Therefore, they are both Raman active.

2.3 Rotational Line Distribution

The rotational levels of a linear molecule are called positive or negative depending on whether the total eigenfunction (describing rotation, vibration, and electronic state) remains unchanged or changes sign by reflection of all nuclei and electrons at the origin. If the electronic and vibrational eigenfunctions are unchanged by such a symmetry operation, the symmetry character positive-negative depends on the rotational function only so that the even rotational levels are positive and the odd ones negative. For the linear CO_2 we have the property symmetric or antisymmetric with respect to an exchange of the oxygen nuclei. For the electronic ground state of CO_2 the positive rotational levels have symmetric wave functions, the negative antisymmetric with respect to a simultaneous exchange of the oxygen nuclei (inversion) for all vibrational levels that are symmetric to this inversion (for example, 10^00, 20^00, 02^00, 02^20, 00^02) while the reverse is the case for all vibrational levels that are antisymmetric with respect to this inversion (for example, 00^01, 02^01, 10^01, 03^10).

Further, since CO_2 has a center of symmetry and the spins of the oxygen (^{16}O) nuclei are zero, the negative rotational levels are missing. Thus for the electronic ground state the molecules with symmetric vibrational levels have only even j numbers (symmetric wave functions) in the rotational spectrum, while for antisymmetric vibrational levels only odd j numbers (antisymmetric wave functions) exist. It should be noted that if one oxygen atom is replaced by a different isotope, the distinction between symmetric and antisymmetric rotational levels no longer exist and therefore none of the rotational levels are missing.

The thermal distribution of rotational levels is according to Boltzmann given by

$$n_{vj} = N_v C g(j) \exp\left[-F(j)\frac{hc}{kT}\right] , \qquad (2.18)$$

where n_{vj} is the number of molecules with rotational number j per unit volume, N_v the total number of molecules per unit volume having a particular vibrational state, C a normalization constant, and $g(j)$ the statistical weight according to (2.9) equal to $2j + 1$; $F(j)$ is given by, see (2.17),

$$F(j) = B_v j(j+1) - D_v j^2(j+1)^2 . \qquad (2.19)$$

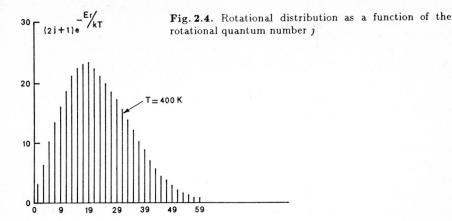

Fig. 2.4. Rotational distribution as a function of the rotational quantum number j

Since, depending on the vibrational state, either the odd or the even j values are missing, the rotational distribution is expressed by

$$n_{vj} \simeq N_v \left(\frac{2hcB}{kT}\right)(2j+1)\exp\left[-F(j)\frac{hc}{kT}\right] \ . \qquad (2.20)$$

good for $\ell \to 0$

The number of molecules in the different rotational levels goes through a maximum for increasing j value (Fig. 2.4). Since the second term in (2.19) is very small compared to the first one, this maximum is found for

$$j_{max} \simeq \sqrt{\frac{kT}{2Bhc}} - \frac{1}{2} \ . \qquad (2.21)$$

For $T = 400\,\text{K}$ we obtain $j_{max} \simeq 19$.

The above expression for n_{vj} only holds as long as the degenerate ν_2 vibration has the angular momentum $l = 0$. This is the case for the levels involving the laser transitions $(00^01)-(10^00)$ and $(00^01)-(02^00)$. However, for the degenerate vibrational levels with $l \neq 0$ the j value is always equal to or larger than $|l|$. Thus for l and j both even or odd we have

$$j = |l| \ , \quad |l|+2 \ , \quad |l|+4 \ , \quad \ldots \ , \qquad (2.22)$$

and for l odd and j even, or l even and j odd we have

$$j = |l|+1 \ , \quad |l|+3 \ , \quad |l|+5 \ , \quad \ldots \ . \qquad (2.23)$$

The levels with j values smaller than l do not occur.

2.4 Fermi Resonance

For CO_2 molecules we also have an accidenta degeneracy because ω_1 is close to $2\omega_2$ so that vibrational levels belonging to different vibrations have nearly the same energy. Such a resonance leads to a perturbation of the energy levels, as was first recognized by *Fermi* [2.7]. The perturbation is produced by the anharmonic forces between the two vibrations and can be described by the anharmonic terms in the potential energy. The result of the interaction is that one level is shifted up and the other shifted down so that the separation of the two levels is much greater than expected. At the same time there is a mixing of the two states, as can be seen in Table 2.1 for the states $(10^00)'$ and $(02^00)'$. The magnitude of this energy perturbation depends strongly on the matrix element $H_{vv'}$ of the anharmonic part of the Hamiltonian H_v given by

$$H_{vv'} = \int \psi_v^* H_v \psi_{v'} d\tau \tag{2.24}$$

where ψ_v and $\psi_{v'}$ are the unperturbed wave functions of the two considered states. Since the unperturbed energies are nearly equal the magnitude of the shift can be obtained according to first-order perturbation theory from the secular determinant

$$\begin{vmatrix} E_v^0 - E & H_{vv'} \\ H_{v'v} & E_{v'}^0 - E \end{vmatrix} = 0 \tag{2.25}$$

where E_v^0 and $E_{v'}^0$ are the unperturbed energies. Solving (2.25) we obtain

$$E = \overline{E}_{vv'} \pm \tfrac{1}{2}[4|H_{vv'}|^2 + \delta^2]^{1/2} \tag{2.26}$$

where $\overline{E}_{vv'} = \tfrac{1}{2}(E_v^0 + E_{v'}^0)$ and $\delta = E_v^0 - E_{v'}^0$.

The wave functions ψ_v' and $\psi_{v'}'$ of the two resulting states are, according to perturbation theory, the following mixtures of the unperturbed wave functions:

$$\psi_v' = a\psi_v - b\psi_{v'} \ , \quad \psi_{v'}' = b\psi_v + a\psi_{v'} \quad \text{where} \tag{2.27}$$

$$a = \left(\frac{(4|H_{vv'}|^2 + \delta^2)^{1/2} + \delta}{2(4|H_{vv'}|^2 + \delta^2)^{1/2}} \right)^{1/2} , \tag{2.28a}$$

$$b = \left(\frac{(4|H_{vv'}|^2 + \delta^2)^{1/2} - \delta}{2(4|H_{vv'}|^2 + \delta^2)^{1/2}} \right)^{1/2} . \tag{2.28b}$$

Using (2.26) the coefficients a and b can also be expressed in terms of δ and the energy-level splitting Δ, including Fermi resonance effects, as

18

$$a = \left(\frac{(\Delta + \delta)}{2\Delta} \right)^{1/2} , \qquad\qquad (2.29a)$$

$$b = (1 - a^2)^{1/2} . \qquad\qquad (2.29b)$$

It is usual to designate these mixed terminal states ψ'_v and $\psi'_{v'}$ by (I) and (II). Applying the above results to the mixing of (10^00) and (02^00) we obtain

$$(I) = 0.73(10^00) - 0.68(02^00) , \qquad\qquad (2.30a)$$

$$(II) = 0.68(10^00) + 0.73(02^00) . \qquad\qquad (2.30b)$$

The corresponding energies are

$$E_I = 1388.3 \, \text{cm}^{-1} \quad \text{and} \quad E_{II} = 1285.5 \, \text{cm}^{-1}$$

which have a much larger energy separation than expected on the basis of $\omega_1 = 1351.2 \, \text{cm}^{-1}$ and $\omega_2 = 672.2 \, \text{cm}^{-1}$.

It should be noted that $H_{vv'}$ vanishes for $\psi_v = (10^00)$ and $\psi_{v'} = (02^00)$ or $(02^{-2}0)$ so that these levels have no Fermi resonance. Similar results can be obtained for the mixing of the terminal states of (11^10) and (03^10) of the so-called hot bands, and (10^01) and (02^01), (10^02) and (02^02) of the sequence bands (Fig. 2.3).

2.5 Transitions of the Regular Bands

The first-order allowed infrared spectrum of CO_2 results from transitions between the vibrational-rotational levels, subject to certain selection rules. Symmetric vibrations have only even j numbers and antisymmetric vibrations only odd j numbers (Sect. 2.3). The selection rule for electric dipole radiation between (00^01)-(I) and (00^01)-(II) leads to allowed transitions for $\Delta j = \pm 1$ (Sect. 3.1). It is customary in spectroscopy to indicate the rotational transition part by the j number of the lower level and any change of j refers to this level. The rotational-vibrational transitions arising from the vibrational bands (00^01)-(I) and (00^01)-(II) have wavelengths near their band edge of 10.4 and 9.4 μm. The transitions with $\Delta j = -1$ are called the *P-branch* transitions and those with $\Delta j = +1$ are called *R-branch* transitions. Figure 2.5 shows the absorption spectra of the two regular band of CO_2. In the most common CO_2 lasers the stimulated emission occurs on one or more lines of these regular bands.

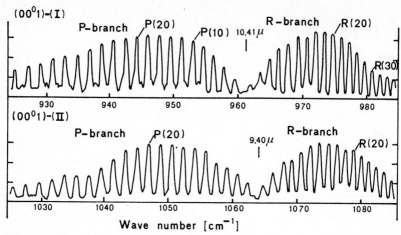

Fig. 2.5. Absorption spectra for CO_2 in the 9.4 and 10.4 μm vibrational bands

2.6 The Gain Ratios of Regular Bands

The population densities of the terminal states of the two regular bands can be regulated in electrical discharges of CO_2 gas mixtures. For laser action the discharge conditions in the mixture are such that the density of the upper level (00^01) is much larger than that of the level (I) or (II). Since the levels (I) and (II) are in close resonance, the population densities of these levels are practically equal so that the inversion density, i.e. the difference between the upper and lower densities, of both bands are nearly equal. In that case the gain ratio of a P or R transition in both bands for the same j values depends on the ratio of the vibrational dipole matrix element which can be calculated from the terminal vibrational-state wave functions. In Chap. 3 it is shown that the gain, as given by (3.19 and 20), is proportional to $|R_{vv'}|^2$ where $R_{vv'}$ is the dipole matrix element of the vibrational states. It is also seen that for the same j values and assuming the same line shape and lower vibrational state densities the gain ratio of the regular bands is simply given by the ratio of the values $\lambda^2|R_{vv'}|^2$ for the two bands.

The dipole moment operator of the vibrational states can be expressed by [2.3]

$$p = d_3\varsigma + d_{13}\sigma\varsigma , \qquad (2.31)$$

where ς and σ are defined by (2.3). The dipole nonlinearity d_{13} is very small and can be neglected in the further calculations. The dipole matrix element can be derived by using the wave functions of the molecular vibrations. For the lower level it is sufficient to use the approximate wavefunction given by (2.27). Neglecting the wavelength dependence the gain ratio of the regular bands is then obtained as [2.8]

Table 2.2. Low-lying infrared and Raman bands of CO_2

Upper State $\nu_1\nu_2\nu_3$	Species	Lower state $\nu_1\nu_2\nu_3$	Species	ν_{vac} Observed [2.6] $[\text{cm}^{-1}]$
01^10	π_u	00^00	Σ_g^+	667.3
02^00	Σ_g^+	00^00	Σ_g^+	1285.8
10^00	Σ_g^+	00^00	Σ_g^+	1388.1
03^10	π_u	00^00	Σ_g^+	1931.9
00^01	Σ_u^+	00^00	Σ_g^+	2349.4
02^00	Σ_g^+	01^10	π_u	618.5
02^20	Δ_g	01^10	π_u	668.1
10^00	Σ_g^+	01^10	π_u	720.8
03^10	π_u	01^10	π_u	1264.6
04^00	Σ_g^+	01^10	π_u	1880.1
20^00	Σ_g^+	01^10	π_u	2131.5
03^10	π_u	02^20	Δ_g	596.5
03^10	π_u	02^00	Σ_g^+	646.1
00^01	Σ_u^+	10^00	Σ_g^+	961.3
00^01	Σ_u^+	02^00	Σ_g^+	1063.6
04^20	Δ_g	02^20	Δ_g	1248
04^20	Δ_g	02^00	Σ_g^+	1297.6
20^00	Σ_g^+	02^00	Σ_g^+	1513

$$\left|\frac{R_{\text{I}-00^01}}{R_{\text{II}-00^01}}\right|^2 = \left|\frac{\langle\text{I}|p|00^01\rangle}{\langle\text{II}|p|00^01\rangle}\right|^2 = \left|\frac{a+0.13b}{b-0.13a}\right|^2 , \qquad (2.32)$$

where a and b are given by (2.29).

Assuming that the tabulated presentation of the wave functions of the energy levels in terms of the unperturbed wave function, as presented in Table 2.1, holds also for isotopic substitutions the gain ratio for CO_2 isotopes is then also given by (2.32). For each isotope the values of a and b are obtained in terms of the unperturbed (δ) and perturbed energy-level splitting (Δ), as expressed by (2.29). The calculated matrix element ratios of the regular bands of some CO_2 isotopes are compared with the j-independent portion of the matrix element ratios inferred from gain and loss measurements. The results are shown in Table 2.3. It is seen that the effects due to

Table 2.3. Vibrational matrix element ratios of the regular bands of some CO_2 isotopes

$\left|\dfrac{R_{\text{I}-00^01}}{R_{\text{II}-00^01}}\right|^2$	$^{12}C^{16}O_2$	$^{12}C^{18}O_2$	$^{13}C^{16}O_2$
Theory	1.4	0.48	3.2
Gain coefficient	1.3	0.4	2.5
Absorption coefficient	1.0	0.3	1.8

Fermi resonance (expressed by a and b) play a major role in the gain of the regular bands.

2.7 Absolute Frequencies of CO_2 Isotopes

Several authors have reported on absolute frequency measurements in CO_2 lasers [2.9]. Very extensive and highly accurate measurements have been performed by *Petersen* et al. [2.10] and by *Freed* et al. [2.11]. In essence, optical heterodyne techniques were used to generate beat frequencies between two lasers oscillating on adjacent rotational transitions. The lasers employed were accurately stabilized by using a $4.3\,\mu$m fluorescence technique [2.12]. Microwave frequency counter measurements of the difference frequencies were then used to calculate the band centers, rotational constants, and transition frequencies. The measured data were fitted to a standard formula that includes a further expansion of the rotational energy in terms of $j(j+1)$ as compared to (2.19); i.e.

$$
\begin{aligned}
T(v,j) = G_v + B_v j(j+1) &- D_v[j(j+1)]^2 \\
&+ H_v[j(j+1)]^3 + L_v[j(j+1)]^4 + \dots \,,
\end{aligned} \tag{2.33}
$$

where the first term on the right-hand side describes the vibrational part of the transition. The reference line in this scheme is the absolute measurement of the $R(30)$ of the (00^01)-(I) band [2.10].

The spectral range of the regular bands can be greatly extended by the nature of the CO_2 isotopes. The isotopes not only have different band centers and rotational constants but some of them contain also the odd P and R transitions. The absolute frequencies, band centers, rotational constants, and vacuum wave numbers (using $c = 299792458$ m/s) of the regular bands in nine CO_2 isotopes, as obtained by *Bradley* et al. [2.11], are listed in Tables 2.4–12.

For some lines one should be somewhat cautious in interpreting the standard deviations given in these tables, because some measurements could be influenced by the proximity of lines belonging to the hot and sequence bands. This is particularly true for the frequencies which are demarcated by horizontal lines drawn in the tables.

Table 2.4 $^{16}O^{12}C^{16}O$

NUMBER	SYMBOL		CONSTANTS (MHZ)	STD.DEV. (MHZ)
1	V(001-I)	=	2.880 881 382 455 D+07	3.6D-03
2	V(001-II)	=	3.188 996 017 636 D+07	3.7D-03
3	B(001)	=	1.160 620 695 034 D+04	2.3D-05
4	B(I)	=	1.169 756 942 611 D+04	2.5D-05
5	B(II)	=	1.170 636 464 791 D+04	2.4D-05
6	D(001)	=	3.988 109 863 D-03	3.2D-08
7	D(I)	=	3.445 940 508 D-03	3.3D-08
8	D(II)	=	4.711 559 114 D-03	3.3D-08
9	H(001)	=	0.481 534 D-09	1.9D-11
10	H(I)	=	5.625 110 D-09	1.8D-11
11	H(II)	=	7.066 300 D-09	1.8D-11
12	L(001)	=	-0.96 936 D-14	3.5D-15
13	L(I)	=	1.06 816 D-14	3.3D-15
14	L(II)	=	-4.31 765 D-14	3.2D-15

BAND I

LINE	FREQUENCY (MHZ)	STD.DEV. (MHZ)	VAC.WAVE NO. (CM-1)
P(60)	2707 7607.5077	0.0246	903.2117 6484
P(58)	2714 6404.4578	0.0154	905.5065 8408
P(56)	2721 4396.1809	0.0097	907.7745 4384
P(54)	2728 1588.8741	0.0064	910.0158 5084
P(52)	2734 7988.4259	0.0049	912.2307 0148
P(50)	2741 3600.4235	0.0043	914.4192 8214
P(48)	2747 8430.1601	0.0040	916.5817 6938
P(46)	2754 2482.6413	0.0038	918.7183 3017

BAND II

LINE	FREQUENCY (MHZ)	STD.DEV. (MHZ)	VAC.WAVE NO. (CM-1)
P(60)	3014 3456.0742	0.0172	1005.4774 6515
P(58)	3021 2223.6949	0.0110	1007.7713 0607
P(56)	3028 0322.1201	0.0072	1010.0428 2503
P(54)	3034 7743.7465	0.0051	1012.2917 6841
P(52)	3041 4481.1364	0.0042	1014.5178 8812
P(50)	3048 0527.0251	0.0039	1016.7209 4183
P(48)	3054 5874.3277	0.0038	1018.9006 9322
P(46)	3061 0516.1462	0.0039	1021.0569 1219

23

Line						Line					
P(44)	2760	5762.5914	0.0037	920.8291	2210	P(44)	3067	4445.7759	0.0039	1023.1893	7509
P(42)	2766	8274.4599	0.0036	922.9142	9359	P(42)	3073	7656.7119	0.0039	1025.2978	6496
P(40)	2773	0022.4271	0.0036	924.9739	8407	P(40)	3080	0142.6555	0.0039	1027.3821	7169
P(38)	2779	1010.4094	0.0036	927.0083	3596	P(38)	3086	1897.5199	0.0038	1029.4420	9223
P(36)	2785	1242.0651	0.0035	929.0174	3596	P(36)	3092	2915.4360	0.0038	1031.4774	3083
P(34)	2791	0720.7986	0.0035	931.0014	3295	P(34)	3098	3190.7583	0.0038	1033.4879	9917
P(32)	2796	9449.7656	0.0035	932.9604	2043	P(32)	3104	2718.0700	0.0038	1035.4736	1655
P(30)	2802	7431.8776	0.0035	934.8944	9550	P(30)	3110	1492.1877	0.0037	1037.4341	1009
P(28)	2808	4669.8055	0.0035	936.8037	4726	P(28)	3115	9508.1671	0.0037	1039.3693	1486
P(26)	2814	1165.9839	0.0035	938.6882	5692	P(26)	3121	6761.3064	0.0037	1041.2790	7402
P(24)	2819	6922.6147	0.0036	940.5480	9793	P(24)	3127	3247.1518	0.0038	1043.1632	3901
P(22)	2825	1941.6703	0.0036	942.3833	3608	P(22)	3132	8961.5006	0.0038	1045.0216	6964
P(20)	2830	6224.8967	0.0036	944.1940	2961	P(20)	3138	3900.4054	0.0038	1046.8542	3425
P(18)	2835	9773.8165	0.0036	945.9802	2931	P(18)	3143	8060.1774	0.0037	1048.6608	0978
P(16)	2841	2589.7314	0.0035	947.7419	7860	P(16)	3149	1437.3897	0.0037	1050.4412	8194
P(14)	2846	4673.7246	0.0035	949.4793	1361	P(14)	3154	4028.8804	0.0037	1052.1955	4524
P(12)	2851	6026.6628	0.0035	951.1922	6324	P(12)	3159	5831.7547	0.0037	1053.9235	0313
P(10)	2856	6649.1983	0.0035	952.8808	4927	P(10)	3164	6843.3878	0.0037	1055.6250	6805
P(8)	2861	6541.7701	0.0035	954.5450	8632	P(8)	3169	7061.4264	0.0037	1057.3001	6151
P(6)	2866	5704.6061	0.0036	956.1849	8202	P(6)	3174	6483.7910	0.0037	1058.9487	1415
P(4)	2871	4137.7235	0.0036	957.8005	3691	P(4)	3179	5108.6771	0.0037	1060.5706	6576
P(2)	2876	1840.9300	0.0036	959.3917	4460	P(2)	3184	2934.5560	0.0037	1062.1659	6536
V(0)	2880	8813.8246	0.0036	960.9585	9171	V(0)	3188	2993.1764	0.0037	1063.7345	7121
R(0)	2883	2026.2225	0.0036	961.7328	7396	R(0)	3191	3172.5743	0.0037	1064.5088	5347
R(2)	2887	7902.4412	0.0036	963.2631	3990	R(2)	3195	8996.0672	0.0036	1066.0373	6066
R(4)	2892	3046.4336	0.0036	964.7689	8140	R(4)	3200	4017.3872	0.0036	1067.5391	1025
R(6)	2896	7457.0695	0.0035	966.2503	6076	R(6)	3204	8236.2544	0.0036	1069.0140	9289
R(8)	2901	1133.0097	0.0035	967.7072	3331	R(8)	3209	1652.6660	0.0036	1070.4623	0849
R(10)	2905	4072.7058	0.0035	969.1395	4739	R(10)	3213	4266.8953	0.0036	1071.8837	6618
R(12)	2909	6274.3988	0.0034	970.5472	4435	R(12)	3217	6079.4907	0.0036	1073.2784	8423
R(14)	2913	7736.1185	0.0034	971.9302	5845	R(14)	3221	7091.2743	0.0037	1074.6464	9008
R(16)	2917	8455.6817	0.0033	973.2885	1688	R(16)	3225	7303.3400	0.0037	1075.9878	2021
R(18)	2921	8430.6909	0.0033	974.6219	3025	R(18)	3229	6717.0518	0.0037	1077.3025	2013
R(20)	2925	7658.5324	0.0032	975.9304	3960	R(20)	3233	5334.0411	0.0038	1078.5906	4423
R(22)	2929	6136.3740	0.0032	977.2139	2224	R(22)	3237	3156.2043	0.0039	1079.8522	5580
R(24)	2933	3861.1629	0.0032	978.4722	8575	R(24)	3241	0185.7000	0.0039	1081.0874	2682
R(26)	2937	0829.6231	0.0031	979.7054	2084	R(26)	3244	6424.9456	0.0039	1082.2962	3794
R(28)	2940	7038.2525	0.0031	980.9132	1071	R(28)	3248	1876.6140	0.0040	1083.4623	7831
R(30)	2944	2483.3197	0.0031	982.0955	3089	R(30)	3251	6543.6298	0.0040	1084.6351	4549

R(32)	2947 7160.8609	0.0031	983.2522 4916
R(34)	2951 1066.6762	0.0031	984.3832 2542
R(36)	2954 4196.3256	0.0032	985.4883 1157
R(38)	2957 6545.1250	0.0033	986.5673 5137
R(40)	2960 8108.1417	0.0033	987.6201 8028
R(42)	2963 8880.1900	0.0034	988.6466 2533
R(44)	2966 8855.8259	0.0035	989.6465 0491
R(46)	2969 8029.3420	0.0035	990.6196 2866
R(48)	2972 6394.7621	0.0036	991.5657 9723
R(50)	2975 3945.8353	0.0045	992.4848 0211
R(52)	2978 0676.0297	0.0070	993.3764 2542
R(54)	2980 6578.5263	0.0117	994.2404 3971
R(56)	2983 1646.2123	0.0193	995.0766 0771
R(58)	2985 5871.6737	0.0307	995.8846 8212

R(32)	3255 0429.1653	0.0039	1085.7654 4528
R(34)	3258 3536.6360	0.0039	1086.8697 9163
R(36)	3261 5869.6965	0.0039	1087.9483 0644
R(38)	3264 7432.2354	0.0038	1089.0011 1941
R(40)	3267 8228.3702	0.0038	1090.0283 6790
R(42)	3270 8262.4421	0.0038	1091.0301 9670
R(44)	3273 7539.0104	0.0038	1092.0067 5790
R(46)	3276 6062.8469	0.0039	1092.9582 1067
R(48)	3279 3838.9297	0.0043	1093.8847 2107
R(50)	3282 0872.4368	0.0055	1094.7864 6180
R(52)	3284 7168.7402	0.0081	1095.6636 1206
R(54)	3287 2733.3987	0.0127	1096.5163 5728
R(56)	3289 7572.1515	0.0199	1097.3448 8889
R(58)	3292 1690.9108	0.0305	1098.1494 0411

Table 2.5 $^{16}O^{13}C^{16}O$

NUMBER	SYMBOL	CONSTANTS (MHZ)	STD.DEV. (MHZ)
15	V(001-I) =	2.738 379 258 341 D+07	4.5D-03
16	V(001-II) =	3.050 865 923 183 D+07	4.6D-03
17	B(001) =	1.161 016 490 148 D+04	1.1D-04
18	B(I) =	1.168 344 168 872 D+04	1.2D-04
19	B(II) =	1.171 936 491 647 D+04	1.2D-04
20	D(001) =	3.984 584 753 D-03	1.4D-07
21	D(I) =	3.604 500 429 D-03	1.5D-07
22	D(II) =	4.747 234 294 D-03	1.5D-07
23	H(001) =	0.495 934 D-09	7.2D-11
24	H(I) =	6.338 964 D-09	7.6D-11
25	H(II) =	8.203 342 D-09	7.8D-11
26	L(001) =	-2.29 763 D-14	1.3D-14
27	L(I) =	5.77 901 D-14	1.4D-14
28	L(II) =	-7.93 174 D-14	1.5D-14

BAND I

LINE	FREQUENCY (MHZ)	STD.DEV. (MHZ)	VAC.WAVE NO. (CM-1)
P(60)	2572 0428.2139	0.1461	857.9411 3653
P(58)	2578 4672.4669	0.0893	860.0840 9414
P(56)	2584 8281.2715	0.0506	862.2058 5547
P(54)	2591 1259.2641	0.0254	864.3065 7519
P(52)	2597 3610.8059	0.0104	866.3863 9875
P(50)	2603 5339.9930	0.0046	868.4454 6280
P(48)	2609 6450.6662	0.0060	870.4838 9544
P(46)	2615 6946.4203	0.0071	872.5018 1658

BAND II

LINE	FREQUENCY (MHZ)	STD.DEV. (MHZ)	VAC.WAVE NO. (CM-1)
P(60)	2872 9056.6508	0.2555	958.2981 7876
P(58)	2879 9935.4314	0.1690	960.6624 4039
P(56)	2887 0077.7355	0.1079	963.0021 3581
P(54)	2893 9475.6922	0.0660	965.3170 0248
P(52)	2900 8121.5973	0.0383	967.6067 8340
P(50)	2907 6007.9210	0.0208	969.8712 2741
P(48)	2914 3127.3153	0.0108	972.1100 8942
P(46)	2920 9472.6220	0.0060	974.3231 3064

P(44)	2621 6830.6129	0.0069	874.4993 3824	2927 5036.8795	0.0046	976.5101 1886	
P(42)	2627 6106.3727	0.0062	876.4765 6475	2933 9813.3301	0.0042	978.6708 2867	
P(40)	2633 4776.6070	0.0055	878.4335 9311	2940 3795.4270	0.0040	980.8050 4170	
P(38)	2639 2844.0093	0.0050	880.3705 1317	2946 6976.8409	0.0037	982.9125 4682	
P(36)	2645 0311.0659	0.0048	882.2874 0784	2952 9351.4664	0.0036	984.9931 4037	
P(34)	2650 7180.0624	0.0047	884.1843 5338	2959 0913.4284	0.0035	987.0466 2638	
P(32)	2656 3453.0895	0.0047	886.0614 1951	2965 1657.0881	0.0034	989.0728 1677	
P(30)	2661 9132.0488	0.0046	887.9186 6968	2971 1577.0489	0.0034	991.0715 3152	
P(28)	2667 4218.6576	0.0045	889.7561 6117	2977 0668.1613	0.0035	993.0425 9887	
P(26)	2672 8714.4542	0.0044	891.5739 4527	2982 8925.5288	0.0037	994.9858 5548	
P(24)	2678 2620.8016	0.0045	893.3720 6747	2988 6344.5126	0.0039	996.9011 4661	
P(22)	2683 5938.8924	0.0046	895.1505 6754	2994 2920.7360	0.0043	998.7883 2629	
P(20)	2688 8669.7518	0.0048	896.9094 7969	2999 8650.0890	0.0048	1000.6472 5741	
P(18)	2694 0814.2416	0.0049	898.6488 3264	3005 3528.7321	0.0050	1002.4778 1190	
P(16)	2699 2373.0626	0.0051	900.3686 4979	3010 7553.1003	0.0050	1004.2798 7085	
P(14)	2704 3346.7581	0.0051	902.0689 4925	3016 0719.9062	0.0050	1006.0533 2460	
P(12)	2709 3735.7157	0.0051	903.7497 4396	3021 3026.1432	0.0050	1007.7980 7286	
P(10)	2714 3540.1700	0.0051	905.4110 4173	3026 4469.0880	0.0050	1009.5140 2480	
P(8)	2719 2760.2041	0.0050	907.0528 4534	3031 5046.3034	0.0049	1011.2010 9911	
P(6)	2724 1395.7513	0.0049	908.6751 5257	3036 4755.6398	0.0048	1012.8592 2409	
P(4)	2728 9446.5964	0.0048	910.2779 5624	3041 3595.2374	0.0048	1014.4883 3771	
P(2)	2733 6912.3769	0.0046	911.8612 4425	3046 1563.5271	0.0047	1016.0883 8762	
V(0)	2738 3792.5834	0.0045	913.4249 9962	3050 8659.2318	0.0046	1017.6593 3124	
R(0)	2740 7012.8973	0.0044	914.1995 4592	3053 1879.5457	0.0046	1018.4338 7754	
R(2)	2745 3013.4681	0.0042	915.7339 5979	3057 7664.6183	0.0045	1019.9611 0317	
R(4)	2749 8426.5523	0.0041	917.2487 7723	3062 2575.1933	0.0044	1021.4591 5870	
R(6)	2754 3251.1293	0.0039	918.7439 6418	3066 6611.0178	0.0042	1022.9280 3569	
R(8)	2758 7486.0315	0.0038	920.2194 8169	3070 9772.1308	0.0042	1024.3677 3546	
R(10)	2763 1129.9444	0.0037	921.6752 8592	3075 2058.8624	0.0041	1025.7782 6899	
R(12)	2767 4181.4046	0.0037	923.1113 2806	3079 3471.8321	0.0040	1027.1596 5697	
R(14)	2771 6638.7995	0.0037	924.5275 5431	3083 4011.9476	0.0040	1028.5119 2966	
R(16)	2775 8500.3648	0.0037	925.9239 0583	3087 3680.4026	0.0040	1029.8351 2689	
R(18)	2779 9764.1836	0.0037	927.3003 1866	3091 2478.6741	0.0040	1031.1292 9793	
R(20)	2784 0428.1832	0.0037	928.6567 2369	3095 0408.5204	0.0040	1032.3945 0141	
R(22)	2788 0490.1338	0.0037	929.9930 4651	3098 7471.9774	0.0040	1033.6308 0526	
R(24)	2791 9947.6446	0.0036	931.3092 0741	3102 3671.3557	0.0039	1034.8382 8655	
R(26)	2795 8798.1618	0.0036	932.6051 2117	3105 9009.2365	0.0039	1036.0170 3137	
R(28)	2799 7038.9645	0.0036	933.8806 9704	3109 3488.4681	0.0040	1037.1671 3474	
R(30)	2803 4667.1610	0.0036	935.1358 3858	3112 7112.1611	0.0040	1038.2887 0042	

R(32)	2807 1679.6853	0.0036	936.3704	4349
R(34)	2810 8073.2921	0.0037	937.5844	0354
R(36)	2814 3844.5522	0.0037	938.7776	0434
R(38)	2817 8989.8473	0.0037	939.9499	2520
R(40)	2821 3505.3647	0.0038	941.1012	3893
R(42)	2824 7387.0908	0.0039	942.2314	1167
R(44)	2828 0630.8053	0.0047	943.3403	0262
R(46)	2831 3232.0738	0.0075	944.4277	6388
R(48)	2834 5186.2410	0.0138	945.4936	4017
R(50)	2837 6488.4223	0.0251	946.5377	6855
R(52)	2840 7133.4961	0.0432	947.5599	7818
R(54)	2843 7116.0949	0.0708	948.5600	9002
R(56)	2846 6430.5956	0.1112	949.5379	1651
R(58)	2849 5071.1098	0.1688	950.4932	6124

R(32)	3115 9883.6840	0.0040	1039.3818	4075
R(34)	3119 1806.6581	0.0041	1040.4466	7655
R(36)	3122 2884.9526	0.0041	1041.4833	3687
R(38)	3125 3122.6789	0.0043	1042.4919	5885
R(40)	3128 2524.1847	0.0047	1043.4726	8752
R(42)	3131 1094.0483	0.0054	1044.4256	7559
R(44)	3133 8837.0719	0.0071	1045.3510	8324
R(46)	3136 5758.2755	0.0112	1046.2490	7794
R(48)	3139 1862.8900	0.0193	1047.1198	3415
R(50)	3141 7156.3502	0.0332	1047.9635	3316
R(52)	3144 1644.2875	0.0555	1048.7803	6283
R(54)	3146 5332.5230	0.0892	1049.5705	1732
R(56)	3148 8227.0596	0.1386	1050.3341	9685
R(58)	3151 0334.0742	0.2089	1051.0716	0749

Table 2.6 $^{16}O^{12}C^{18}O$

NUMBER	SYMBOL	CONSTANTS (MHZ)	STD.DEV. (MHZ)
29	v(001-I) =	2.896 801 233 901 D+07	1.0D-02
30	v(001-II) =	3.215 835 064 653 D+07	2.3D-02
31	B(001) =	1.095 102 264 016 D+04	2.8D-04
32	B(I) =	1.104 772 438 281 D+04	3.0D-04
33	B(II) =	1.103 600 443 963 D+04	2.8D-04
34	D(001) =	3.550 909 355 D-03	5.2D-07
35	D(I) =	3.064 795 497 D-03	5.5D-07
36	D(II) =	4.096 110 317 D-03	6.3D-07
37	H(001) =	0.074 060 060 D-09	4.0D-10
38	H(I) =	2.945 673 D-09	4.2D-10
39	H(II) =	4.419 934 D-09	6.2D-10
40	L(001) =	5.56 243 D-14	1.1D-13
41	L(I) =	7.69 066 D-14	1.1D-13
42	L(II) =	64.89 108 D-14	2.1D-13

BAND I

LINE	FREQUENCY (MHZ)	STD.DEV. (MHZ)	VAC.WAVE NO. (CM-1)
P(60)	2729 6371.2432	2.7312	910.5089 3759
P(59)	2733 0160.6851	2.2947	911.6360 3206
P(58)	2736 3739.9515	1.9182	912.7561 1582
P(57)	2739 7109.7225	1.5948	913.8692 1156
P(56)	2743 0270.6628	1.3182	914.9753 4147
P(55)	2746 3223.4212	1.0828	916.0745 2717
P(54)	2749 5968.6313	0.8834	917.1667 8981
P(53)	2752 8506.9114	0.7153	918.2521 5000

BAND II

LINE	FREQUENCY (MHZ)	STD.DEV. (MHZ)	VAC.WAVE NO. (CM-1)
P(60)	3054 3244.7021	15.2104	1018.8129 7835
P(59)	3057 4756.3872	12.8533	1019.8640 9502
P(58)	3060 6123.1746	10.8119	1020.9103 7843
P(57)	3063 7344.4579	9.0504	1021.9518 0834
P(56)	3066 8419.6268	7.5364	1022.9883 6440
P(55)	3069 9348.0678	6.2405	1024.0200 2614
P(54)	3073 0129.1655	5.1363	1025.0467 7304
P(53)	3076 0762.3030	4.1999	1026.0685 8452

P(52)	2756	0838.8646	0.5746	919.3306	2788
P(51)	2759	2965.0793	0.4575	920.4022	4305
P(50)	2762	4886.1287	0.3606	921.4670	1465
P(49)	2765	6602.5715	0.2812	922.5249	6130
P(48)	2768	8114.9518	0.2165	923.5761	0116
P(47)	2771	9423.7991	0.1644	924.6204	5190
P(46)	2775	0529.6287	0.1229	925.6580	3069
P(45)	2778	1432.9417	0.0901	926.6888	5425
P(44)	2781	2134.2250	0.0647	927.7129	3883
P(43)	2784	2633.9515	0.0453	928.7303	0020
P(42)	2787	2932.5802	0.0308	929.7409	5366
P(41)	2790	3030.5565	0.0203	930.7449	1409
P(40)	2793	2928.3119	0.0130	931.7421	9586
P(39)	2796	2626.2643	0.0084	932.7328	1292
P(38)	2799	2124.8184	0.0059	933.7167	7877
P(37)	2802	1424.3652	0.0050	934.6941	0645
P(36)	2805	0525.2826	0.0048	935.6648	0857
P(35)	2807	9427.9351	0.0047	936.6288	9729
P(34)	2810	8132.6743	0.0045	937.5863	8432
P(33)	2813	6639.8385	0.0044	938.5372	8096
P(32)	2816	4949.7533	0.0043	939.4815	9808
P(31)	2819	3062.7312	0.0042	940.4193	4608
P(30)	2822	0979.0720	0.0041	941.3505	3498
P(29)	2824	8699.0627	0.0041	942.2751	7434
P(28)	2827	6222.9778	0.0041	943.1932	7332
P(27)	2830	3551.0828	0.0041	944.1048	4065
P(26)	2833	0683.6152	0.0041	945.0098	8464
P(25)	2835	7620.8235	0.0040	945.9084	1320
P(24)	2838	4362.9282	0.0040	946.8004	3379
P(23)	2841	0910.1411	0.0040	947.6859	5350
P(22)	2843	7262.6619	0.0039	948.5649	7897
P(21)	2846	3420.6781	0.0039	949.4375	1647
P(20)	2848	9384.3647	0.0040	950.3035	7184
P(19)	2851	5153.8849	0.0040	951.1631	5050
P(18)	2854	0729.3895	0.0039	952.0162	5751
P(17)	2856	6111.0174	0.0039	952.8628	9749
P(16)	2859	1298.8955	0.0039	953.7030	7466
P(15)	2861	6293.1385	0.0039	954.5367	9287
P(14)	2864	1093.8494	0.0039	955.3640	5553

Right block:

P(52)	3079	1246.8631	3.4099	1027.0854	3999
P(51)	3082	1582.2290	2.7471	1028.0973	1888
P(50)	3085	1767.7849	2.1943	1029.1042	0064
P(49)	3088	1802.9170	1.7363	1030.1060	6481
P(48)	3091	1687.0141	1.3596	1031.1028	9099
P(47)	3094	1419.4680	1.0521	1032.0946	5890
P(46)	3097	0999.6746	0.8032	1033.0813	4838
P(45)	3100	0427.0340	0.6039	1034.0629	3943
P(44)	3102	9700.9514	0.4459	1035.0394	1221
P(43)	3105	8820.8376	0.3222	1036.0107	4706
P(42)	3108	7786.1094	0.2268	1036.9769	2453
P(41)	3111	6596.1900	0.1545	1037.9379	2538
P(40)	3114	5250.5098	0.1008	1038.8937	3060
P(39)	3117	3748.5064	0.0620	1039.8443	2145
P(38)	3120	2089.6253	0.0350	1040.7896	7942
P(37)	3123	0273.3202	0.0177	1041.7297	8628
P(36)	3125	8299.0533	0.0093	1042.6646	2411
P(35)	3128	6166.2956	0.0090	1043.5941	7526
P(34)	3131	3874.5277	0.0104	1044.5184	2240
P(33)	3134	1423.2391	0.0107	1045.4373	4850
P(32)	3136	8811.9297	0.0101	1046.3509	3688
P(31)	3139	6040.1090	0.0088	1047.2591	7118
P(30)	3142	3107.2971	0.0075	1048.1620	3539
P(29)	3145	0013.0245	0.0065	1049.0595	1385
P(28)	3147	6756.8324	0.0058	1049.9515	9126
P(27)	3150	3338.2731	0.0055	1050.8382	5268
P(26)	3152	9756.9100	0.0054	1051.7194	8355
P(25)	3155	6012.3177	0.0053	1052.5952	6968
P(24)	3158	2104.0824	0.0052	1053.4655	9727
P(23)	3160	8031.8019	0.0050	1054.3304	5290
P(22)	3163	3795.0859	0.0049	1055.1898	2355
P(21)	3165	9393.5558	0.0048	1056.0436	9660
P(20)	3168	4826.8453	0.0048	1056.8920	5981
P(19)	3171	0094.5998	0.0050	1057.7349	0138
P(18)	3173	5196.4775	0.0052	1058.5722	0989
P(17)	3176	0132.1485	0.0056	1059.4039	7435
P(16)	3178	4901.2955	0.0060	1060.2301	8416
P(15)	3180	9503.6138	0.0065	1061.0508	2917
P(14)	3183	3938.8112	0.0071	1061.8658	9961

P(13)	2866	5701.1190	0.0040	956.1848	6570
P(12)	2869	0115.0266	0.0041	956.9992	2600
P(11)	2871	4335.6392	0.0044	957.8071	3867
P(10)	2873	6363.0123	0.0047	958.6086	0557
P(9)	2876	2197.1895	0.0052	959.4036	2814
P(8)	2878	5838.2025	0.0057	960.1922	0745
P(7)	2880	9286.0713	0.0063	960.9743	4417
P(6)	2883	2540.8044	0.0069	961.7500	3857
P(5)	2885	5602.3982	0.0075	962.5192	9054
P(4)	2887	8470.8376	0.0081	963.2820	9957
P(3)	2890	1146.0958	0.0087	964.0384	6476
P(2)	2892	3628.1341	0.0093	964.7883	8484
P(1)	2894	5916.9025	0.0097	965.5318	5813
V(0)	2896	8012.3390	0.0101	966.2688	8256
R(0)	2898	9914.3701	0.0104	966.9994	5567
R(1)	2901	1622.9105	0.0106	967.7235	7464
R(2)	2903	3137.8634	0.0108	968.4412	3622
R(3)	2905	4459.1202	0.0108	969.1524	3679
R(4)	2907	5586.5606	0.0108	969.8571	7235
R(5)	2909	6520.0529	0.0107	970.5554	3848
R(6)	2911	7259.4532	0.0107	971.2472	3042
R(7)	2913	7804.6065	0.0105	971.9325	4296
R(8)	2915	8155.3456	0.0103	972.6113	7055
R(9)	2917	8311.4918	0.0100	973.2837	0722
R(10)	2919	8272.8546	0.0096	973.9495	4661
R(11)	2921	8039.2319	0.0093	974.6088	8199
R(12)	2923	7610.4096	0.0089	975.2617	0620
R(13)	2925	6986.1619	0.0084	975.9080	1173
R(14)	2927	6166.2511	0.0080	976.5477	9064
R(15)	2929	5150.4278	0.0075	977.1810	3461
R(16)	2931	3938.4306	0.0071	977.8077	3493
R(17)	2933	2529.9861	0.0066	978.4278	8247
R(18)	2935	0924.8091	0.0062	979.0414	6772
R(19)	2936	9122.6024	0.0057	979.6484	8076
R(20)	2938	7123.0568	0.0053	980.2489	1129
R(21)	2940	4925.8509	0.0050	980.8427	4858
R(22)	2942	2530.6513	0.0047	981.4299	8152
R(23)	2943	9937.1125	0.0044	982.0105	9856
R(24)	2945	7144.8766	0.0041	982.5845	8779

P(13)	3185	8206.6080	0.0079	1062.6753	8618
P(12)	3188	2306.7374	0.0090	1063.4792	7997
P(11)	3190	6238.9453	0.0102	1064.2775	7250
P(10)	3193	0002.9904	0.0116	1065.0702	5572
P(9)	3195	3598.6444	0.0132	1065.8573	2201
P(8)	3197	7025.6917	0.0148	1066.6387	6420
P(7)	3200	0283.9296	0.0164	1067.4145	7551
P(6)	3202	3373.1687	0.0180	1068.1847	4962
P(5)	3204	6293.2322	0.0195	1068.9492	8064
P(4)	3206	9043.9565	0.0208	1069.7081	6312
P(3)	3209	1625.1911	0.0218	1070.4613	9203
P(2)	3211	4036.7984	0.0226	1071.2089	6278
P(1)	3213	6278.6540	0.0230	1071.9508	7123
V(0)	3215	8350.6465	0.0231	1072.6871	1365
R(0)	3218	0252.6776	0.0228	1073.4176	8677
R(1)	3220	1984.6620	0.0223	1074.1425	8774
R(2)	3222	3546.5277	0.0213	1074.8618	1416
R(3)	3224	4938.2155	0.0201	1075.5753	6406
R(4)	3226	6159.6795	0.0186	1076.2832	3590
R(5)	3228	7210.8868	0.0169	1076.9854	2859
R(6)	3230	8091.8175	0.0150	1077.6819	4147
R(7)	3232	8802.4647	0.0131	1078.3727	7430
R(8)	3234	9342.8347	0.0112	1079.0579	2729
R(9)	3236	9712.9467	0.0094	1079.7374	0109
R(10)	3238	9912.8328	0.0078	1080.4111	9676
R(11)	3240	9942.5380	0.0065	1081.0793	1581
R(12)	3242	9802.1204	0.0056	1081.7417	6018
R(13)	3244	9491.6508	0.0051	1082.3985	3222
R(14)	3246	9011.2129	0.0050	1083.0496	3472
R(15)	3248	8360.9032	0.0051	1083.6950	7091
R(16)	3250	7540.8306	0.0052	1084.3348	4443
R(17)	3252	6551.1172	0.0052	1084.9692	5933
R(18)	3254	5391.8971	0.0052	1085.5974	2010
R(19)	3256	4063.3174	0.0051	1086.2202	3164
R(20)	3258	2565.5374	0.0050	1086.8373	9927
R(21)	3260	0898.7287	0.0049	1087.4498	2871
R(22)	3261	9063.0753	0.0050	1088.0548	2609
R(23)	3263	7058.7733	0.0051	1088.6550	9797
R(24)	3265	4886.0308	0.0053	1089.2497	5127

R(25)	2947 4153.5738	0.0041	983.1519 3686
R(26)	2949 0962.8216	0.0040	983.7126 3301
R(27)	2950 7572.2255	0.0040	984.2666 6309
R(28)	2952 3981.3785	0.0039	984.8140 1352
R(29)	2954 0189.8609	0.0039	985.3546 7029
R(30)	2955 6197.2409	0.0039	985.8886 1901
R(31)	2957 2003.0737	0.0040	986.4158 4485
R(32)	2958 7606.9022	0.0042	986.9363 3254
R(33)	2960 3008.2564	0.0044	987.4500 6642
R(34)	2961 8206.6535	0.0047	987.9572 0788
R(35)	2963 3201.5979	0.0049	988.4572 3038
R(36)	2964 7992.5812	0.0052	988.9505 8198
R(37)	2966 2579.0817	0.0058	989.4371 3526
R(38)	2967 6960.5648	0.0074	989.9168 4990
R(39)	2969 1136.4828	0.0109	990.3897 0763
R(40)	2970 5106.2746	0.0168	990.8556 8972
R(41)	2971 8869.3657	0.0256	991.3147 7703
R(42)	2973 2425.1684	0.0381	991.7669 4994
R(43)	2974 5773.0814	0.0549	992.2121 8839
R(44)	2975 8912.4896	0.0772	992.6504 7187
R(45)	2977 1842.7644	0.1060	993.0817 7941
R(46)	2978 4563.2633	0.1428	993.5060 8958
R(47)	2979 7073.3299	0.1892	993.9233 8048
R(48)	2980 9372.2936	0.2469	994.3336 2975
R(49)	2982 1459.4700	0.3181	994.7368 1456
R(50)	2983 3334.1601	0.4052	995.1329 1159
R(51)	2984 4995.6509	0.5108	995.5218 9705
R(52)	2985 6443.2145	0.6381	995.9037 4667
R(53)	2986 7676.1087	0.7903	996.2784 3569
R(54)	2987 8693.5765	0.9714	996.6459 3886
R(55)	2988 9494.8460	1.1856	997.0062 3042
R(56)	2990 0079.1304	1.4377	997.3592 8415
R(57)	2991 0445.6275	1.7331	997.7050 7327
R(58)	2992 0593.5203	2.0776	998.0435 7054
R(59)	2993 0521.9760	2.4777	998.3747 4817

R(25)	3267 2545.0679	0.0055	1089.8387 9334
R(26)	3269 0036.1164	0.0056	1090.4222 3192
R(27)	3270 7359.4198	0.0057	1091.0000 7512
R(28)	3272 4515.2331	0.0061	1091.5723 3145
R(29)	3274 1503.8227	0.0068	1092.1390 0980
R(30)	3275 8325.4660	0.0079	1092.7001 1943
R(31)	3277 4980.4516	0.0093	1093.2556 6995
R(32)	3279 1469.0786	0.0106	1093.8056 7134
R(33)	3280 7791.6570	0.0114	1094.3501 3395
R(34)	3282 3948.5069	0.0113	1094.8890 6845
R(35)	3283 5979.9585	0.0101	1095.4224 8586
R(36)	3285 5766.3518	0.0106	1095.9503 9752
R(37)	3287 1428.0366	0.0187	1096.4728 1509
R(38)	3288 6925.3717	0.0360	1096.9897 5055
R(39)	3290 2258.7249	0.0634	1097.5072 1615
R(40)	3291 7428.4725	0.1029	1098.0072 2447
R(41)	3293 2434.9992	0.1577	1098.5077 8832
R(42)	3294 7278.6976	0.2316	1099.0029 2080
R(43)	3296 1959.9675	0.3291	1099.4926 3525
R(44)	3297 6479.2160	0.4554	1099.9769 4525
R(45)	3299 0836.8567	0.6169	1100.4558 6459
R(46)	3300 5033.3092	0.8207	1100.9294 0728
R(47)	3301 9068.9988	1.0750	1101.3975 8749
R(48)	3303 2944.3560	1.3893	1101.8604 1958
R(49)	3304 6659.8155	1.7743	1102.3179 1807
R(50)	3306 0215.8163	2.2423	1102.7700 9758
R(51)	3307 3612.8006	2.8069	1103.2169 7288
R(52)	3308 6851.2130	3.4839	1103.6585 5878
R(53)	3309 9931.5003	4.2906	1104.0948 7020
R(54)	3311 2854.1107	5.2465	1104.5259 2209
R(55)	3312 5619.4927	6.3736	1104.9517 2939
R(56)	3313 8228.0944	7.6959	1105.3723 0708
R(57)	3315 0680.3629	9.2405	1105.7876 7005
R(58)	3316 2976.7434	11.0373	1106.1978 3315
R(59)	3317 5117.6781	13.1190	1106.6028 1114

Table 2.7 $^{18}O\ ^{12}C\ ^{18}O$

NUMBER	SYMBOL	CONSTANTS (MHZ)	STD.DEV. (MHZ)
43	V(001-I) =	2.898 859 706 882 D+07	3.6D-03
44	V(001-II) =	3.248 919 295 228 D+07	3.9D-03
45	B(001) =	1.031 555 954 654 D+04	3.7D-05
46	B(I) =	1.041 489 423 454 D+04	3.4D-05
47	B(II) =	1.038 852 773 874 D+04	3.6D-05
48	D(001) =	3.150 554 563 D-03	5.9D-08
49	D(I) =	2.768 029 928 D-03	5.4D-08
50	D(II) =	3.518 236 015 D-03	5.8D-08
51	H(001) =	0.267 825 D-09	3.9D-11
52	H(I) =	2.501 922 D-09	3.5D-11
53	H(II) =	4.744 762 D-09	3.7D-11
54	L(001) =	0.52 648 D-14	9.2D-15
55	L(I) =	-0.28 458 D-14	7.8D-15
56	L(II) =	-3.21 944 D-14	8.3D-15

BAND I

LINE	FREQUENCY (MHZ)	STD.DEV. (MHZ)	VAC.WAVE NO. (CM-1)
P(60)	2738 4653.4520	0.0476	913.4537 1510
P(58)	2744 9958.7473	0.0308	915.6320 6528
P(56)	2751 4420.9601	0.0193	917.7822 9325
P(54)	2757 8044.1062	0.0117	919.9045 3296
P(52)	2764 0832.0154	0.0071	921.9989 1218
P(50)	2770 2788.3340	0.0047	924.0655 5251
P(48)	2776 3916.5272	0.0038	926.1045 6955
P(46)	2782 4219.8818	0.0036	928.1160 7295

BAND II

LINE	FREQUENCY (MHZ)	STD.DEV. (MHZ)	VAC.WAVE NO. (CM-1)
P(60)	3099 1695.4754	0.0052	1033.7716 8599
P(58)	3104 9479.9052	0.0041	1035.6991 6710
P(56)	3110 6752.5492	0.0039	1037.6095 7686
P(54)	3116 3509.5006	0.0038	1039.5027 8498
P(52)	3121 9746.9218	0.0038	1041.3786 6343
P(50)	3127 5461.0484	0.0038	1043.2370 8665
P(48)	3133 0648.1932	0.0038	1045.0779 3166
P(46)	3138 5304.7512	0.0038	1046.9010 7818

Right block:

P(44)	3143 9427.2028	0.0038	1048.7064	0885
P(42)	3149 3012.1187	0.0038	1050.4938	0924
P(40)	3154 6056.1633	0.0038	1052.2631	6812
P(38)	3159 8556.0987	0.0037	1054.0143	7746
P(36)	3165 0508.7885	0.0037	1055.7473	3266
P(34)	3170 1911.2012	0.0037	1057.4619	3259
P(32)	3175 2760.4139	0.0037	1059.1580	7975
P(30)	3180 3053.6153	0.0037	1060.8356	8037
P(28)	3185 2788.1092	0.0037	1062.4946	4452
P(26)	3190 1961.3173	0.0037	1064.1348	8618
P(24)	3195 0570.7820	0.0037	1065.7563	2339
P(22)	3199 8614.1691	0.0037	1067.3588	7829
P(20)	3204 6089.2707	0.0037	1068.9424	7722
P(18)	3209 2994.0069	0.0037	1070.5070	5081
P(16)	3213 9326.4282	0.0037	1072.0525	3403
P(14)	3218 5084.7176	0.0037	1073.5788	6627
P(12)	3223 0267.1923	0.0037	1075.0859	9140
P(10)	3227 4872.3053	0.0038	1076.5738	5781
P(8)	3231 8898.6464	0.0038	1078.0424	1848
P(6)	3236 2344.9440	0.0038	1079.4916	3097
P(4)	3240 5210.0656	0.0039	1080.9214	5753
P(2)	3244 7493.0190	0.0039	1082.3318	6503
V(0)	3248 9192.9523	0.0039	1083.7228	2508
R(0)	3250 9824.0588	0.0039	1084.4110	0472
R(2)	3255 0648.1734	0.0039	1085.7727	5061
R(4)	3259 0887.7557	0.0039	1087.1149	9859
R(6)	3263 0542.4476	0.0039	1088.4377	3674
R(8)	3266 9612.0317	0.0039	1089.7409	5778
R(10)	3270 8096.4309	0.0039	1091.0246	5916
R(12)	3274 5995.7070	0.0039	1092.2888	4294
R(14)	3278 3310.0604	0.0039	1093.5335	1579
R(16)	3282 0039.8288	0.0039	1094.7586	8899
R(18)	3285 6185.4856	0.0039	1095.9643	7832
R(20)	3289 1747.6385	0.0040	1097.1506	0405
R(22)	3292 6727.0276	0.0040	1098.3173	9088
R(24)	3296 1124.5237	0.0040	1099.4647	6785
R(26)	3299 4941.1261	0.0040	1100.5927	6828
R(28)	3302 8177.9601	0.0040	1101.7014	2973
R(30)	3306 0836.2746	0.0040	1102.7907	9384

Left block:

P(44)	2788 3701.5085	0.0036	930.1001	6644
P(42)	2794 2364.3439	0.0035	932.0569	4801
P(40)	2800 0211.1532	0.0035	933.9865	0987
P(38)	2805 7244.5316	0.0036	935.8889	3859
P(36)	2811 3466.9070	0.0036	937.7643	1517
P(34)	2816 8880.5416	0.0036	939.6127	1506
P(32)	2822 3487.5337	0.0036	941.4342	0825
P(30)	2827 7289.8196	0.0036	943.2288	5923
P(28)	2833 0289.1753	0.0036	944.9967	2755
P(26)	2838 2487.2180	0.0036	946.7378	6683
P(24)	2843 3885.4075	0.0035	948.4523	2589
P(22)	2848 4485.0477	0.0035	950.1403	4821
P(20)	2853 4287.2879	0.0035	951.8013	7213
P(18)	2858 3293.1241	0.0035	953.4360	3087
P(16)	2863 1503.3996	0.0035	955.0441	5256
P(14)	2867 8918.8066	0.0036	956.6257	6030
P(12)	2872 5539.8870	0.0036	958.1808	7215
P(10)	2877 1367.0327	0.0036	959.7095	0119
P(8)	2881 6400.4870	0.0036	961.2116	5553
P(6)	2886 0640.3445	0.0036	962.6873	3834
P(4)	2890 4086.5522	0.0036	964.1365	4783
	2898 6738.9096	0.0036	965.5592	7733
V(0)	2898 8597.0688	0.0036	966.9555	1523
R(0)	2900 9228.1753	0.0035	967.6436	9487
R(2)	2904 9894.0639	0.0035	969.0001	6290
R(4)	2908 9764.2422	0.0034	970.3300	8890
R(6)	2912 8837.8481	0.0034	971.6334	4410
R(.8)	2916 7113.8724	0.0033	972.9101	9484
R(10)	2920 4591.1584	0.0033	974.1603	0254
R(12)	2924 1268.4016	0.0032	975.3837	2368
R(14)	2927 7144.1495	0.0032	976.5837	0982
R(16)	2931 2216.8003	0.0032	977.7503	0752
R(18)	2934 6484.6028	0.0032	978.8933	5838
R(20)	2937 9945.6557	0.0031	980.0094	9896
R(22)	2941 2597.9062	0.0031	981.0986	6080
R(24)	2944 4439.1493	0.0031	982.1607	7034
R(26)	2947 5467.4893	0.0031	983.1957	4893
R(28)	2950 5679.0262	0.0032	984.2035	1276
R(30)	2953 5072.4789	0.0032	985.1839	7280

R(32)	2956 3644.5594	0.0032	986.1370 3482	
R(34)	2959 1392.2837	0.0033	987.0625 9928	
R(36)	2961 8312.5075	0.0034	987.9605 6129	
R(38)	2964 4401.9249	0.0035	988.8308 1058	
R(40)	2966 9657.0662	0.0038	989.6732 3141	
R(42)	2969 4074.2966	0.0045	990.4877 0255	
R(44)	2971 7649.8141	0.0063	991.2740 9717	
R(46)	2974 0379.6472	0.0100	992.0322 8279	
R(48)	2976 2259.6531	0.0162	992.7621 2122	
R(50)	2978 3285.5157	0.0260	993.4634 6851	
R(52)	2980 3452.7430	0.0404	994.1361 7480	
R(54)	2982 2756.6650	0.0610	994.7800 8433	
R(56)	2984 1192.4311	0.0898	995.3950 3529	
R(58)	2985 8755.0078	0.1292	995.9808 5979	

R(32)	3309 2917.4396	0.0040	1103.8609 0632	
R(34)	3312 4422.9433	0.0041	1104.9118 1680	
R(36)	3315 5354.3890	0.0041	1105.9435 7877	
R(38)	3318 5713.4920	0.0042	1106.9562 4945	
R(40)	3321 5502.0763	0.0044	1107.9498 8966	
R(42)	3324 4722.0714	0.0051	1108.9245 6379	
R(44)	3327 3375.5084	0.0069	1109.8803 3957	
R(46)	3330 1464.5166	0.0104	1110.8172 8802	
R(48)	3332 8991.3192	0.0163	1111.7354 8333	
R(50)	3335 5958.2301	0.0252	1112.6350 0265	
R(52)	3338 2367.6494	0.0382	1113.5159 2605	
R(54)	3340 8222.0593	0.0564	1114.3783 3634	
R(56)	3343 3524.0201	0.0812	1115.2223 1891	
R(58)	3345 8276.1658	0.1145	1116.0479 6161	

Table 2.8 $^{18}O^{13}C^{18}O$

NUMBER	SYMBOL	CONSTANTS (MHZ)	STD.DEV. (MHZ)
57	V(001-I) =	2.783 855 114 188 D+07	3.8D-03
58	V(001-II) =	3.078 588 436 561 D+07	4.1D-03
59	B(001) =	1.031 909 579 289 D+04	4.3D-05
60	B(I) =	1.040 347 357 162 D+04	4.5D-05
61	B(II) =	1.039 898 242 205 D+04	4.2D-05
62	D(001) =	3.148 155 581 D-03	7.2D-08
63	D(I) =	2.717 913 129 D-03	7.5D-08
64	D(II) =	3.657 050 669 D-03	6.7D-08
65	H(001) =	0.317 872 D-09	3.9D-11
66	H(I) =	3.111 016 D-09	4.2D-11
67	H(II) =	5.173 276 D-09	3.6D-11
68	L(001) =	-0.13 805 D-14	6.7D-15
69	L(I) =	0.34 339 D-14	7.5D-15
70	L(II) =	-1.95 866 D-14	6.0D-15

BAND I

LINE	FREQUENCY (MHZ)	STD.DEV. (MHZ)	VAC.WAVE NO. (CM-1)
P(60)	2628 8254.3691	0.1107	876.8817 7830
P(58)	2635 0100.2532	0.0715	878.9447 3493
P(56)	2641 1213.9553	0.0444	880.9832 6861
P(54)	2647 1600.1418	0.0263	882.9975 3498
P(52)	2653 1263.2613	0.0148	884.9876 8242
P(50)	2659 0207.5479	0.0083	886.9538 5218
P(48)	2664 8437.0249	0.0053	888.8961 7847
P(46)	2670 5955.5086	0.0046	890.8147 8856

BAND II

LINE	FREQUENCY (MHZ)	STD.DEV. (MHZ)	VAC.WAVE NO. (CM-1)
P(60)	2926 4508.5336	0.0293	976.1589 3104
P(58)	2932 3720.0883	0.0169	978.1340 1591
P(56)	2938 2386.9924	0.0092	980.0909 3319
P(54)	2944 0503.8821	0.0051	982.0295 0396
P(52)	2949 8065.5045	0.0038	983.9495 5301
P(50)	2955 5066.7231	0.0036	985.8509 0900
P(48)	2961 1502.5235	0.0035	987.7334 0467
P(46)	2966 7368.0181	0.0035	989.5968 7699

Line				Line			
P(44)	2676 2766.6111	0.0045	892.7098 0296	P(44)	991.4411 6733	2972 2658.4516	0.0036
P(42)	2681 8873.7438	0.0046	894.5813 3546	P(42)	993.2661 2164	2977 7369.2056	0.0037
P(40)	2687 4280.1203	0.0048	896.4294 9324	P(40)	995.0715 9062	2983 1495.8037	0.0040
P(38)	2692 8988.7594	0.0050	898.2543 7701	P(38)	996.8574 2980	2988 5033.9156	0.0043
P(36)	2698 3002.4878	0.0051	900.0546 8106	P(36)	998.6234 9979	2993 7979.3620	0.0045
P(34)	2703 6323.9427	0.0052	901.8346 9334	P(34)	1000.3696 6634	2999 0328.1180	0.0047
P(32)	2708 8955.5742	0.0052	903.5902 9560	P(32)	1002.0958 0049	3004 2076.3181	0.0048
P(30)	2714 0899.6478	0.0051	905.3229 6339	P(30)	1003.8017 7873	3009 3220.2591	0.0049
P(28)	2719 2158.2464	0.0049	907.0327 6620	P(28)	1005.4874 8308	3014 3756.4042	0.0049
P(26)	2724 2733.2727	0.0048	908.7197 6748	P(26)	1007.1528 0123	3019 3681.3863	0.0048
P(24)	2729 2626.4506	0.0046	910.3840 2476	P(24)	1008.7976 2663	3024 2992.0111	0.0047
P(22)	2734 1839.3276	0.0044	912.0255 8964	P(22)	1010.4218 5859	3029 1685.2604	0.0046
P(20)	2739 0373.2761	0.0042	913.6445 0790	P(20)	1012.0254 0240	3033 9758.2945	0.0044
P(18)	2743 8229.4949	0.0041	915.2408 1953	P(18)	1013.6081 6939	3038 7208.4549	0.0043
P(16)	2748 5409.0108	0.0039	916.8145 5878	P(16)	1015.1700 7699	3043 4033.2670	0.0042
P(14)	2753 1912.6797	0.0039	918.3657 5421	P(14)	1016.7110 4887	3048 0230.4418	0.0041
P(12)	2757 7741.1880	0.0038	919.8944 2870	P(12)	1018.2310 1494	3052 5797.8781	0.0041
P(10)	2762 2895.0531	0.0038	921.4005 9952	P(10)	1019.7299 1142	3057 0733.6641	0.0040
P(8)	2766 7374.6251	0.0037	922.8842 7833	P(8)	1021.2076 8092	3061 5036.0793	0.0040
P(6)	2771 1180.0865	0.0037	924.3454 7124	P(6)	1022.6642 7246	3065 8703.5950	0.0040
P(4)	2775 4311.4538	0.0038	925.7841 7879	P(4)	1024.0996 4150	3070 1734.8761	0.0040
P(2)	2779 6768.5773	0.0038	927.2003 9599	P(2)	1025.5137 4997	3074 4128.7817	0.0041
V(0)	2783 8551.1419	0.0038	928.5941 1233	V(0)	1026.9065 6633	3078 5884.3656	0.0041
R(0)	2785 9189.3209	0.0038	929.2825 2788	R(0)	1027.5949 8188	3080 6522.5446	0.0041
R(2)	2789 9959.0945	0.0038	930.6424 6114	R(2)	1030.2953 0584	3084 7319.2989	0.0040
R(4)	2794 0052.7980	0.0037	931.9798 4314	R(4)	1031.6134 4527	3088 7476.2204	0.0040
R(6)	2797 9469.5379	0.0037	933.2946 4405	R(6)	1032.9102 3115	3092 6993.0464	0.0039
R(8)	2801 8208.2548	0.0036	934.5868 2856	R(8)	1034.1856 6769	3096 5869.7090	0.0039
R(10)	2805 6267.7235	0.0036	935.8563 5578	R(10)	1035.4397 6556	3100 4106.3345	0.0039
R(12)	2809 3646.5527	0.0036	937.1031 7932	R(12)	1036.6725 4184	3104 1703.2428	0.0039
R(14)	2813 0343.1839	0.0036	938.3272 4718	R(14)	1037.8840 1999	3107 8980.1471	0.0039
R(16)	2816 6355.8909	0.0037	939.5285 0178	R(16)	1039.0742 2978	3111 0661.7389	0.0039
R(18)	2820 1682.7789	0.0037	940.7068 7992	R(18)	1040.2432 0726	3115 5706.8021	0.0039
R(20)	2823 6321.7838	0.0037	941.8623 1275	R(20)	1041.3909 9467	3118 0116.6032	0.0040
R(22)	2827 0270.6704	0.0037	942.9947 2572	R(22)	1042.5176 4040	3122 3892.5924	0.0040
R(24)	2830 3527.0319	0.0037	944.1040 3853	R(24)	1043.6231 9888	3125 7036.4017	0.0040
R(26)	2833 6088.2880	0.0037	945.1901 6513	R(26)	1044.7077 3049	3128 9549.8415	0.0040
R(28)	2836 7951.6837	0.0037	946.2530 1360	R(28)	1045.7713 0151	3131 1434.8987	0.0040
R(30)	2839 9114.2873	0.0037	947.2924 8617	R(30)		3135	

R(32)	2842 9572.9890	0.0037	948.3084 7909
R(34)	2845 9324.4986	0.0037	949.3008 8263
R(36)	2848 8365.3438	0.0038	950.2695 8096
R(38)	2851 6691.8679	0.0038	951.2144 5210
R(40)	2854 4300.2272	0.0038	952.1353 6783
R(42)	2857 1186.3889	0.0039	953.0321 9366
R(44)	2859 7346.1285	0.0039	953.9047 8864
R(46)	2862 2775.0267	0.0040	954.7530 0539
R(48)	2864 7468.4669	0.0052	955.5766 8989
R(50)	2867 1421.6320	0.0090	956.3756 8147
R(52)	2869 4629.5011	0.0166	957.1498 1266
R(54)	2871 7086.8467	0.0292	957.8989 0907
R(56)	2873 8788.2305	0.0486	958.6227 8932
R(58)	2875 9728.0006	0.0773	959.3212 6487

R(32)	3138 2693.7329	0.0039	1046.8139 8399
R(34)	3141 3328.6739	0.0039	1047.8358 5563
R(36)	3144 3342.2180	0.0039	1048.8369 9969
R(38)	3147 2737.0241	0.0038	1049.8175 0489
R(40)	3150 1515.9106	0.0039	1050.7774 6521
R(42)	3152 9681.8507	0.0039	1051.7169 7984
R(44)	3155 7237.9690	0.0039	1052.6361 5301
R(46)	3158 4187.5362	0.0041	1053.5350 9381
R(48)	3161 0533.9655	0.0045	1054.4139 1609
R(50)	3163 6280.8072	0.0061	1055.2727 3829
R(52)	3166 1431.7443	0.0100	1056.1116 8325
R(54)	3168 5990.5869	0.0172	1056.9308 7806
R(56)	3170 9961.2676	0.0288	1057.7304 5390
R(58)	3173 3347.8356	0.0462	1058.5105 4584

16 14 16
O C O

NUMBER	SYMBOL		CONSTANTS (MHZ)	STD.DEV. (MHZ)
71	ν(001-I)	=	2.596 591 761 827 D+07	1.3D-02
72	ν(001-II)	=	2.946 000 239 110 D+07	4.9D-03
73	B(001)	=	1.161 366 771 367 D+04	1.4D-04
74	B(I)	=	1.167 472 474 237 D+04	1.6D-04
75	B(III)	=	1.172 705 299 921 D+04	1.5D-04
76	D(001)	=	3.979 815 697 D-03	3.6D-07
77	D(I)	=	3.776 614 494 D-03	4.2D-07
78	D(II)	=	4.821 389 232 D-03	4.0D-07
79	H(001)	=	-0.922 033 D-09	3.4D-10
80	H(I)	=	2.658 350 D-09	3.6D-10
81	H(II)	=	11.449 041 D-09	4.1D-10
82	L(001)	=	37.29 216 D-14	1.1D-13
83	L(I)	=	41.84 717 D-14	1.0D-13
84	L(II)	=	45.06 891 D-14	1.4D-13

BAND I

LINE	FREQUENCY (MHZ)	STD.DEV. (MHZ)	VAC.WAVE NO. (CM-1)
P(60)	2434 9337.5625	1.0707	812.2064 7527
P(58)	2441 0368.8707	0.6772	814.2422 6058
P(56)	2447 0894.8065	0.4100	816.2611 8848
P(54)	2453 0917.6246	0.2378	818.2633 3418
P(52)	2459 0439.4169	0.1379	820.2487 6746
P(50)	2464 9462.1194	0.0923	822.2175 5290
P(48)	2470 7987.5190	0.0790	824.1697 5009
P(46)	2476 6017.2592	0.0755	826.1054 1387

BAND II

LINE	FREQUENCY (MHZ)	STD.DEV. (MHZ)	VAC.WAVE NO. (CM-1)
P(60)	2766 5459.5295	4.1144	922.8203 9762
P(58)	2773 7264.0710	2.9602	925.2155 3931
P(56)	2780 8303.6083	2.0915	927.5851 6321
P(54)	2787 8570.5221	1.4473	929.9290 1516
P(52)	2794 8057.2612	0.9777	932.2468 4329
P(50)	2801 6756.3611	0.6420	934.5383 9860
P(48)	2808 4660.4634	0.4074	936.8034 3564
P(46)	2815 1762.3327	0.2478	939.0417 1307

Left block

Label	value 1	prob	value 2
P(44)	2482 3552.8462	0.0722	828.0245 9447
P(42)	2488 0595.6546	0.0676	829.9273 3775
P(40)	2493 7146.9321	0.0624	831.8136 8532
P(38)	2499 3207.8048	0.0570	833.6836 7475
P(36)	2504 8779.2812	0.0514	835.5373 3967
P(34)	2510 3862.2568	0.0453	837.3747 0997
P(32)	2515 8457.5180	0.0389	839.1865 1186
P(30)	2521 2565.7453	0.0322	841.0006 6805
P(28)	2526 6187.5174	0.0257	842.7892 9784
P(26)	2531 9323.3137	0.0197	844.5617 1722
P(24)	2537 1973.5173	0.0148	846.0579 3897
P(22)	2542 4138.4181	0.0114	848.0579 7276
P(20)	2547 5818.2143	0.0096	849.7818 2521
P(18)	2552 7013.0155	0.0090	851.4894 9996
P(16)	2557 7722.8440	0.0088	853.1809 9777
P(14)	2562 7947.6370	0.0087	854.8563 1653
P(12)	2567 7687.2477	0.0086	856.5154 5136
P(10)	2572 6941.4471	0.0089	858.1583 9460
P(8)	2577 5709.9249	0.0096	859.7851 3592
P(6)	2582 3992.2904	0.0107	861.3956 6227
P(4)	2587 1788.0735	0.0117	862.9899 5799
P(2)	2591 9096.7253	0.0125	864.5680 0475
V(0)	2596 5917.6183	0.0128	866.1297 8163
R(0)	2598 9144.9378	0.0128	866.9045 6161
R(2)	2603 5232.8452	0.0123	868.4418 8873
R(4)	2608 0831.0836	0.0101	869.9628 8224
R(6)	2612 5938.7519	0.0089	871.4675 1210
R(8)	2617 0554.8709	0.0082	872.9557 4563
R(10)	2621 4678.3819	0.0079	874.4275 4754
R(12)	2625 8308.1471	0.0079	875.8828 7985
R(14)	2630 1442.9480	0.0080	877.3217 0194
R(16)	2634 4081.4848	0.0079	878.7439 7043
R(18)	2638 6222.3752	0.0083	880.1496 3923
R(20)	2642 7864.1533	0.0099	881.5386 5943
R(22)	2646 9005.2678	0.0135	882.9109 7929
R(24)	2650 9644.0807	0.0186	884.2665 4418
R(26)	2654 9778.8656	0.0249	885.6052 9650
R(28)	2658 9407.8053	0.0318	886.9271 7564
R(30)	2662 8528.9901		888.2321 1790

Right block

Label	value 1	prob	value 2
P(44)	2821 8054.8734	0.1428	941.2529 9421
P(42)	2828 3531.1460	0.0764	943.4370 4757
P(40)	2834 8184.3816	0.0368	945.5936 4737
P(38)	2841 2007.9961	0.0152	947.7225 7400
P(36)	2847 4995.6042	0.0059	949.8236 1445
P(34)	2853 7141.0312	0.0049	951.8965 6276
P(32)	2859 8438.3257	0.0051	953.9412 2042
P(30)	2865 8881.7700	0.0049	955.9573 9670
P(28)	2871 8465.8912	0.0048	957.9449 0905
P(26)	2877 7185.4704	0.0048	959.9035 8338
P(24)	2883 5035.5520	0.0048	961.8332 5439
P(22)	2889 2108.7652	0.0046	963.7337 6584
P(20)	2894 8108.7652	0.0044	965.6049 7080
P(18)	2900 3323.3734	0.0042	967.4467 3188
P(16)	2905 7651.4505	0.0041	969.2589 2147
P(14)	2911 1089.4690	0.0040	971.0414 2189
P(12)	2916 3634.2047	0.0041	972.7941 2562
P(10)	2921 5282.7418	0.0042	974.5169 3537
P(8)	2926 6032.4761	0.0043	976.2097 6429
P(6)	2931 5881.1188	0.0045	977.8725 3603
P(4)	2936 4826.6991	0.0047	979.5051 8485
P(2)	2941 2867.5662	0.0048	981.1076 5569
V(0)	2946 0002.3911	0.0049	982.6799 0421
R(0)	2948 3229.7106	0.0048	983.4546 8419
R(2)	2952 9003.6861	0.0047	984.9815 3967
R(4)	2957 3869.7091	0.0044	986.4781 0910
R(6)	2961 7827.5803	0.0041	987.9443 8586
R(8)	2966 0877.4220	0.0038	989.3803 7401
R(10)	2970 3019.6766	0.0037	990.7860 8831
R(12)	2974 4255.1041	0.0037	992.1615 5412
R(14)	2978 4584.7800	0.0039	993.5068 0730
R(16)	2982 4010.0913	0.0040	994.8218 9413
R(18)	2986 2532.7332	0.0042	996.1068 7115
R(20)	2990 0154.7042	0.0044	997.3618 0502
R(22)	2993 6878.3016	0.0045	998.5867 7237
R(24)	2997 2706.1154	0.0047	999.7818 5960
R(26)	3000 7641.0223	0.0049	1000.9471 6266
R(28)	3004 1686.1791	0.0050	1002.0827 8686
R(30)	3007 4845.0148	0.0051	1003.1888 4656

R(32)	2666 7140.4154	0.0387	889.5200 5642	
R(34)	2670 5239.9791	0.0452	890.7909 2107	
R(36)	2674 2825.4789	0.0514	892.0446 3839	
R(38)	2677 9894.6097	0.0587	893.2811 3150	
R(40)	2681 6444.9602	0.0707	894.5003 1996	
R(42)	2685 2474.0097	0.0941	895.7021 1969	
R(44)	2688 7979.1248	0.1369	896.8864 4285	
R(46)	2692 2957.5553	0.2076	898.0531 9770	
R(48)	2695 7406.4307	0.3157	899.2022 8849	
R(50)	2699 1322.7556	0.4740	900.3336 1532	
R(52)	2702 4703.4056	0.6993	901.4470 7395	
R(54)	2705 7545.1223	1.0137	902.5425 5570	
R(56)	2708 9844.5086	1.4453	903.6199 4726	
R(58)	2712 1598.0232	2.0293	904.6791 3049	

R(32)	3010 7121.2231	0.0053	1004.2654 6498	
R(34)	3013 8518.7535	0.0054	1005.3127 7386	
R(36)	3016 9041.8019	0.0055	1006.3309 1316	
R(38)	3019 8694.8011	0.0072	1007.3200 3075	
R(40)	3022 7482.4096	0.0154	1008.2802 8201	
R(42)	3025 5409.5012	0.0343	1009.2118 2951	
R(44)	3028 2481.1520	0.0692	1010.1148 4258	
R(46)	3030 8702.6288	0.1280	1010.9894 9690	
R(48)	3033 4079.3751	0.2216	1011.8359 7404	
R(50)	3035 8616.9973	0.3644	1012.6544 6102	
R(52)	3038 2321.2499	0.5755	1013.4451 4977	
R(54)	3040 5198.0199	0.8788	1014.2082 3668	
R(56)	3042 7253.3104	1.3049	1014.9439 2198	
R(58)	3044 8493.2235	1.8916	1015.6524 0922	

Table 2.10 $^{18}O\ ^{14}C\ ^{18}O$

NUMBER	SYMBOL		CONSTANTS (MHZ)	STD.DEV. (MHZ)
85	v(001-I)	=	2.665 794 012 522 D+07	5.1D-01
86	v(001-II)	=	2.945 628 336 884 D+07	4.3D-03
87	B(001)	=	1.032 222 210 875 D+04	1.5D-04
88	B(I)	=	1.039 139 637 214 D+04	2.8D-03
89	B(II)	=	1.041 017 017 924 D+04	1.4D-04
90	D(001)	=	3.145 791 088 D-03	3.1D-07
91	D(I)	=	2.762 860 020 D-03	4.9D-06
92	D(II)	=	3.717 259 657 D-03	2.9D-07
93	H(001)	=	0.074 897 D-09	2.5D-10
94	H(I)	=	-3.624 392 D-09	3.3D-09
95	H(II)	=	4.503 216 D-09	2.2D-10
96	L(001)	=	9.77 438 D-14	6.6D-14
97	L(I)	=	176.77 976 D-14	7.3D-13
98	L(II)	=	7.33 773 D-14	5.7D-14

BAND I

LINE	FREQUENCY (MHZ)	STD.DEV. (MHZ)	VAC.WAVE NO. (CM-1)
P(60)	2516 3563.1916	24.9722	839.3661 1880
P(58)	2522 1736.2103	16.1045	841.3065 6183
P(56)	2527 9300.5125	9.9374	843.2267 0027
P(54)	2533 6261.8905	5.7850	845.1267 2732
P(52)	2539 2625.5971	3.1031	847.0068 1820
P(50)	2544 8396.4026	1.4677	848.8671 3203
P(48)	2550 3578.6456	0.5603	850.7078 1352
P(46)	2555 8176.2776	0.1907	852.5289 9449

BAND II

LINE	FREQUENCY (MHZ)	STD.DEV. (MHZ)	VAC.WAVE NO. (CM-1)
P(60)	2790 5884.6278	0.1963	930.8401 1566
P(58)	2796 6918.2463	0.1219	932.8759 7803
P(56)	2802 7353.5817	0.0718	934.8918 8383
P(54)	2808 7184.5030	0.0394	936.8876 2854
P(52)	2814 6405.0223	0.0198	938.8630 1243
P(50)	2820 5009.2992	0.0093	940.8178 4069
P(48)	2826 2991.6460	0.0056	942.7519 2360
P(46)	2832 0346.5324	0.0052	944.6650 7668

P(44)	2561	2192.9029	0.1915	854.3307	9517
P(42)	2566	5631.8132	0.1816	856.1766	2535
P(40)	2571	8496.0177	0.1304	857.8766	8540
P(38)	2577	0788.2691	0.1036	859.6209	6715
P(36)	2582	2511.0860	0.1289	861.3462	5461
P(34)	2587	3666.7717	0.1595	863.0526	2462
P(32)	2592	4257.4305	0.1718	864.7305	4735
P(30)	2597	4284.9803	0.1655	866.4088	8679
P(28)	2602	3751.1638	0.1484	868.0589	0106
P(26)	2607	2657.5569	0.1306	869.6902	4274
P(24)	2612	1005.5759	0.1194	871.3029	5906
P(22)	2616	8796.4821	0.1155	872.8970	9210
P(20)	2621	6031.3862	0.1159	874.4726	7890
P(18)	2626	2711.2509	0.1210	876.0297	5159
P(16)	2630	8836.8926	0.1375	877.5683	3738
P(14)	2635	4408.9825	0.1719	879.0884	5867
P(12)	2639	9428.0471	0.2234	880.5901	3303
P(10)	2644	3894.4684	0.2855	882.0733	7319
P(8)	2648	7808.4833	0.3503	883.5381	8705
P(6)	2653	1170.1836	0.4107	884.9845	7768
P(4)	2657	3979.5153	0.4603	886.4125	4328
P(2)	2661	6236.2781	0.4944	887.8220	7717
V(0)	2665	7940.1252	0.5095	889.2131	6777
R(0)	2667	8584.5569	0.5096	889.9017	9189
R(2)	2671	9458.0588	0.4944	891.2651	8516
R(4)	2675	9777.1350	0.4603	892.6100	8477
R(6)	2679	9540.9244	0.4107	893.9364	6202
R(8)	2683	8748.4193	0.3503	895.2442	8327
R(10)	2687	7398.4657	0.2855	896.5335	1005
R(12)	2691	5489.7640	0.2234	897.8040	9899
R(14)	2695	3020.8692	0.1719	899.0560	0191
R(16)	2698	9990.1917	0.1375	900.2891	6577
R(18)	2702	6395.9971	0.1211	901.5035	3273
R(20)	2706	2236.4068	0.1160	902.6990	4011
R(22)	2709	7509.3968	0.1156	903.8756	2041
R(24)	2713	2212.7973	0.1195	905.0332	0124
R(26)	2716	6344.2906	0.1307	906.1717	0531
R(28)	2719	9901.4088	0.1485	907.2910	5029
R(30)	2723	2881.5295	0.1656	908.3911	4870

P(44)	946.5571	2085	0.0050	2837	7068.5896
P(42)	948.4278	8258	0.0047	2843	3152.6154
P(40)	950.2771	9403	0.0045	2848	8593.5779
P(38)	952.1048	9318	0.0045	2854	3386.6201
P(36)	953.9108	2398	0.0045	2859	7527.0634
P(34)	955.6948	3645	0.0046	2865	1010.4118
P(32)	957.4567	8683	0.0047	2870	3832.3553
P(30)	959.1965	3768	0.0047	2875	5988.7736
P(28)	960.9139	5798	0.0047	2880	7475.7389
P(26)	962.6089	2326	0.0047	2885	8289.5197
P(24)	964.2813	1568	0.0046	2890	8426.5831
P(22)	965.9310	2412	0.0046	2895	7883.5981
P(20)	967.5579	4429	0.0045	2900	6657.4375
P(18)	969.1619	7875	0.0043	2905	4745.1809
P(16)	970.7430	3706	0.0042	2910	2144.1167
P(14)	972.3010	3579	0.0042	2914	8851.7437
P(12)	973.8358	9861	0.0042	2919	4865.7733
P(10)	975.3475	5631	0.0042	2924	0184.1309
P(8)	976.8359	4687	0.0043	2928	4804.9574
P(6)	978.3010	1551	0.0043	2932	8726.6103
P(4)	979.7427	1469	0.0044	2937	1947.6644
P(2)	981.1610	0416	0.0044	2941	4466.9132
V(0)	982.5558	5098	0.0043	2945	6283.3688
R(0)	983.2444	7510	0.0043	2947	6927.8005
R(2)	984.6041	1215	0.0042	2951	7688.6939
R(4)	985.9402	5618	0.0041	2955	7745.2842
R(6)	987.2528	9984	0.0040	2959	7097.3511
R(8)	988.5420	4309	0.0040	2963	5744.8935
R(10)	989.8076	9317	0.0041	2967	3688.1283
R(12)	991.0498	6457	0.0041	2971	0927.4901
R(14)	992.2685	7903	0.0043	2975	7463.6304
R(16)	993.4638	6545	0.0044	2978	3297.4158
R(18)	994.6357	5989	0.0046	2981	8429.9272
R(20)	995.7843	0549	0.0049	2985	2862.4581
R(22)	996.9095	5243	0.0051	2988	6596.5128
R(24)	998.0115	5787	0.0053	2991	9633.8045
R(26)	999.0903	8584	0.0053	2995	1976.2534
R(28)	1000.1461	0721	0.0055	2998	3625.9839
R(30)	1001.1787	9960	0.0058	3001	4585.3228

R(32)	2726 5281.8713	0.1720	909.4719 0777
R(34)	2729 7099.4864	0.1597	910.5332 2917
R(36)	2732 8331.2527	0.1290	911.5750 0876
R(38)	2735 8973.8628	0.1037	912.5971 3621
R(40)	2738 9023.8112	0.1307	913.5994 9460
R(42)	2741 8477.3784	0.1828	914.5819 5985
R(44)	2744 7330.6123	0.1953	915.5444 0013
R(46)	2747 5579.3060	0.1998	916.4866 7512
R(48)	2750 3218.9719	0.5659	917.4086 3514
R(50)	2753 0244.8124	1.4700	918.3101 2014
R(52)	2755 6651.6845	3.1024	919.1909 5858
R(54)	2758 2434.0616	5.7807	920.0509 6611
R(56)	2760 7585.9881	9.9284	920.8899 4407
R(58)	2763 2101.0296	16.0893	921.7076 7783

R(32)	3004 4856.7961	0.0063	1002.1885 4726
R(34)	3007 4443.1264	0.0068	1003.1754 4101
R(36)	3010 3347.2301	0.0069	1004.1395 7812
R(38)	3013 1572.2138	0.0063	1005.0810 6224
R(40)	3015 9121.3714	0.0078	1006.0000 0322
R(42)	3018 5998.1806	0.0173	1006.8965 1708
R(44)	3021 2206.2990	0.0371	1007.7707 2581
R(46)	3023 7749.5608	0.0703	1008.6227 5731
R(48)	3026 2631.9724	0.1219	1009.4527 4522
R(50)	3028 6857.7089	0.1988	1010.2608 2881
R(52)	3031 0431.1097	0.3096	1011.0471 5282
R(54)	3033 3356.6741	0.4650	1011.8118 6733
R(56)	3035 5639.0573	0.6783	1012.5551 2763
R(58)	3037 7283.0656	0.9658	1013.2770 9404

Table 2.11 $^{16}O\ ^{13}C\ ^{18}O$

NUMBER	SYMBOL		CONSTANTS (MHZ)	STD.DEV. (MHZ)
99	V(001-I)	=	2.769 166 220 212 D+07	4.9D-02
100	V(001-II)	=	3.061 096 273 608 D+07	9.9D-02
101	B(001)	=	1.095 417 207 032 D+04	5.1D-04
102	B(I)	=	1.103 309 137 756 D+04	8.1D-04
103	B(II)	=	1.104 838 028 891 D+04	8.8D-04
104	D(001)	=	3.553 496 537 D-03	1.7D-06
105	D(I)	=	3.099 388 259 D-03	2.8D-06
106	D(II)	=	4.206 493 331 D-03	2.5D-06
107	H(001)	=	5.818 906 D-09	2.2D-09
108	H(I)	=	6.110 893 D-09	3.8D-09
109	H(II)	=	17.600 347 D-09	3.0D-09
110	L(001)	=	-201.16 254 D-14	1.0D-12
111	L(I)	=	7.84 555 D-14	1.8D-12
112	L(II)	=	-507.39 778 D-14	1.2D-12

BAND I

LINE	FREQUENCY (MHZ)	STD.DEV. (MHZ)	VAC.WAVE NO. (CM-1)
P(60)	2608 4932.6048	98.3718	870.0996 9426
P(59)	2611 6598.0182	84.2242	871.1559 3876
P(58)	2614 8086.3664	71.8523	872.2062 7700
P(57)	2617 9398.6843	61.0659	873.2507 4350
P(56)	2621 0535.9568	51.6916	874.2893 7111
P(55)	2624 1499.1212	43.5718	875.3221 9110
P(54)	2627 2289.0702	36.5633	876.3492 3325
P(53)	2630 2906.6539	30.5366	877.3705 2591

BAND II

LINE	FREQUENCY (MHZ)	STD.DEV. (MHZ)	VAC.WAVE NO. (CM-1)
P(60)	2896 3467.3195	86.8068	966.1172 7702
P(59)	2899 5949.0536	73.6194	967.2007 5105
P(58)	2902 8278.6677	62.1727	968.2791 5090
P(57)	2906 0454.6740	52.2716	969.3524 2694
P(56)	2909 2475.6481	43.7390	970.4205 3166
P(55)	2912 4340.2262	36.4145	971.4834 1958
P(54)	2915 6047.1010	30.1535	972.5410 4708
P(53)	2918 7595.0185	24.8251	973.5933 7234

P(n)				
P(52)	25.3746	2633	3352.6819	878.3860 9609
P(51)	20.9716	2636	3627.9259	879.3959 6952
P(50)	17.2329	2639	3733.1212	880.4001 7075
P(49)	14.0733	2642	3668.9688	881.3987 2314
P(48)	11.4166	2645	3436.1373	882.3916 4900
P(47)	9.1949	2648	3035.2642	883.3789 6960
P(46)	7.3480	2651	2466.9580	884.3607 0523
P(45)	5.8222	2654	1731.7991	885.3368 7526
P(44)	4.5705	2657	0830.3416	886.3074 9816
P(43)	3.5512	2659	9763.1145	887.2725 9158
P(42)	2.7280	2662	8530.6229	888.2321 7237
P(41)	2.0692	2665	7133.3489	889.1862 5661
P(40)	1.5471	2668	5571.7531	890.1348 5467
P(39)	1.1381	2671	3846.2752	891.0779 9621
P(38)	0.8216	2674	1957.3353	892.0156 8024
P(37)	0.5802	2676	9905.3343	892.9479 2514
P(36)	0.3991	2679	7690.6549	893.8747 4367
P(35)	0.2658	2682	5313.6624	894.7961 4982
P(34)	0.1701	2685	2774.7054	895.7121 6014
P(33)	0.1031	2688	0074.1161	896.6227 8956
P(32)	0.0581	2690	7212.2113	897.5279 4813
P(31)	0.0293	2693	4189.2925	898.4278 4542
P(30)	0.0127	2696	1005.6468	899.3223 9054
P(29)	0.0066	2698	7661.5470	900.2114 9213
P(28)	0.0071	2701	4157.2519	901.0952 5837
P(27)	0.0075	2704	0493.0072	901.9737 9701
P(26)	0.0071	2706	6669.0452	902.8468 1539
P(25)	0.0064	2709	2685.5857	903.7147 2040
P(24)	0.0059	2711	8584.8354	904.5772 1854
P(23)	0.0057	2714	4240.9891	905.4344 1589
P(22)	0.0057	2716	9780.2293	906.2863 1815
P(21)	0.0056	2719	5160.7266	907.1329 5818
P(20)	0.0055	2722	0382.6396	907.9742 3061
P(19)	0.0054	2724	5446.1155	908.8102 5818
P(18)	0.0054	2727	0351.2897	909.6410 0537
P(17)	0.0055	2729	5098.2865	910.4664 7633
P(16)	0.0057	2731	9687.2184	911.2866 7481
P(15)	0.0057	2734	4118.1871	912.1016 0421
P(14)	0.0059	2736	8391.2828	912.9112 6753

	P(n)			
974.6403 5521	P(52)	20.3120	2921	8982.7755
975.6819 5715	P(51)	16.5089	2925	0209.2160
976.7181 4108	P(50)	13.3216	2928	1273.2288
977.7488 7135	P(49)	10.6662	2931	2173.7450
978.7741 1364	P(48)	8.4680	2934	2909.7356
979.7938 3487	P(47)	6.6610	2937	3480.2089
980.8080 0315	P(46)	5.1870	2940	3884.2089
981.8165 8770	P(45)	3.9946	2943	4120.8131
982.8195 5881	P(44)	3.0391	2946	4189.1307
983.8168 8777	P(43)	2.2814	2949	4088.3007
984.8085 4681	P(42)	1.6877	2952	3817.4909
985.7945 0907	P(41)	1.2287	2955	3375.8957
986.7747 4853	P(40)	0.8797	2958	2762.7353
987.7492 3999	P(39)	0.6192	2961	1977.2545
988.7179 5904	P(38)	0.4293	2964	1018.7210
989.6808 8200	P(37)	0.2950	2966	9886.4250
990.6379 8589	P(36)	0.2032	2969	8579.6778
991.5892 4843	P(35)	0.1430	2972	7097.8114
992.5346 4799	P(34)	0.1043	2975	5440.1771
993.4741 6356	P(33)	0.0790	2978	3606.1454
994.4077 7476	P(32)	0.0608	2981	1595.1050
995.3354 6178	P(31)	0.0463	2983	9406.4623
996.2572 0541	P(30)	0.0342	2986	7039.6411
997.1729 8698	P(29)	0.0242	2989	4494.0818
998.0827 8838	P(28)	0.0166	2992	1769.2416
998.9865 9203	P(27)	0.0118	2994	8864.5933
999.8843 8087	P(26)	0.0099	2997	5779.6258
1000.7761 3839	P(25)	0.0096	3000	2513.8436
1001.6618 4856	P(24)	0.0096	3002	9066.7664
1002.5414 9587	P(23)	0.0092	3005	5437.9293
1003.4150 6530	P(22)	0.0086	3008	1626.8822
1004.2825 4236	P(21)	0.0082	3010	7633.1902
1005.1439 1303	P(20)	0.0081	3013	3456.4332
1005.9991 6380	P(19)	0.0083	3015	9096.2061
1006.8482 8163	P(18)	0.0085	3018	4552.1183
1007.6912 5401	P(17)	0.0084	3020	9823.7944
1008.5280 6889	P(16)	0.0081	3023	4910.8735
1009.3587 1475	P(15)	0.0081	3025	9813.0098
1010.1831 8054	P(14)	0.0097	3028	4529.8723

P(13)	2739 2506.5846	913.7156 6741	0.0066
P(12)	2741 6464.1606	914.5148 0613	0.0081
P(11)	2744 0264.0678	915.3086 8558	0.0105
P(10)	2746 3906.3522	916.0973 0730	0.0136
P(9)	2748 7391.0486	916.8806 7245	0.0174
P(8)	2751 0718.1811	917.6587 8183	0.0216
P(7)	2753 3887.7626	918.4316 3588	0.0260
P(6)	2755 6899.7953	919.1992 3467	0.0305
P(5)	2757 9754.2701	919.9615 7789	0.0349
P(4)	2760 2451.1674	920.7186 6489	0.0389
P(3)	2762 4990.4564	921.4704 9464	0.0424
P(2)	2764 7372.0954	922.2170 6576	0.0453
P(1)	2766 9596.0318	922.9583 7648	0.0474
V(0)	2769 1662.0021	923.6944 2470	0.0486
R(0)	2771 3570.5320	924.4252 0792	0.0489
R(1)	2773 5320.9363	925.1507 2331	0.0484
R(2)	2775 6913.3186	925.8709 6766	0.0469
R(3)	2777 8347.5719	926.5859 3739	0.0445
R(4)	2779 9623.5784	927.2956 2858	0.0414
R(5)	2782 0741.2091	928.0000 3692	0.0376
R(6)	2784 1700.3243	928.6991 5775	0.0333
R(7)	2786 2500.7734	929.3929 8605	0.0287
R(8)	2788 3142.3950	930.0815 1643	0.0239
R(9)	2790 3625.0166	930.7647 4314	0.0192
R(10)	2792 3948.4550	931.4426 6007	0.0148
R(11)	2794 4112.5159	932.1152 6075	0.0109
R(12)	2796 4116.9943	932.7825 3832	0.0078
R(13)	2798 3961.6742	933.4444 8559	0.0059
R(14)	2800 3646.3284	934.1010 9498	0.0053
R(15)	2802 3170.7189	934.7523 5854	0.0055
R(16)	2804 2534.5966	935.3982 6798	0.0058
R(17)	2806 1737.7013	936.0388 1460	0.0057
R(18)	2808 0779.7615	936.6739 8936	0.0054
R(19)	2809 9660.4946	937.3037 8283	0.0049
R(20)	2811 8379.6066	937.9281 8519	0.0045
R(21)	2813 6936.7919	938.5471 8626	0.0044
R(22)	2815 5331.7334	939.1607 7546	0.0046
R(23)	2817 3564.1022	939.7689 4183	0.0050
R(24)	2819 1633.5576	940.3716 7398	0.0054

P(13)	3030 9061.1448	1011.0014 5571	0.0134
P(12)	3033 3406.5264	1011.8135 3022	0.0191
P(11)	3035 7565.7308	1012.6193 9454	0.0261
P(10)	3038 1538.4873	1013.4190 3962	0.0341
P(9)	3040 5324.5400	1014.2124 5694	0.0428
P(8)	3042 8923.6485	1014.9996 3847	0.0518
P(7)	3045 2335.5874	1015.7805 7669	0.0607
P(6)	3047 5560.1470	1016.5552 6461	0.0693
P(5)	3049 8597.1328	1017.3236 9574	0.0773
P(4)	3052 1446.3659	1018.0858 6412	0.0844
P(3)	3054 4107.6828	1018.8417 6429	0.0904
P(2)	3056 6580.9357	1019.5913 9131	0.0949
P(1)	3058 8865.9923	1020.3347 4079	0.0980
V(0)	3061 0962.7361	1021.0718 0882	0.0994
R(0)	3063 2871.0660	1021.8025 9205	0.0992
R(1)	3065 4590.8968	1022.5270 8762	0.0972
R(2)	3067 6122.1589	1023.2452 9321	0.0937
R(3)	3069 7464.7984	1023.9572 0703	0.0888
R(4)	3071 8618.7769	1024.6628 2780	0.0825
R(5)	3073 9584.0718	1025.3621 5477	0.0752
R(6)	3076 0360.6761	1026.0551 8769	0.0670
R(7)	3078 0948.5983	1026.7419 2685	0.0582
R(8)	3080 1347.8624	1027.4223 7306	0.0493
R(9)	3082 1558.5081	1028.0965 2763	0.0403
R(10)	3084 1580.5901	1028.7643 9240	0.0318
R(11)	3086 1414.1789	1029.4259 6971	0.0239
R(12)	3088 1059.3601	1030.0812 6242	0.0169
R(13)	3090 0516.2344	1030.7302 7389	0.0113
R(14)	3091 9784.9179	1031.3730 0799	0.0074
R(15)	3093 8865.5416	1032.0094 6908	0.0059
R(16)	3095 7758.2517	1032.6396 6206	0.0061
R(17)	3097 6463.2092	1033.2635 9228	0.0068
R(18)	3099 4980.5901	1033.8812 6562	0.0072
R(19)	3101 3310.5852	1034.4926 8845	0.0072
R(20)	3103 1453.4002	1035.0978 6761	0.0075
R(21)	3104 9409.2555	1035.6968 1047	0.0080
R(22)	3106 7178.3862	1036.2895 2488	0.0087
R(23)	3108 4761.0423	1036.8760 1915	0.0095
R(24)	3110 2157.4886	1037.4563 0214	0.0102

R(25)	2820 9539.7466	0.0056	940.9689 6015	R(25)	3111 9368.0045	0.0107	1038.0303 8316
R(26)	2822 7282.3040	0.0058	941.5607 8816	R(26)	3113 6392.8845	0.0115	1038.5982 7203
R(27)	2824 4860.8519	0.0062	942.1471 4541	R(27)	3115 3232.4380	0.0135	1039.1599 7907
R(28)	2826 2274.9998	0.0069	942.7280 1885	R(28)	3116 9886.9894	0.0180	1039.7155 1511
R(29)	2827 9524.3435	0.0078	943.3033 9503	R(29)	3118 6356.8784	0.0254	1040.2648 9147
R(30)	2829 6608.4660	0.0097	943.8732 6001	R(30)	3120 2642.4602	0.0363	1040.8081 2000
R(31)	2831 3526.9357	0.0148	944.4375 9942	R(31)	3121 8744.1055	0.0520	1041.3452 1308
R(32)	2833 0279.3072	0.0268	944.9963 9838	R(32)	3123 4662.2009	0.0748	1041.8761 8359
R(33)	2834 6865.1200	0.0490	945.5496 4155	R(33)	3125 0397.1494	0.1084	1042.4010 4497
R(34)	2836 3283.8986	0.0853	946.0973 1305	R(34)	3126 5949.3704	0.1581	1042.9198 1122
R(35)	2837 9535.1515	0.1405	946.6393 9649	R(35)	3128 1319.3005	0.2308	1043.4324 9691
R(36)	2839 5618.3707	0.2211	947.1758 7494	R(36)	3129 6507.3936	0.3348	1043.9391 1716
R(37)	2841 1533.0311	0.3349	947.7067 3087	R(37)	3131 1514.1219	0.4804	1044.4396 8773
R(38)	2842 7278.5901	0.4914	948.2319 4619	R(38)	3132 6339.9758	0.6798	1044.9342 2499
R(39)	2844 2854.4862	0.7021	948.7515 0215	R(39)	3134 0985.4654	0.9478	1045.4227 4594
R(40)	2845 8260.1385	0.9808	949.2653 7940	R(40)	3135 5451.1207	1.3018	1045.9052 6826
R(41)	2847 3494.9459	1.3437	949.7735 5787	R(41)	3136 9737.4926	1.7624	1046.3818 1033
R(42)	2848 8558.2859	1.8099	950.2760 1681	R(42)	3138 3845.1539	2.3539	1046.8523 9126
R(43)	2850 3449.5139	2.4017	950.7727 3471	R(43)	3139 7774.7001	3.1045	1047.3170 3091
R(44)	2851 8167.9616	3.1450	951.2636 8928	R(44)	3141 1526.7506	4.0470	1047.7757 4994
R(45)	2853 2712.9361	4.0697	951.7488 5741	R(45)	3142 5101.9501	5.2193	1048.2285 6985
R(46)	2854 7083.7186	5.2100	952.2282 1511	R(46)	3143 8500.9695	6.6649	1048.6755 1303
R(47)	2856 1279.5630	6.6054	952.7017 3751	R(47)	3145 1724.5076	8.4334	1049.1166 0278
R(48)	2857 5299.6945	8.3004	953.1693 9876	R(48)	3146 4773.2927	10.5813	1049.5518 6340
R(49)	2858 9143.3078	10.3459	953.6311 7200	R(49)	3147 7648.0840	13.1728	1049.9813 2021
R(50)	2860 2809.5658	12.7992	954.0870 2930	R(50)	3149 0349.6734	16.2804	1050.4049 9963
R(51)	2861 6297.5977	15.7249	954.5369 4161	R(51)	3150 2878.8878	19.9855	1050.8229 2923
R(52)	2862 9606.4969	19.1957	954.9808 7870	R(52)	3151 5236.5905	24.3797	1051.2351 3783
R(53)	2864 2735.3193	23.2928	955.4188 0908	R(53)	3152 7423.6840	29.5654	1051.6416 5551
R(54)	2865 5683.0813	28.1068	955.8506 9993	R(54)	3153 9441.1120	35.6568	1052.0425 1376
R(55)	2866 8448.7571	33.7388	956.2765 1704	R(55)	3155 1289.8621	42.7811	1052.4377 4552
R(56)	2868 1031.2769	40.3009	956.6962 2472	R(56)	3156 2970.9683	51.0791	1052.8273 8528
R(57)	2869 3429.5242	47.9175	957.1097 8574	R(57)	3157 4485.5139	60.7073	1053.2114 6918
R(58)	2870 5642.3331	56.7257	957.5171 6119	R(58)	3158 5834.6344	71.8380	1053.5900 3509
R(59)	2871 7668.4857	66.8772	957.9183 1046	R(59)	3159 7019.5211	84.6613	1053.9631 2275

Table 2.12 $^{17}O^{12}C^{17}O$

NUMBER	SYMBOL		CONSTANTS (MHZ)	STD.DEV. (MHZ)
113	V(001-I)	=	2.894 953 846 245 732 D+07	1.0D-02
114	V(001-II)	=	3.214 846 174 949 D+07	6.1D-03
115	B(001)	=	1.092 153 560 997 D+04	1.2D-04
116	B(I)	=	1.101 834 696 644 D+04	1.5D-04
117	B(II)	=	1.100 609 965 843 D+04	1.1D-04
118	D(001)	=	3.530 964 947 D-03	2.1D-07
119	D(I)	=	3.049 684 815 D-03	3.1D-07
120	D(II)	=	4.069 984 195 D-03	1.9D-07
121	H(001)	=	-0.086 853 D-09	1.5D-10
122	H(I)	=	3.143 802 D-09	3.0D-10
123	H(II)	=	5.840 864 D-09	1.4D-10
124	L(001)	=	10.84 024 D-14	3.8D-14
125	L(I)	=	-1.29 247 D-14	1.0D-13
126	L(II)	=	6.45 892 D-14	3.3D-14

BAND I

LINE	FREQUENCY (MHZ)	STD.DEV. (MHZ)	VAC.WAVE NO. (CM-1)
P(60)	2728 1083.6929	7.6946	909.9989 9981
P(59)	2731 4822.5368	6.5234	911.1244 0650
P(58)	2734 8351.3738	5.5069	912.2428 0812
P(57)	2738 1670.8532	4.6277	913.3542 2632
P(56)	2741 4781.6111	3.8702	914.4586 8232
P(55)	2744 7684.2700	3.2200	915.5561 9688
P(54)	2748 0379.4394	2.6644	916.6467 9034
P(53)	2751 2867.7152	2.1917	917.7304 8257

BAND II

LINE	FREQUENCY (MHZ)	STD.DEV. (MHZ)	VAC.WAVE NO. (CM-1)
P(60)	3053 8359.5142	0.4634	1018.6500 2602
P(59)	3056 9761.4634	0.3777	1019.6974 8230
P(58)	3060 1020.0398	0.3054	1020.7401 5617
P(57)	3063 2134.5330	0.2448	1021.7780 2395
P(56)	3066 3104.2389	0.1942	1022.8110 6214
P(55)	3069 3928.4602	0.1524	1023.8392 4749
P(54)	3072 4606.5061	0.1182	1024.8625 5695
P(53)	3075 5137.6927	0.0904	1025.8809 6772

Left table:

P				
P(52)	918.8072 9302	1.7914	2754	5149.6802
P(51)	919.8772 4067	1.4542	2757	7225.9038
P(50)	920.9403 4407	1.1718	2760	9096.9421
P(49)	921.9966 2134	0.9366	2764	0763.3380
P(48)	923.0460 9014	0.7420	2767	2225.6209
P(47)	924.0887 6767	0.5821	2770	3484.3071
P(46)	925.1246 7074	0.4519	2773	4539.8997
P(45)	926.1538 1567	0.3466	2776	5392.8885
P(44)	927.1762 1836	0.2623	2779	6043.7501
P(43)	928.1918 9428	0.1955	2782	6492.9481
P(42)	929.2008 5844	0.1433	2785	6740.9328
P(41)	930.2031 2544	0.1029	2788	6788.1415
P(40)	931.1987 0943	0.0723	2791	6634.9985
P(39)	932.1876 2412	0.0495	2794	6281.9151
P(38)	933.1698 8280	0.0330	2797	5729.2897
P(37)	934.1454 9833	0.0214	2800	4977.5075
P(36)	935.1144 8314	0.0136	2803	4026.9412
P(35)	936.0768 4922	0.0088	2806	2877.9504
P(34)	937.0326 0815	0.0063	2809	1530.8823
P(33)	937.9817 7107	0.0052	2811	9986.0709
P(32)	938.9243 4872	0.0047	2814	8243.8379
P(31)	939.8603 5139	0.0044	2817	6304.4921
P(30)	940.7897 8898	0.0042	2820	4168.3301
P(29)	941.7126 7095	0.0040	2823	1835.6355
P(28)	942.6290 0636	0.0040	2825	9306.6799
P(27)	943.5388 0384	0.0039	2828	6581.0086
P(26)	944.4420 7161	0.0039	2831	3661.0086
P(25)	945.3388 1749	0.0039	2834	0544.7738
P(24)	946.2290 4889	0.0039	2836	7233.2396
P(23)	947.1127 7279	0.0039	2839	3726.6158
P(22)	947.9899 9579	0.0039	2842	0025.0999
P(21)	948.8607 2408	0.0039	2844	6128.8772
P(20)	949.7249 6344	0.0040	2847	2038.1213
P(19)	950.5827 1923	0.0041	2849	7752.9932
P(18)	951.4339 9646	0.0042	2852	3273.6422
P(17)	952.2787 9968	0.0043	2854	8600.2058
P(16)	953.1171 3309	0.0043	2857	3732.8091
P(15)	953.9490 0047	0.0044	2859	8671.5658
P(14)	954.7744 0521	0.0045	2862	3416.5773

Right table:

P				
P(52)	1026.8944 5722	0.0680	3078	5521.3436
P(51)	1027.9030 0313	0.0503	3081	5756.7892
P(50)	1028.9065 8337	0.0366	3084	5843.3680
P(49)	1029.9051 7612	0.0261	3087	5780.4257
P(48)	1030.8987 5984	0.0184	3090	5567.3161
P(47)	1031.8873 1323	0.0131	3093	5203.4011
P(46)	1032.8708 1527	0.0098	3096	4688.0508
P(45)	1033.8492 4526	0.0080	3099	4020.6437
P(44)	1034.8225 8272	0.0071	3102	3200.5669
P(43)	1035.7908 0753	0.0066	3105	2227.2163
P(42)	1036.7538 9982	0.0063	3108	1099.9967
P(41)	1037.7118 4004	0.0059	3110	—
P(40)	1038.6646 0896	0.0056	3113	9818.3221
P(39)	1039.6121 8765	0.0052	3116	8381.6157
P(38)	1040.5545 5750	0.0049	3119	6789.3101
P(37)	1041.4917 0024	0.0047	3122	5040.8476
P(36)	1042.4235 9791	0.0045	3125	3135.6800
P(35)	1043.3502 3290	0.0044	3127	1073.2694
P(34)	1044.2715 8793	0.0044	3130	8853.0875
P(33)	1045.1876 4608	0.0044	3133	6474.6165
P(32)	1046.0983 9076	0.0043	3136	3937.3489
P(31)	1047.0038 0574	0.0043	3138	1240.7875
P(30)	1047.9038 7516	0.0043	3141	8384.4458
P(29)	1048.7985 8351	0.0042	3144	5367.8482
P(28)	1049.6879 1566	0.0042	3146	2190.5298
P(27)	1050.5718 5682	0.0042	3149	8852.0367
P(26)	1051.4503 9262	0.0042	3152	5351.9262
P(25)	1052.3235 0902	0.0042	3154	7865.1382
P(24)	1053.1911 9241	0.0042	3157	3877.6318
P(23)	1054.0534 2954	0.0042	3159	9726.8505
P(22)	1054.9102 0754	0.0042	3162	5412.4088
P(21)	1055.7615 1395	0.0042	3165	0933.9330
P(20)	1056.6073 3671	0.0043	3167	6291.0614
P(19)	1057.4476 6414	0.0043	3170	1483.4440
P(18)	1058.2824 8498	0.0043	3172	6510.7431
P(17)	1059.1117 8836	0.0043	3175	1372.6330
P(16)	1059.9355 6383	0.0044	3177	6068.8002
P(15)	1060.7538 0134	0.0044	3180	0598.9435
P(14)	1061.5664 9125	0.0045	3182	4962.7742

P(13)	2864 7967.9336	0.0046	955.5933 5030		
P(12)	2867 2325.7126	0.0048	956.4058 3835		
P(11)	2869 6489.9806	0.0051	957.2118 7157		
P(10)	2872 0460.7920	0.0055	958.0114 5178		
P(9)	2874 4238.1899	0.0060	958.8045 8040		
P(8)	2876 7822.2052	0.0066	959.5912 5847		
P(7)	2879 1212.8575	0.0073	960.3714 8665		
P(6)	2881 4410.1546	0.0079	961.1452 6519		
P(5)	2883 7414.0928	0.0086	961.9125 9397		
P(4)	2886 0224.6567	0.0091	962.6734 7248		
P(3)	2888 2841.8194	0.0096	963.4278 9982		
P(2)	2890 5265.5424	0.0100	964.1758 7471		
P(1)	2892 7495.7756	0.0103	964.9173 9547		
V(0)	2894 9532.4573	0.0104	965.6524 6005		
R(0)	2897 1375.5144	0.0104	966.3810 6601		
R(1)	2899 3024.8621	0.0103	967.1032 1052		
R(2)	2901 4480.4042	0.0100	967.8188 9037		
R(3)	2903 5742.0327	0.0096	968.5281 0195		
R(4)	2905 6809.6283	0.0090	969.2308 4130		
R(5)	2907 7683.0600	0.0084	969.9271 0404		
R(6)	2909 8362.1852	0.0077	970.6168 8540		
R(7)	2911 8846.8498	0.0070	971.3001 8027		
R(8)	2913 9136.8882	0.0063	971.9769 8310		
R(9)	2915 9232.1228	0.0057	972.6472 8797		
R(10)	2917 9132.3648	0.0051	973.3110 8859		
R(11)	2919 8837.4136	0.0046	973.9683 7827		
R(12)	2921 8347.0567	0.0042	974.6191 4992		
R(13)	2923 7661.0702	0.0040	975.2633 9606		
R(14)	2925 6779.2184	0.0038	975.9011 0884		
R(15)	2927 5701.2537	0.0037	976.5322 8000		
R(16)	2929 4426.9168	0.0037	977.1569 0089		
R(17)	2931 2955.9368	0.0036	977.7749 6246		
R(18)	2933 1288.0305	0.0036	978.3864 5529		
R(19)	2934 9422.9032	0.0035	978.9913 6953		
R(20)	2936 7360.2482	0.0035	979.5896 9496		
R(21)	2938 5099.7467	0.0034	980.1814 2093		
R(22)	2940 2641.0680	0.0034	980.7665 3643		
R(23)	2941 9983.8693	0.0034	981.3450 3001		
R(24)	2943 7127.7959	0.0035	981.9168 8985		

Label					
P(13)	1062.3736 2435	0.0045	3184 9160.0159		
P(12)	1063.1751 9184	0.0046	3187 3190.4045		
P(11)	1063.9711 8532	0.0048	3189 7053.6889		
P(10)	1064.7615 9684	0.0049	3192 0749.6301		
P(9)	1065.5464 1886	0.0051	3194 4278.0022		
P(8)	1066.3256 4425	0.0053	3196 7638.5917		
P(7)	1067.0992 6632	0.0055	3199 0831.1981		
P(6)	1067.8672 7881	0.0056	3201 3855.6333		
P(5)	1068.6296 7588	0.0058	3203 6711.7224		
P(4)	1069.3864 5211	0.0059	3205 9399.3031		
P(3)	1070.1376 0253	0.0060	3208 1918.2262		
P(2)	1070.8831 2259	0.0061	3210 4268.3552		
P(1)	1071.6230 0816	0.0061	3212 6449.5665		
V(0)	1072.3572 5555	0.0061	3214 8461.7495		
R(0)	1073.0858 6151	0.0060	3217 0304.8066		
R(1)	1073.8088 2320	0.0059	3219 1978.6530		
R(2)	1074.5261 3824	0.0058	3221 3483.2169		
R(3)	1075.2378 0466	0.0056	3223 4818.4395		
R(4)	1075.9438 2093	0.0054	3225 5984.2747		
R(5)	1076.6441 8594	0.0052	3227 6980.6896		
R(6)	1077.3388 9903	0.0051	3229 7807.6639		
R(7)	1078.0279 5994	0.0049	3231 8465.1904		
R(8)	1078.7113 6887	0.0047	3233 8953.2747		
R(9)	1079.3891 2643	0.0046	3235 9271.9352		
R(10)	1080.0612 3366	0.0045	3237 9421.2029		
R(11)	1080.7276 9202	0.0045	3239 9401.1219		
R(12)	1081.3885 0340	0.0044	3241 9211.7486		
R(13)	1082.0436 7011	0.0044	3243 8853.1525		
R(14)	1082.6931 9488	0.0044	3245 8325.4153		
R(15)	1083.3370 8086	0.0044	3247 7628.6314		
R(16)	1083.9753 3162	0.0044	3249 6762.9079		
R(17)	1084.6079 5114	0.0045	3251 5728.3640		
R(18)	1085.2349 4381	0.0045	3253 4525.1314		
R(19)	1085.8563 1444	0.0044	3255 3153.3541		
R(20)	1086.4720 6823	0.0044	3257 1613.1883		
R(21)	1087.0822 1080	0.0044	3258 0028.3769		
R(22)	1087.6867 4817	0.0044	3260 8028.3769		
R(23)	1088.2856 8676	0.0043	3262 5984.1040		
R(24)	1088.8790 3338	0.0043	3264 3772.1880		

R				
R(25)	2945 4072.4807	0.0035	982.4821 0369	
R(26)	2947 0817.5447	0.0035	983.0406 5891	
R(27)	2948 7362.5966	0.0036	983.5925 4243	
R(28)	2950 3707.2326	0.0037	984.1377 4080	
R(29)	2951 9851.0371	0.0038	984.6762 4016	
R(30)	2953 5793.5817	0.0040	985.2080 2620	
R(31)	2955 1534.4257	0.0043	985.7330 8424	
R(32)	2956 7073.1163	0.0046	986.2513 9917	
R(33)	2958 2409.1878	0.0050	986.7629 5545	
R(34)	2959 7542.1621	0.0059	987.2677 3714	
R(35)	2961 2471.5487	0.0082	987.7657 2787	
R(36)	2962 7196.8443	0.0126	988.2569 1086	
R(37)	2964 1717.5329	0.0206	988.7412 6890	
R(38)	2965 6033.0860	0.0322	989.2187 8435	
R(39)	2967 0142.9621	0.0489	989.6894 3916	
R(40)	2968 4046.6072	0.0718	990.1532 1483	
R(41)	2969 7743.4542	0.1026	990.6100 9247	
R(42)	2971 1232.9234	0.1431	991.0600 5273	
R(43)	2972 4514.4219	0.1957	991.5030 7584	
R(44)	2973 7587.3442	0.2628	991.9391 4159	
R(45)	2975 0451.0716	0.3476	992.3682 2934	
R(46)	2976 3104.9725	0.4534	992.7903 1804	
R(47)	2977 5548.4022	0.5844	993.2053 8618	
R(48)	2978 7780.7031	0.7450	993.6134 1182	
R(49)	2979 9801.2044	0.9406	994.0143 7258	
R(50)	2981 1609.2224	1.1769	994.4082 4567	
R(51)	2982 3204.0601	1.4606	994.7950 0782	
R(52)	2983 4585.0076	1.7992	995.1746 3537	
R(53)	2984 5751.3418	2.2012	995.5471 0418	
R(54)	2985 6702.3264	2.6758	995.9123 8971	
R(55)	2986 7437.2121	3.2337	996.2704 6696	
R(56)	2987 7955.2367	3.8863	996.6213 1049	
R(57)	2988 8255.6246	4.6466	996.9648 9445	
R(58)	2989 8337.5873	5.5289	997.3011 9252	
R(59)	2990 8200.3233	6.5490	997.6301 7799	

R				
R(25)	3266 1392.8451	0.0043	1089.4667 9523	
R(26)	3267 8846.3029	0.0043	1090.0489 7991	
R(27)	3269 6132.8008	0.0043	1090.6255 9542	
R(28)	3271 3252.5895	0.0044	1091.1966 5010	
R(29)	3273 0205.9314	0.0045	1091.7621 5272	
R(30)	3274 6993.0998	0.0046	1092.3221 1238	
R(31)	3276 3614.3794	0.0048	1092.8765 3859	
R(32)	3278 0070.0659	0.0050	1093.4254 4121	
R(33)	3279 6360.4658	0.0052	1093.9688 3046	
R(34)	3281 2485.8964	0.0054	1094.5067 1692	
R(35)	3282 8446.6858	0.0056	1095.0391 1155	
R(36)	3284 4243.1725	0.0059	1095.5660 2563	
R(37)	3285 9875.7055	0.0061	1096.0874 7080	
R(38)	3287 5344.6439	0.0065	1096.6034 5905	
R(39)	3289 0650.3571	0.0071	1097.1140 0268	
R(40)	3290 5793.2244	0.0083	1097.6191 1437	
R(41)	3292 0773.6349	0.0105	1098.1188 0707	
R(42)	3293 5591.9873	0.0142	1098.6130 9411	
R(43)	3295 0248.6901	0.0196	1099.1019 8909	
R(44)	3296 4744.1609	0.0271	1099.5855 0595	
R(45)	3297 9078.8267		1100.0636 5893	
R(46)	3299 3253.1235	0.0372	1100.5364 6258	
R(47)	3300 7267.4961	0.0502	1101.0039 3173	
R(48)	3302 1122.3983	0.0668	1101.4660 8152	
R(49)	3303 4818.2921	0.0876	1101.9229 2736	
R(50)	3304 8355.6482	0.1135	1102.3744 8496	
R(51)	3306 1734.9456	0.1452	1102.8207 7028	
R(52)	3307 4956.6710	0.1838	1103.2617 9957	
R(53)	3308 8021.3193	0.2304	1103.6975 8933	
R(54)	3310 0929.3931	0.2863	1104.1281 5632	
R(55)	3311 3681.4023	0.3529	1104.5535 1756	
R(56)	3312 6277.8646	0.4317	1104.9736 9032	
R(57)	3313 8719.3043	0.5246	1105.3886 9208	
R(58)	3315 1006.2533	0.6335	1105.7985 4058	
R(59)	3316 3139.2500	0.7606	1106.2032 5379	

3. Laser Processes in CO$_2$

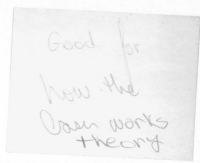

Like any other laser the basic structure of a CO$_2$ laser consists of an amplifying medium with inverted population between two mirrors [3.1]. The mirrors form a stable or an unstable resonator between which the radiation oscillates. For the CO$_2$ laser the inverted population is between molecular vibrational-rotational transitions of the electronic ground state level of the CO$_2$ molecule. The populations of the upper and lower states are obtained during an electrical discharge in a gas mixture containing CO$_2$. Other gases like N$_2$, He, H$_2$O, and Xe are added to CO$_2$ because of their favorable effects on the homogeneity of the discharge or the energy transfer processes so that a higher production rate of the inverted medium is obtained. There are many vibrational-rotational transitions in the CO$_2$ molecule for which laser action can be observed. In this chapter we shall treat the process of stimulated emission, the gain, power extraction, and the molecular energy transfer processes.

3.1 Spontaneous Emission

From the quantum theory of radiation, the transition rate for spontaneous emission from a molecule in the initial vibrational-rotational state (v', j', m') to the state (v, j, m) is given by the Einstein A-coefficient

$$A_{v'j'm' \to vjm} = \frac{16\pi^3\nu^3}{\varepsilon hc^3}|\langle v'j'm'|P|vjm\rangle|^2 \ , \qquad (3.1)$$

where P is the dipole operator, and $|\langle v'j'm'|P|vjm\rangle|$ is the electronic dipole matrix element of the full vibrational rotational wave functions. v' and v indicate the initial and final vibrational state, respectively, of the molecule. Similarly the rotational and magnetic quantum numbers of the initial state, described by j' and m', respectively, change to j and m. The rotational wave functions are given by (2.7) and for the vibrational parts we must use the expansions listed in Table 2.1. The total spontaneous emission rate for $v' \to v$ and $j' \to j$ can be obtained from (3.1) by averaging over the initial m' states and summig over all final m states. Each initial m' state has an equal probability of occurrence. Since a priori knowledge of the initial m' number is not possible, the total emission rate from an initial m' state is obtained by summing (3.1) over all possible m' states and simultaneously dividing by the number of possible m' states, i.e. $g(j')$. The total spontaneous emission rate then becomes

$$A_{v'j' \to vj} = \sum_{m'} \sum_{m} \frac{1}{2j'+1} A_{v'j'm' \to vjm} \ . \tag{3.2}$$

The matrix elements are calculated by substituting the rotational-vibrational wave functions and integrating over the rotational angles and normal vibrational coordinates. The dependence of P on the rotational co-ordinates of the molecules is found in the cosine of the angle between the E field and the molecular axis. P can be written as the product of $\cos \theta$ and a term containing the effective charge and the normal vibrational coordinates. The matrix element then becomes a product of a part $R_{v'v}$ that describes the vibrational electronic dipole element and a part that is the sum of the rotational transitions over all m' and m states. So we may write

$$\sum_{m'} \sum_{m} |\langle v'j'm'|P|vjm\rangle|^2 = |R_{v'v}|^2 \sum_{m'} \sum_{m} |\langle j'm'|\cos\theta|jm\rangle|^2 \ . \tag{3.3a}$$

Using (2.7) for the rotational functions we get for the rotational element

$$\langle j'm'|\cos\theta|jm\rangle$$
$$= \frac{1}{2\pi} \int_0^{2\pi} \int_{-1}^{+1} N_{j'm'}^{1/2} N_{jm}^{1/2} P_{j'}^{|m'|}(u) P_j^{|m|}(u) e^{-i(m'-m)\phi} du\, d\phi \ , \tag{3.3b}$$

where $u = \cos\theta$. Since the integral is non-vanishing for $m' = m$ we find by using (2.7)

$$\langle j'm|\cos\theta|jm\rangle = \left(\frac{2j'+1}{2} \frac{(j'-|m|)!}{(j'+|m|)!} \right)^{1/2}$$
$$\times \left(\frac{2j+1}{2} \frac{(j-|m|)!}{(j+|m|)!} \right)^{1/2} \int_{-1}^{+1} P_j^{|m|} P_{j'}^{|m|} u\, du \ . \tag{3.3c}$$

Next we use the recurrence relation

$$u P_j^m(u) = \frac{1}{2j+1}[(j-|m|+1)P_{j+1}^m(u) + (j+|m|)P_{j-1}^m(u)]$$

and the orthogonality relation

$$\int_{-1}^{+1} P_j^m(u) P_{j'}^m(u) du = \frac{2(j+|m|)!}{(2j+1)(j-|m|)!} \delta_{jj'} \ .$$

It is seen that the integral in (3.3c) vanishes unless

$$j' = j - 1 \quad (P \text{ transition}) \quad \text{or}$$

$$j' = j + 1 \quad (R \text{ transition}) .$$

In the following we continue with a P transition. Using the orthogonality and recurrence relations we find

$$\langle j'm | \cos \theta | jm \rangle = \left(\frac{(j^2 - m^2)}{(2j + 1)(2j - 1)} \right)^{1/2} . \tag{3.3d}$$

Next we have to sum the absolute value squared of the matrix element over all m states. Since the m state is identical to the initial m' state we have to sum over $2j - 1$ possible m states, i.e.,

$$\sum_m |\langle j'm | \cos \theta | jm \rangle|^2 = \frac{1}{(2j + 1)(2j - 1)} \sum_{m=-j+1}^{m=j-1} (j^2 - m^2) . \tag{3.3e}$$

Using $\sum_{m=1}^{j-1} m^2 = \frac{1}{6} j(j - 1)(2j - 1)$ we get

$$\sum_m |\langle j'm | \cos \theta | jm \rangle|^2 = \frac{1}{3} j . \tag{3.3f}$$

Substituting (3.1) into (3.2) and using (3.3a and f) we finally obtain for a P transition

$$A_{v'j-1 \rightarrow vj} = \frac{16\pi^3 \nu^3}{3\varepsilon h c^3} |R_{v'v}|^2 \frac{j}{2j - 1} . \tag{3.4a}$$

Similarly one obtains for a R transition

$$A_{v'j+1 \rightarrow vj} = \frac{16\pi^3 \nu^3}{3\varepsilon h c^3} |R_{v'v}|^2 \frac{j + 1}{2j + 3} . \tag{3.4b}$$

It should be noted that the difference between a P and R transition starting from the j' level is not only determined by the j values in (3.4) but also by the difference of the factor ν^3/c^3 for these transitions. These transition probabilities and the broadening of the lines can be determined from measurements of the absorption coefficients in CO_2 gas. In Table 3.1 the values of A for the transitions (00^01)-(I) and (00^01)-(II) are given for the P and R branches [3.3].

It is seen that the probabilities A for all the lines of (00^01)-(I) transition are smaller than those of the appropriate lines of the (00^01)-(II) transition. Table 3.1 contains also the collisional line-broadening per torr for each transition. The broadening for the lines of the (00^01)-(I) transition are larger than those for the (00^01)-(II) transition.

Table 3.1

j for lower level	A_P [s^{-1}]	$2\Delta\nu_P$ [MHz/ torr]	A_R [s^{-1}]	$2\Delta\nu_P$ [MHz/ torr]	A_P [s^{-1}]	$2\Delta\nu_P$ [MHz/ torr]	A_R [s^{-1}]	$2\Delta\nu_P$ [MHz/ torr]
	(00^01)-(I) transition				**(00^01)-(II) transition**			
	P branch		**R branch**		**P branch**		**R branch**	
6	0.185	8.72	0.161	8.41	0.198	8.66	0.165	7.88
8	0.183	8.61	0.163	8.30	0.196	8.45	0.168	7.78
10	0.181	8.42	0.166	8.17	0.194	8.31	0.171	7.68
12	0.179	8.28	0.168	8.09	0.192	8.15	0.174	7.58
14	0.178	8.10	0.170	7.94	0.190	8.01	0.177	7.48
16	0.176	8.02	0.172	7.85	0.188	7.84	0.179	7.42
18	0.174	7.86	0.174	7.76	0.186	7.70	0.182	7.30
20	0.173	7.70	0.177	7.62	0.185	7.56	0.185	7.21
22	0.171	7.54	0.179	7.51	0.183	7.38	0.188	7.10
24	0.169	7.43	0.181	7.39	0.181	7.24	0.191	6.99
26	0.167	7.25	0.183	7.29	0.179	7.10	0.194	6.91
28	0.165	7.09	0.186	7.17	0.177	6.94	0.196	6.84
30	0.163	7.00	0.188	7.06	0.175	6.77	0.199	6.72
32	0.162	6.81	0.190	6.94	0.173	6.61	0.202	6.63
34	0.160	6.67	0.193	6.84	0.171	6.47	0.205	6.54

3.2 Stimulated Emission

In the presence of external radiation, the rates for stimulated emission or absorption of this radiation by the excited molecules can be expressed in terms of the Einstein B coefficient. The probability per second that the incident monochromatic radiation at frequency ν induces a molecular transition from the initial state $(v'j'm')$ to the state (vjm) is given by

$$\Gamma_{v'j'm'\rightarrow vjm} = \frac{I(\nu)}{c} S(\nu,\nu_{12}) B_{v'j'm'\rightarrow vjm} \ , \tag{3.5}$$

where $I(\nu)$ is the incident radiant power per unit area, $S(\nu,\nu_{12})$ is the normalized line shape function for molecular resonance centered at the frequency, $\nu_{12} = \nu(v'j'\rightarrow vj)$, between the two states. The Einstein B coefficient is related to the Einstein A coefficient by

$$B_{v'j'm'\rightarrow vjm} = \frac{c^3}{8\pi h\nu_{12}^3} A_{v'j'm'\rightarrow vjm} \ . \tag{3.6}$$

According to quantum theory each induced transition probability is equal to its inverse. This can also be obtained from the so-called "principle of detailed balancing", which says that the rates for forward and reverse elementary processes are equal for a system in equilibrium. Thus we have

$$B_{v'j'm'\rightarrow vjm} = B_{vjm\rightarrow v'j'm'} \ . \tag{3.7}$$

Similar to the previous section the total stimulated rates for the processes $(v'j')\rightarrow(vj)$ and the reverse $(vj)\rightarrow(v'j')$ can be obtained by averaging over all initial m' or m states and summing over the final states. We obtain

$$B_{v'j'\rightarrow vj} = \sum_{m'}\sum_{m} \frac{1}{2j'+1} B_{v'j'm'\rightarrow vjm} \ , \tag{3.8}$$

$$B_{vj\rightarrow v'j'} = \sum_{m}\sum_{m'} \frac{1}{2j+1} B_{vjm\rightarrow v'j'm'} \ . \tag{3.9}$$

Using (3.7) it follows that the reduced Einstein B coefficients satisfy the relation

$$(2j+1)B_{vj\rightarrow v'j'} = (2j'+1)B_{v'j'\rightarrow vj} \tag{3.10}$$

and with (3.6)

$$B_{v'j'\rightarrow vj} = \frac{c^3}{8\pi h\nu_{12}^3} A_{v'j'\rightarrow vj} \ . \tag{3.11}$$

The probability per second that the incident monochromatic radiation induces a P transition is found by substituting $j = j'+1$ into (3.11). Making use of (3.4a) we find

$$B_{v'j-1\rightarrow vj} = \frac{2\pi^3 j|R_{v'v}|^2}{3\varepsilon h^2(2j-1)} \ ; \tag{3.12}$$

and similarly for the R transition with $j = j' - 1$

$$B_{v'j+1\rightarrow vj} = \frac{2\pi^3(j+1)|R_{v'v}|^2}{3\varepsilon h^2(2j+3)} \ . \tag{3.13}$$

The probability per second of an induced molecular transition can also be expressed in terms of the product of the incident photon density and a cross section $\sigma_{v'j'\rightarrow vj}$ for stimulated emission or absorption, so that

$$\frac{I(\nu)}{c} S(\nu,\nu_{12}) B_{v'j'\rightarrow vj} = \frac{I(\nu)}{h\nu} \sigma_{v'j'\rightarrow vj} \ . \tag{3.14}$$

Using (3.11) we find for the stimulated emission cross section

$$\sigma_{v'j'\rightarrow vj} = \frac{\lambda^2}{8\pi} S(\nu,\nu_{12}) A_{v'j'\rightarrow vj} \ . \tag{3.15}$$

3.3 Laser Gain

The stimulated emission and absorption of incident monochromatic radiation leads to a change of the radiant power. The net increase dI of the incident monochromatic beam at frequency ν travelling a distance dx of the active medium is given by

$$dI = \frac{h\nu}{c}(n_{v'j'}B_{v'j'\to vj} - n_{vj}B_{vj\to v'j'})IS(\nu,\nu_{12})dx \; , \tag{3.16}$$

where $n_{v'j'}$ and n_{vj} are, respectively, the number of molecules per unit volume in the upper and lower state. Expressions for $n_{v'j'}$ and n_{vj} can be obtained from (2.20). The gain α is defined as the relative increase of radiant power of energy per unit length, or

$$\alpha = \frac{1}{I}\frac{dI}{dx} \; . \tag{3.17}$$

From (3.17) we obtain

$$\alpha = \frac{h\nu}{c}(n_{v'j'}B_{v'j'\to vj} - n_{vj}B_{vj\to v'j'})S(\nu,\nu_{12}) \; . \tag{3.18}$$

The gain of a P transition as a function of j can be obtained by substituting (3.10, 11) and (2.18) into (3.18), i.e.,

$$\alpha_{\mathrm{P}}(j) = \frac{\lambda^2 hcB}{4\pi kT}(2j-1)A_{\mathrm{P}}S(\nu,\nu_{12})$$
$$\left\{ N_{v'}\exp\left[-F(j-1)\frac{hc}{kT}\right] - N_v\exp\left[-F(j)\frac{hc}{kT}\right]\right\} \; ; \tag{3.19}$$

and similarly for a R transition as a function of j

$$\alpha_{\mathrm{R}}(j) = \frac{\lambda^2 hcB}{4\pi kT}(2j+3)A_{\mathrm{R}}S(\nu,\nu_{12})$$
$$\left\{ N_{v'}\exp\left[-F(j+1)\frac{hc}{RT}\right] - N_v\exp\left[-F(j)\frac{hc}{kT}\right]\right\} \; , \tag{3.20}$$

where A_{P} and A_{R}, the spontaneous transition probability of the P and R branch, respectively, are given in Table 3.1 as a function of j.

3.4 Line Shape

A very important parameter that determines the gain of the laser medium is the line shape function S, as mentioned in Sect. 3.3. This function describes the spectral distribution of the radiative spontaneous decay. In general, this radiative transition has a spectral distribution that is broadened due to lifetime effects, collisions or inhomogeneities. The probability density that the observed emission will be in the frequency interval between ν and $\nu + d\nu$ is given by $S(\nu, \nu_{12})$ which is normalized to unity, i.e.

$$\int\limits_{0}^{\infty} S(\nu, \nu_{12}) d\nu = 1 \ . \tag{3.21}$$

At low gas pressures (below about 10 torr) medium inhomogeneities due to the Doppler shifts of the transition frequencies of the molecules determine the line shape. If v_x is the component of the velocity of the molecule in the direction of observation then the transition frequency is Doppler-shifted to

$$\nu = \nu_{12}\left(1 + \frac{v_x}{c}\right) \ , \tag{3.22}$$

where c is the velocity of light.

If it is assumed that the molecular velocities are distributed according to the Maxwell distribution of a gas with molecular mass M and temperature T we have the following distribution function

$$f(v_x, v_y, v_z) = \left(\frac{M}{2\pi kT}\right)^{3/2} \exp\left[-\frac{M}{2kT}(v_x^2 + v_y^2 + v_z^2)\right] \ . \tag{3.23}$$

Integration of $f(v_x, v_y, v_z)$ over all velocity space gives unity. The probability $S(\nu, \nu_{12}) d\nu$ that the transition frequency is between ν and $\nu + d\nu$ is equal to the probability that v_x will be found between $v_x = (\nu - \nu_{12})c/\nu_{12}$ and $(\nu + d\nu - \nu_{12})c/\nu_{12}$ irrespective of the values of v_y and v_z.

Although an emitting molecule at the Doppler-shifted frequency ν has its own spectrum due to its lifetime, its width [about a few tenths of one Hz] is much less than the Doppler width of about 150 MHz at 400 K so that the broadening by the natural lifetime is negligible. The probability $S(\nu, \nu_{12}) d\nu$ is thus obtained by substituting $v_x = (\nu - \nu_{12})c/\nu_{12}$ into $f(v_x, v_y, v_z) dv_x dv_y dv_z$, and then integrating over all values of v_y and v_z. The result gives

$$S(\nu, \nu_{12}) = \frac{c}{\nu_{12}}\left(\frac{M}{2\pi kT}\right)^{1/2} \exp\left[-\frac{Mc^2}{2kT\nu_{12}^2}(\nu - \nu_{12})^2\right] \tag{3.24}$$

for the Doppler-broadened line shape. The width of $S(\nu, \nu_{12})$ is taken as the frequency separation between the points where $S(\nu, \nu_{12})$ has half its maximum value. This Doppler width obtained from (3.24) is given by

$$\Delta\nu_D = 2\nu_{12}\sqrt{\frac{2kT}{Mc^2}\ln 2} \ . \tag{3.25}$$

In terms of $\Delta\nu_D$ the Doppler line shape is written as

$$S(\nu,\nu_{12}) = \frac{2}{\Delta\nu_D}\sqrt{\frac{\ln 2}{\pi}}\exp\left(-\frac{4(\ln 2)(\nu-\nu_{12})^2}{\Delta\nu_D^2}\right) \ . \tag{3.26}$$

This Doppler line shape is applicable for continuously operating gas discharge CO_2 lasers (Chap. 4).

At much higher gas pressures (above 50 torr CO_2) the line shape results from collisional processes that limit the lifetime of excited states, modulate their energy levels, or interrupt the phases of the radiating molecules and thereby modulate the emission frequencies. The effects of the collisional processes have been calculated in the past [3.4] and result to a collision-broadened line shape given by

$$S(\nu,\nu_{12}) = \frac{\Delta\nu_P}{2\pi}\left[(\nu-\nu_{12})^2 + \left(\frac{\Delta\nu_P}{2}\right)^2\right]^{-1} , \tag{3.27}$$

where $\Delta\nu_P$ is the pressure-broadened linewidth and corresponds to the frequency separation between the points where $S(\nu,\nu_{12})$ has half its maximum value. It is seen from (3.27) that pressure broadening leads to a Lorentzian line shape.

The broadening mechanism and the transition probability can be obtained from absorption measurements in CO_2 and mixtures of CO_2 and other gases used in CO_2 lasers [3.3, 5, 6]. At low densities where the line is Doppler broadened the line-center absorption coefficient for the $P(j)$ line of a laser beam passing through an absorption cell can be obtained from (3.19) by using the Doppler profile given by (3.24)

$$\alpha_P(j) = \frac{\lambda^2 hcB}{4\pi kT}(2j-1)A_P\frac{c}{\nu_{21}}\left(\frac{M}{2\pi kT}\right)^{1/2}$$
$$\left\{N_v\exp\left[-F(j)\frac{hc}{kT}\right] - N_{v'}\exp\left[-F(j-1)\frac{hc}{kT}\right]\right\} , \tag{3.28}$$

where again N_v and $N_{v'}$ are the densities of the vibrational states (I) and (00^01), respectively, and A_P is the transition probability. It is seen that the absorption coefficient is proportional to the density. At thermal equilibrium all quantities except A_P are known for the absorption cell so that the measured absorption yields the quantity A_P. At high densities where the line is pressure broadened the absorption at line center is expressed by

$$\alpha_P(j) = \frac{\lambda^2 hcB}{4\pi kT}(2j-1)A_P\frac{2}{\pi\Delta\nu_P}$$
$$\left\{N_v\exp\left[-F(j)\frac{hc}{kT}\right] - N_{v'}\exp\left[-F(j-1)\frac{hc}{kT}\right]\right\} \qquad (3.29)$$

where we have used (3.27).

In thermal equilibrium both the populations and line width $\Delta\nu_P$ are proportional to the density so that the absorption coefficient is independent of the density in the pressure-broadened limit. Measurements of the absorption coefficient in this limit yield the ratio $A_P/\Delta\nu_P$ since all other quantities in (3.29) are known at equilibrium. The absorption measurements in pure CO_2 over a range of temperatures from 298 to 404 K and over a range of densities from 0.4 to 760 torr were first obtained by *Gerry* and *Leonard* [3.5] and are shown in Fig. 3.1. More accurate measurements for various P and R lines were performed by *Nevdakh* [3.3] and are shown in Table 3.1. In most practical cases the small dependence of $\Delta\nu_P$ on the transition is not important.

The pressure broadened line width in the laser gas mixture can then be simply obtained as the sum of the contributions of the individual components and is usually written in the following form [3.7, 8]

$$\Delta\nu_P = 7.58(\psi_{CO_2} + 0.73\psi_{N_2} + 0.64\psi_{He})p\left(\frac{300}{T}\right)^{1/2} \text{MHz} \qquad (3.30)$$

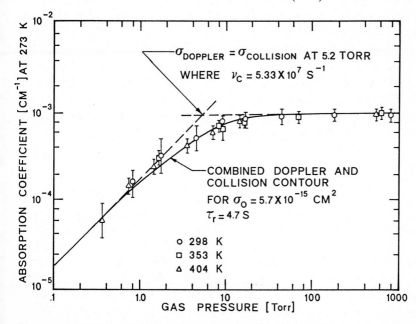

Fig. 3.1. Absorption coefficient of CO_2 gas for the $P(20)$ transition of the CO_2 laser as a function of CO_2 density. The absorption coefficients are scaled to 273 K [3.5]

where ψ_x is the fraction of the gas component x and p is the total gas pressure in torr.

3.5 Gain Saturation

Substituting (3.10, 11 and 15) into (3.16) we can write for the gain

$$\frac{dI}{I\,dx} = \left(n_{v'j'} - \frac{2j'+1}{2j+1} n_{vj} \right) \sigma_{v'j' \to vj} \ . \tag{3.31}$$

Each quantum added to the radiation field decreases the inversion density by two particles provided that the relaxation rate of the lower level is much slower than the process of stimulated emission. Therefore the inversion density obeys the following rate equation

$$\frac{d}{dt}\left(n_{v'j'} - \frac{2j'+1}{2j+1} n_{vj} \right)$$
$$= -\left(n_{v'j'} - \frac{2j'+1}{2j+1} n_{vj} \right) \left(\frac{2I\sigma_{v'j' \to vj}}{h\nu} + \frac{1}{\tau} \right) + P \ , \tag{3.32}$$

where τ is a time constant for the loss of inversion by collisional relaxation processes and spontaneous decay and, P is a source term describing the production rate of inversion per unit volume. If we consider the inversion density as stationary and we substitute this value for the inversion obtained from (3.32) into (3.31) we obtain

$$\frac{dI}{I\,dx} = \frac{P\tau\sigma_{v'j' \to vj}}{1 + I/I_0} \ , \tag{3.33}$$

where the saturation intensity I_0 is given by

$$I_0 = \frac{h\nu}{2\tau\sigma_{v'j' \to vj}} \ . \tag{3.34}$$

When the population of the lower level is much faster than the stimulated emission process or in the case of a stationary radiation intensity, we have for each quantum produced by stimulated emission a decrease of only one particle in the inversion. In the final result for the gain we find a saturation intensity given by

$$I_0 = \frac{h\nu}{\tau\sigma_{v'j' \to vj}} \ . \tag{3.35}$$

The term $P\tau$ in (3.33) is for the stationary case equal to the inversion density in the absence of stimulated emission. The small-signal gain α_0 is thus

$$\alpha_0 = \left(n_{v'j'}^0 - \frac{2j'+1}{2j+1} n_{vj}^0 \right) \sigma_{v'j' \to vj} \; , \tag{3.36}$$

where n_{vj}^0 is the density in the absence of stimulated emission. The gain can be written as

$$\alpha = \frac{\alpha_0}{1 + I/I_0} \; . \tag{3.37}$$

The stimulated emission cross section (3.15) depends on the line shape. Since the laser line width is much smaller than the emission line of the medium one might expect for low gas pressure a depletion of those molecules having a Doppler frequency close to the laser frequency because the beam interacts only with the molecules that have a Doppler shift within the pressure broadened line width around the laser frequency. This "hole" in the normal distribution will be restored by collisional processes of other molecules taking the place of those that were lost. Even for low-pressure systems the lifetime of the upper level of about 10^{-3} s [3.9] is much larger than the collisional time of about 10^{-8} s at 10 torr, so that the lost molecules will be replaced very fast; all excited molecules will be shifted during their life to the laser frequency. This means that the distortion of the line shape by the radiation field will be relatively small, and secondly the radiation field interacts with all excited molecules so that the interaction can be considered as homogeneous. The medium is called homogeneously broadened. The same is even more true for high-pressure systems with a pressure-broadened profile because the radiation field interacts directly with all excited molecules belonging to the same profile.

Finally we shall make two remarks. First the saturation intensity I_0 increases with increasing frequency shift of the laser line from the line center of the medium because I_0 is inversely proportional to $S(\nu, \nu_{12})$. Second for a Doppler width of 150 MHz the stimulated emission cross section at line center of the $P(20)$ is about 0.5×10^{-16} cm^2. The saturation intensity calculated for a lifetime τ of 10^{-3} s is about 0.4 W/cm^2, which is much smaller than the measured values of 5 W/cm^2 [3.10]. The main reason for this discrepancy is the fact that the CO_2 laser has many rotational levels closely coupled by collisions. Similar to the collisional effect on the Doppler shift there is also a continuous exchange between the rotational energies of the excited molecules during collisions. The rotational energy exchange is also very fast because the energy difference between the rotational levels is small compared to the average kinetic energy. An excited molecule will therefore always arrive at the lasing transition during its life. This means effectively that all rotational levels are coupled to the radiation field. The observed saturation intensity that gives to all excited molecules a stimulated emission lifetime equal to the spontaneous decay is therefore much larger than the calculated one for which only one level was considered. The difference must be roughly a factor equal to the ratio of the total number of excited molecules and the number of the lasing rotational transitions. This is in good agreement with the observed value of $5\,W/cm^2$.

3.6 The Temperature Model of the Laser Process

Laser action in CO_2 molecules involves, in fact, transitions between different degrees of freedom, i.e., different vibrational modes. Within each degree of freedom the thermalization is much faster than among the different vibrations. This is because there is – apart from a small anharmonic defect – no excess energy to be exchanged with translation. Each vibrational mode can therefore be described by a temperature. Among the vibrational modes the excess energy to be exchanged with translation during collisions in a relaxation process can be considerable compared with kT (T being the translational temperature). Collisions with an excess energy ΔE, much larger than kT have very small probabilities. The larger $\Delta E/kT$ the slower the energy exchange.

The excitation process by an electrical discharge is different for the vibrational modes. Usually the discharge conditions are most favorable for the vibrational excitation of N_2 and much less effective for the ν_1 and ν_2 vibrations of CO_2. Each vibrational temperature may then be different. The differences tend to decrease by relaxation processes. The relaxation between translation and rotation is very fast and is comparable with the thermalization with a vibrational mode. The translation and rotation can therefore be described by the same temperature.

The relaxation process between the various degrees of freedom can be elegantly described in terms of temperatures provided all vibrational modes may be treated as perfectly harmonic. The laser process based on population inversion between different vibrational modes can also be easily incorporated into a temperature model [3.11, 12]. In the "five-temperature" model we assign three vibrational temperatures for CO_2, one for N_2, and one for translation and rotation. Under certain conditions the temperatures of the ν_1 and ν_2 vibration are practically equal, so that a "four temperature" model is adequate.

3.7 Vibrational Excitation of the Upper Laser Level

Practically all efficient CO_2-laser systems use N_2 as an additional channel for energy transfer to the ν_3 vibration of CO_2. It has been observed that an electric discharge in nitrogen leads to a very effective formation of vi-

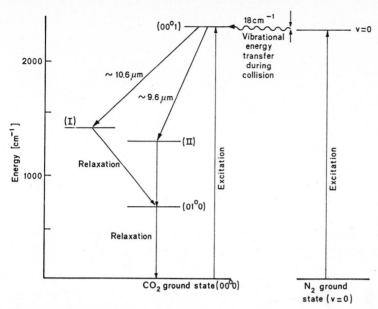

Fig. 3.2. (a) Some of the low-lying vibrational levels of the CO_2 molecule. **(b)** Ground state $(v = 0)$ and the first excited state $(v = 1)$ of the N_2 molecule, which plays an important role in the selective excitation of the (00^01) level of CO_2

brationally excited N_2 molecules up to 50 % of all N_2 molecules [3.13,14]. Since the N_2 molecule has two identical nuclei, its dipole radiation is forbidden. It can only decay by collisions with the wall of the containing vessel or by collisions with other molecules. In the presence of CO_2 the vibrational energy of N_2 can be easily transfered to CO_2 because of the close resonance between the N_2 vibrations and the ν_3 vibration of CO_2. The (00^01) level of CO_2 is only $\Delta E = 18\,\mathrm{cm}^{-1}$ higher than the vibrational level $\nu = 1$ of nitrogen, as can be seen in Fig. 3.2. This energy difference is much smaller than the average kinetic energy so that during collisions the CO_2 molecules can easily draw the vibrational energy of N_2 to excite the ν_3 vibration. The rate constant is $1.9 \times 10^4\,\mathrm{torr}^{-1}\,\mathrm{s}^{-1}$ [3.15]. It should be emphasized that the efficient transfer is not limited to the first excited states of N_2 but also up to values of $\nu = 4$ in N_2 because even at this level the anharmonicity of the N_2 molecule still does not lead to quantum values differing from that of the (00^01) level by more than the average kinetic energy.

Accompanying to the energy resonance between CO_2 and N_2 there is a similar effect between CO and CO_2. During the discharge an appreciable number of CO molecules can be produced. This is especially the case in laser gas mixtures without H_2O or H_2 as will be discussed later for sealed-off systems. In these dry mixtures the dissociation of CO_2 can be as high as 50 %. (The dissociation energy of CO_2 is only 2.8 eV). These CO molecules can transfer a considerable amount of energy to the ν_3 vibration because

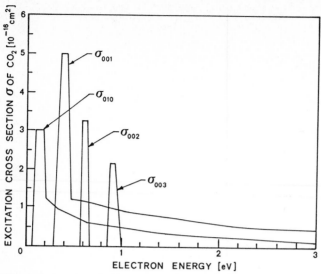

Fig. 3.3. Cross sections for vibrational excitation of CO_2 by electron impact [3.16]

the cross section for vibrational excitation of CO is rather large as we shall see further on, and secondly because the difference between the energies of the vibrational level of CO and the (00^01) level of CO_2 is $170\,cm^{-1}$ which is smaller than the average kinetic energy kT. However, CO is not as effective as N_2 in exciting the ν_3 vibration of CO_2. This is understandable since the excess energy coming from the translation is for CO larger than for N_2 and furthermore since the CO molecule unlike N_2 has a dipole moment it has also spontaneous decay.

Next we shall turn our attention to the vibrational excitation cross section of the molecules. In Fig. 3.3 we show the experimentally obtained cross sections by *Hake* and *Phelps* [3.16] for the vibrational excitation of CO_2 by electrons as a function of their energy. The figure shows that there are four resonances at 0.08, 0.3, 0.6, and 0.9 eV, two of which have high energy tails. The 0.3, 0.6 and 0.9 eV energy loss processes are, according to *Hake* and *Phelps*, most likely associated with the three lowest levels of the ν_3 vibration, i.e. (00^01), (00^02) and (00^03). The 0.08 eV energy loss is associated with the lowest ν_2 vibration. *Schulz* [3.17, 18, 19] investigated experimentally the cross sections for CO and N_2. The observed cross section of the individual vibrational levels are unusually large, which is related to the resonance effect of short-lived negative ions N_2^- and CO^-. *Schulz* found that the cross sections of the levels from the first to the fourth are comparable with one another, and that the cross section of the seventh and eighth are smaller by an order of magnitude. The total cross section for the excitation from the first to the eighth level are shown in Figs. 3.4, 5 for N_2 and CO, respectively. It is seen from these figures that the total cross sections are very

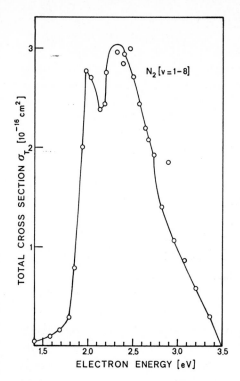

Fig. 3.4. Total effective cross section for vibrational excitation of N_2 ($v = 1-8$) by electron impact [3.17, 19]

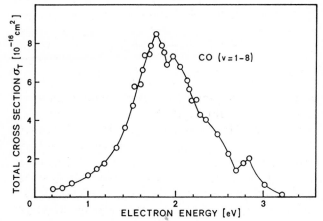

Fig. 3.5. Total effective cross section for vibrational excitation of CO ($v = 1-8$) by electron impact [3.17, 19]

large at electron energies from 1.7 to 3.5 eV in the case of N_2 and at electron energies from 1.0 to 3.0 eV in the case of CO. Fortunately, the experimental conditions for the discharge of a continuous CO_2 laser such as given by gas composition, total pressure, and tube diameter can be adjusted in such a way that the energy range of the electrons is from 1 to 3 eV [3.20, 21, 22]. For pulsed systems too the energy distribution of the electrons can be optimized [3.23].

Finally, we remark that in a system optimized with respect to electron energy distribution the cross section for the lower laser level is negligible. However, at a much higher electron energy of about 4 eV the cross section for the (00^01) level is much smaller whereas the cross section for the lower level is greatly enhanced. Laser action is then likely to be suppressed since population inversion is reduced [3.15].

3.8 Relaxation Phenomena and Vibrational Temperatures

Within each of the vibrational modes the levels are strongly coupled by collisions to reach equilibrium with each other. This process is very rapid because the near-resonant vibration-vibration $(V \rightarrow V)$ energy transfer has only a very small energy exchange with translation due to the anharmonicity of the vibration. For instance

$$2CO_2(00^01) \rightleftharpoons CO_2(00^02) + CO_2(00^00) + \Delta E = 25\,cm^{-1}\ .$$

This means that each vibration can be described by its own temperature. It is also expected that the ν_1 and ν_2 vibrations are in rapid equilibrium with each other because of the near-resonant processes. For instance

$$CO_2(I) + CO_2(00^00) \rightleftharpoons 2CO_2(01^10) + \Delta E = 54\,cm^{-1}\ ,$$

$$CO_2(I) + M \rightleftharpoons CO_2(II) + M + \Delta E = 103\,cm^{-1}\ ,$$

$$CO_2(I) + M \rightleftharpoons CO_2(02^00) + M + \Delta E = 52\,cm^{-1}\ ,$$

$$CO_2(II) + CO_2(00^00) \rightleftharpoons 2CO_2(01^10) + \Delta E = -46\,cm^{-1}\ ,$$

$$CO_2(02^20) + CO_2(00^00) \rightleftharpoons 2CO_2(01^10) + \Delta E = 1\,cm^{-1}\ ,$$

where M means any colliding particle.

Apart from the small energy exchange with translation there is also the additional coupling by Fermi resonance, as we have seen. For cw systems the ν_1 and ν_2 vibrations can therefore be described by the same temperature T_2. In pulsed systems, however, there may be large energy transfer during the gain switched peak and T_1 and T_2 are not practically equal (Sect. 6.5). The temperature of the ν_3 vibration will be indicated by T_3. Similarly the vibrational energy of the N_2 molecule can be described by T_4.

As well as vibrational temperatures, the translational and rotational energies can also be described by their temperatures. However, the rotational relaxation time is also very small, about 10^{-7} s at one torr, so that rotation and translation can be considered in equilibrium, and their energy is described with the same temperature T. The CO_2 laser is, depending on the conditions, a five- or four-temperature system because the coupling between the energy groups 0,1,2,3, and 4 with temperatures T, T_1, T_2, T_3, and T_4, respectively, is much weaker than that within each group. In the following we consider $T_1 = T_2$ so that we are dealing with a four-temperature model.

Since the laser gain depends on the inversion, efficient depopulation of the lower laser level is required. This occurs by energy relaxation from group 2 to group 0. For continuous systems the photon production rate by stimulated emission is equal to the relaxation rate of energy quanta of group 2. The output of the laser is limited by this relaxation rate. It has been determined [3.25] that the relaxation phenomena in CO_2 occur by a direct relaxation of the ν_2 vibration to the translational degrees of freedom and an indirect relaxation of the ν_1 vibration to the ν_2 vibration. The latter means that because of the near resonance between the two vibrations, the energy quanta of the ν_1 vibration are first transferred, as twice as many quanta, to the ν_2 vibration, which in their turn are transferred to the translation. This process is much faster than a direct exchange with translation because the energy quanta to be exchanged with translation are for the indirect case only half as large.

The energy relaxation of the lower laser level can therefore be described by

$$\frac{dE_2(T_2)}{dt} = -\frac{1}{\tau_2}[E_2(T_2) - E_2(T)] \ , \qquad (3.38)$$

where $E_2(T)$ is the energy of the ν_2 vibration for the temperature T.

In order to make an estimate of the temperatures involved we consider a sealed-off continuous system of 130 cm discharge length and 10 mm internal diameter. The tube is filled with 2.5 torr CO_2, 3.5 torr N_2, 12 torr He, 0.2 torr Xe, and 0.2 torr H_2O. Having an outcoupling of 78 % an output power of 70 W can be obtained for a discharge current of about 30 mA. This corresponds to roughly 2 W/cm^3 in the case of single-Gaussian-mode operation. Using the experimentally obtained relaxation rates k_2 (Table 3.3) for the gas components, we find for this laser mixture $1/\tau_2 = 1.25 \times 10^5\,\mathrm{s}^{-1}$. The energy relaxation of group 2 is equal to

$$\frac{h\nu_1}{h(\nu_3 - \nu_1)} \cdot 2\,\mathrm{W} \sim 3\,\mathrm{W/cm}^3 \ .$$

From this we find $\Delta E = [E_2(T_2) - E_2(T)] = \tau_2 \cdot 3\,\mathrm{W} = 2.4 \times 10^{-5}\,\mathrm{J}$. It is easy to verify that this small value of ΔE corresponds to a slight increase of T_2 with respect to T. For $T = 400\,\mathrm{K}$ we find $T_2 = 415\,\mathrm{K}$. Even in the absence of water vapour ΔE would be $9 \times 10^{-5}\,\mathrm{J}$ corresponding to a temperature difference of $T_2 - T = 56\,\mathrm{K}$.

For the lasing system, T_3 is coupled to T_2 by the radiation field. Since the total radiation field in the laser cavity is in equilibrium with the inversion, the inversion density is only determined by the gain which in turn must be equal to all losses such as outcoupling, absorption and diffraction. The gain of our system is about 0.1 % per cm and the Doppler width 150 MHz. Using (3.19) we find that for the lasing $P(20)$ transition the difference between upper and lower level $\Delta n = 4.24 \times 10^{13}\,\mathrm{cm}^{-3}$. Knowing T_2 we calculate N_{I}, the density of all particles in state (I) and obtain $1.1 \times 10^{15}\,\mathrm{cm}^{-3}$. From this we find for the density of the lower state $n_{\mathrm{I}}(20) = 6.94 \times 10^{13}\,\mathrm{cm}^{-3}$. The density N_{00^01} of the upper level is then given by

$$N_{00^01} = [\Delta n + n_{\mathrm{I}}(20)]\left(\frac{kT}{2hcB}\right)\frac{\exp[F(j-1)\frac{hc}{kT}]}{2j-1}$$
$$= 1.74 \times 10^{15}\,\mathrm{cm}^{-3} \ .$$

The total inversion $\Delta N = N_{00^01} - N_{\mathrm{I}} = 6.4 \times 10^{14}\,\mathrm{cm}^{-3}$. Knowing N_{00^01} we calculate the temperature of the ν_3 vibration. We find $T_3 = 857\,\mathrm{K}$.

Since each vibration is itself in equilibrium and can be described by its temperature, the energy transfer between the vibration of N_2 having the temperature T_4 and the ν_3 vibration of CO_2 also obeys a simple relationship. The energy transfer to the ν_3 vibration is then

$$\frac{dE_3(T_3)}{dt} = \frac{1}{\tau_{43}}[E_3(T_4) - E_3(T_3)] \ , \tag{3.39}$$

where E_3 is the energy of the ν_3 vibration. The time constant τ_{43} is inversely proportional to the N_2 density. For our density of 3.5 torr N_2 we find $\tau_{43} = 1.4 \times 10^{-5}\,\mathrm{s}$ [3.26].

The vibrational energy transfer from the nitrogen, if we assume that by far most of the vibrational excitation occurs by N_2, is equal to

$$\frac{h\nu_3}{h(\nu_3 - \nu_1)} \cdot 2\,\mathrm{W} \sim 5\,\mathrm{W/cm}^3 \ .$$

From this we find $[E_3(T_4) - E_3(T_3)] = \tau_{43} \cdot 5\,\mathrm{W} = 7 \times 10^{-5}\,\mathrm{J}$. Knowing T_3 we can now calculate T_4. In our example we find $T_4 = 1012\,\mathrm{K}$. In Table 3.2 we show the calculated values of T_2, T_3, and T_4 of the above described laser system for various gas temperatures T. The deduced temperatures are obtained for a lasing system. For a non-lasing discharge the temperatures T_3 and T_4 must be considerably higher because T_3 and T_2 are then no longer coupled by the radiation field.

It is seen that T_3 and T_4 increase considerably with T. The larger T_4 the more difficult the excitation by the hot electrons, because this energy transfer process is also expected to depend on the difference between the

Table 3.2. Vibrational temperatures [K] of nitrogen and the groups 0, 2, 3, respectively

T	T_2	T_3	T_4
400	415	857	1012
500	514	1025	1147
600	613	1203	1305
700	712	1387	1476

average energies. Furthermore the collisional deexcitation of the ν_3 vibration increases considerably with T. Efficient pumping thus requires low T values and this is obtained by cooling or flowing the gas.

3.9 Gain Measurements and Vibrational Temperatures

Quantitative information on the vibrational temperatures of the excited CO_2 molecules in a discharge can be obtained by measuring the gain of regular, sequence, and hot bands. (See the energy level diagram of Fig. 2.3). Since we may characterize the CO_2 discharge by a five-temperature model, the population density of all levels can be conveniently expressed in terms of temperatures. For reasons of simplicity, we introduce

$$q = \exp(-h\nu_1/kT_1) \ , \tag{3.40a}$$

$$r = \exp(-h\nu_2/kT_2) \ , \tag{3.40b}$$

$$s = \exp(-h\nu_3/kT_3) \ , \tag{3.40c}$$

and the partition function Q given by

$$Q(q,r,s) = [(1-q)(1-r)^2(1-s)]^{-1} \ . \tag{3.41}$$

If the vibrational modes of CO_2 may be approximated by simple harmonic oscillators, the population density of the upper level (00^01) is given by

$$N_{00^01} = N_{CO_2}\frac{s}{Q} \tag{3.42}$$

and the densities of the lower levels (I) and (II) by

$$N_I = N_{CO_2}\frac{q}{Q} \ , \tag{3.43a}$$

$$N_{II} = N_{CO_2}\frac{r^2}{Q} \ . \tag{3.43b}$$

Looking at (3.19 or 20) it is seen that the gain depends on the gas temperature T, j value, line shape S, and the transition probability A. For two bands we may assume the same T and S values. With respect to j we are dealing with even j numbers for the regular band and with odd values for the sequence band. However, the gain that should correspond with an equal j value, on the other band, can be easily estimated by averaging the measured gain values for the two adjacent j values. In this procedure we take the gain ratio for equal j values. Substituting (3.4) into (3.19 or 20) we obtain for the gain ratio of "corresponding" lines in the sequence and regular band, respectively,

$$\frac{\alpha_s}{\alpha_r} = \left|\frac{R_{00^02-Is}}{R_{00^01-Ir}}\right|^2 \frac{N_{00^02} - uN_{Is}}{N_{00^01} - uN_{Ir}} \tag{3.44}$$

where $u = \exp\{[F(j-1) - F(j)]hc/kT\}$. The additional subscripts s and r refer to sequence and regular bands, respectively. The above expression holds for both P and R transitions.

For a harmonic oscillator one derives for the matrix element between the states n and $n+1$ by using the appropriate wave functions the following relation

$$|R_{n-n+1}|^2 = (n+1)|R_{0-1}|^2 . \tag{3.45}$$

Considering the vibrational modes as harmonic oscillators we obtain

$$|R_{00^02-10^01}|^2 = 2|R_{00^01-10^00}|^2 = 2M_1 \tag{3.46a}$$

$$|R_{00^02-02^01}|^2 = 2|R_{00^01-02^00}|^2 = 2M_2 . \tag{3.46b}$$

If we now assume that the perturbed lower laser levels (I) and (II) of the two bands can be described with the same coefficients a and b in terms of the unperturbed wave functions, as presented by (2.27), we derive

$$|R_{00^01-Ir}|^2 = a^2M_1 + b^2M_2 - 2ab\sqrt{M_1M_2} \quad \text{and} \tag{3.47a}$$

$$|R_{00^02-Is}|^2 = 2(a^2M_1 + b^2M_2 - 2ab\sqrt{M_1M_2}) . \tag{3.47b}$$

For the ratio of the inversion densities we derive by using (3.40, 43)

$$\frac{N_{00^02} - uN_{Is}}{N_{00^01} - uN_{Ir}} = \frac{s^2 - uqs}{s - uq} = s . \tag{3.48}$$

We finally arrive at a very simple relationship for the gain ratio of the two bands

$$\frac{\alpha_s}{\alpha_r} = 2\exp(-h\nu_3/kT_3) , \tag{3.49}$$

where we have substituted (3.40c, 47, 48) into (3.44). Similarly the gain ratio of a line of the hot band and a corresponding line of the regular band is, within the same approximations, given by

$$\frac{\alpha_h}{\alpha_r} = \exp(-h\nu_2/kT_2) \ . \tag{3.50}$$

The gain is measured by transmitting a weak probe laser beam through the discharge of a CO_2 gas mixture. From the increased beam intensity the amplification per unit intensity and per unit length, i.e. the gain, can be deduced. The physics and construction details of regular, sequence, and hot band lasers and their tunability will be discussed in Chap. 4.

After measuring the gain ratios, the temperatures T_3 and T_2 are then simply obtained by means of (3.49, 50). The gas temperature which is assumed to be equal to the rotational temperature can be determined by measuring the gain of a wide range of rotational lines in the P and R branch of a band. By fitting the relative gain values of the lines to the theoretical expression as given by (3.19 or 20) the temperature T is obtained. In this way the gas and vibrational temperatures of CO_2 in a gas discharge were measured by *Siemsen* et al. [3.27]. Their results are displayed in Fig. 3.6 as a function of the discharge current. The measurements were performed on a 11.5 mm internal diameter discharge tube of 80 cm length that was cooled by a water jacket. The gas mixture of 1 torr CO_2, 1 torr N_2, and 3 torr He is flowing through the discharge tube. It is seen that T and T_2 increase linearly with the discharge current. This behaviour of T can be expected because the heat dissipation by the discharge is almost linear with the cur-

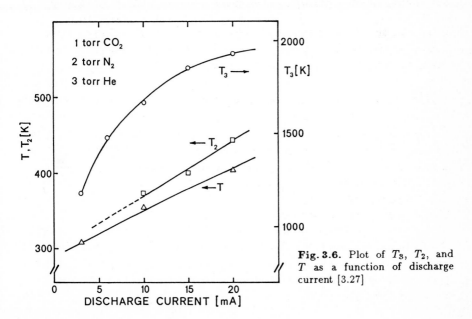

Fig. 3.6. Plot of T_3, T_2, and T as a function of discharge current [3.27]

73

rent and the cooling is almost linear with temperature difference between gas and water jacket. The linear increase of $T_2 - T$ with the current can be explained by considering the energy or temperature coupling as described by (3.38). At relatively low temperatures the cooling of the ν_2 vibration is proportional to $T_2 - T$. Since the energy transfer to the ν_2 vibration is expected to be proportional to the discharge current we conclude that $T_2 - T$ is proportional to the discharge current.

The situation seems to be different for T_3 as it shows a decreasing slope at higher current. The reason is most likely related to the increased deexcitation rate of the ν_3 vibration with increasing gas temperature. During the deexcitation of the ν_3 vibration the energy quantum $h\nu_3$ is partly transferred as two quanta $h\nu_1$ and $h\nu_2$ whereas the excess energy $h(\nu_3 - \nu_1 - \nu_2)$ is transferred to the translational motion of the colliding particles [3.25]. The transfer probability of the $h\nu_3$ quantum turns out to be very sensitive to the factor $h(\nu_3 - \nu_1 - \nu_2)/kT$ and increases with increasing T. On the other hand, the excitation rate of E_3 is expected to be proportional to the current. Then in the steady state where there is a balance between excitation and deexcitation E_3 or T_3 must show a decreasing slope with the discharge current.

Finally we mention that T_3 of the non-lasing discharge is much higher than that of the lasing system discussed in Sect. 3.8.

3.10 The Role of He, H_2O, and Xe in the Laser Mixture

The cooling of the discharge gas is effectively obtained by the addition of He. The energy levels of He are all above $20\,eV$ so that for electron energies in the range of 1 to $3\,eV$ this gas component will not influence the discharge itself, except for relatively small energy transfer by inelastic collisions with translation. However, the thermal conductivity is about six times as large as that of CO_2 and N_2. The difference between the temperature of the gas and the wall of the tube, as caused by the discharge heat, is inversely proportional to the conductivity of the gas. Although the gas conductivity does not depend on its pressure it is seen that for the described system the heat conductivity is practically determined by He. The considerable increase of heat transfer obtained by the addition of He means that the radiation production of the system saturates at a higher discharge current. This is the main advantage of adding He to the CO_2 laser system. For pulsed systems where the conductivity of the gas is not relevant, He has a favorable influence on the discharge, especially for those systems that operate near the region of instability. Although He also contributes to the relaxation of group 2 it is easy to verify that this function can be easily taken over by the other gas components without loss of performance. The relaxation phenomena in CO_2 and CO_2 mixtures have been studied extensively in the past [3.28].

Table 3.3. Relaxation rate constants k_2 of group 2 caused by the various gas components in CO_2 lasers at 300 K [3.26]

gas	k_2 [torr^{-1} s^{-1}]
CO_2	194
N_2	650
He	3 270
H_2O	450 000

This is also the case for the internal relaxation among the normal vibrations [3.25]. The measured relaxation rate constants k_2 of the main constituents in the laser gas that determine the relaxation of group 2 are shown in Table 3.3 [3.26].

As far as the relaxation phenomena are concerned we find with the above relaxation rates that in the presence of a high He concentration there is no need for the addition of H_2O. However, the water vapor is mainly used to reduce the degree of dissociation of CO_2 into CO and O. It is observed that with a content of about 0.2 torr H_2O there is practically no dissociation, as will be discussed in Sect. 4.6 for sealed-off systems. The addition of H_2O increases the lifetime of the gas and maintains the pumping rate by the optimized CO_2 concentration. In such a system the percentage of CO is negligible.

The gas mixture also contains a low concentration Xe. A concentration of about 0.5 torr is found to increase the output power and efficiency compared with a mixture without Xe [3.29, 30, 31]. The influence of Xe is mainly due to its effect on the discharge conditions. It is observed that Xe changes the electron energy distribution; the number of electrons with energy smaller than 4 eV increases whereas the number of those with larger energies decreases [3.31]. This change of electron energy distribution has a favorable effect on the vibrational excitation of CO_2 and N_2 as can be understood from the Figs. 3.3, 4.

Xe has an ionization potential of 12.1 eV which is 2–3 eV less than that of the other gas components. This means that the longitudinal electric field of the discharge tube establishes itself at a lower level to maintain the discharge than in the absence of Xe. (Chap. 4). The low ionization of Xe facilitates the production of new electrons for maintaining the discharge. Therefore with the current remaining the same, the longitudinal electric field decreases which, in turn, reduces the mean energy of the electrons in the discharge. In addition, the Xe atom has metastable states $6S_5$ and $6S_3$ with energies of 8.3 and 9.4 eV, respectively. These levels can be populated from the $B^3\pi_u$ state of nitrogen by collisions [3.32]. It is likely that these levels take part in a cascade ionization process of Xe.

The constituents of the laser gas also effect the lifetime of the upper laser level. The relaxation times of the (00^01) level in various gas mixtures have been measured by an induced fluorescence technique [3.9, 24] by means of so-called Q-switching. In this technique one uses a laser with a fast rotating mirror. During the short period the rotating mirror forms a cavity with the other mirror, so a short pulse, typically of about 0.2 to $0.5 \mu s$, is produced. This pulse pumps a passive gas cell containing a gas mixture of CO_2 with another gas. The CO_2 molecules of this cell excited to the (00^01) level decay spontaneously to the ground state and can be observed spectroscopically. The results are shown in Table 3.4.

Table 3.4. Relaxation rate constants k_3 of group 3 caused by various gas components in CO_2 lasers at 300 K [3.24]

gas	k_3 [torr^{-1} s^{-1}]
CO_2	350
N_2	106
He	85
H_2O	24 000

It is seen that H_2O has a large effect on the relaxation of the upper laser level. This is in agreement with the observation that the gain of a system depends critically on the H_2O partial pressure [3.11, 33]. The saturation intensity which is proportional to the relaxation rate, see (3.34), will therefore increase in proportion to the H_2O pressure. Thus for an optimized system the maximum water vapor pressure is set by the intensity of the radiation inside the cavity. As we shall see in Chap. 4 the density of the radiation production inside the discharge tube is inversely proportional to the cross section of the tube. Therefore the water vapor pressure in an optimized continuous discharge system is also inversely proportional to the tube cross section (for example, 0.1 torr H_2O at 20 mm and 0.4 torr H_2O at 10 mm internal diameter).

3.11 Power Extraction

This section deals with the laser power extractable from a CO_2 system. The maximum laser power depends on many parameters such as outcoupling, mirror system, small-signal gain α_0, absorption losses, saturation intensity I_0, and the length of the active medium. The question is what is the maximum extractable power and the corresponding relation between those parameters. Although the CO_2 laser can be considered as a homogeneously broadened system, the problem is still complicated and only numerical solutions can be obtained. However, an exact solution is found if a stable optical resonator for a one dimensional stationary laser system with constant small signal gain α_0 and constant loss coefficient γ_0 is considered. The losses may be due to diffraction, absorption, and scattering and are non-saturable. This problem was originally solved by *Schindler* [3.34]. The irradiances I_+ and I_- in z_+ and z_- direction, respectively, are determined according to (3.37) from the following differential equations

$$\frac{1}{\beta_+}\frac{d\beta_+}{dz} = -\frac{1}{\beta_-}\frac{d\beta_-}{dz} = \frac{\alpha_0}{1+\beta_++\beta_-} - \gamma_0 \, , \tag{3.51}$$

where $\beta_+ = I_+/I_0$ and $\beta_- = I_-/I_0$. The term describing the saturation in the denominator of (3.37) contains for a laser system the sum of the two irradiances. The complete solution of (3.51) is obtained as

$$\beta_+ \cdot \beta_- = \beta_0^2 = \text{const} \, , \tag{3.52}$$

$$\gamma_0 z + \ln\beta_+ + \frac{\alpha_0}{\sqrt{(\alpha_0-\gamma_0)^2 - (2\gamma_0\beta_0)^2}}\ln F(\beta_+) = \text{const} \, , \tag{3.53}$$

where

$$F(\beta_+) = \frac{\sqrt{(\alpha_0-\gamma_0)^2 - (2\gamma_0\beta_0)^2} + (\alpha_0 - \gamma_0 - 2\gamma_0\beta_+)}{\sqrt{(\alpha_0-\gamma_0)^2 - (2\gamma_0\beta_0)^2} - (\alpha_0 - \gamma_0 - 2\gamma_0\beta_+)} \, . \tag{3.54}$$

In the cavity (Fig. 3.7) extending from $z = 0$ to $z = L$, the boundary conditions are

$$\beta_+(0) = \beta_1 \quad \text{and} \quad \beta_+(L) = \beta_2 \, . \tag{3.55}$$

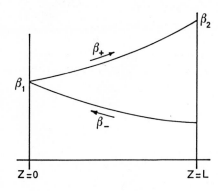

Fig. 3.7. Schematic representation of the irradiances $\beta_- = I_-/I_0$ and $\beta_+ = I_+/I_0$ of the two running waves in the cavity between $z = 0$ and $z = L$. A perfectly reflecting mirror is located at $z = 0$ and the outcoupling mirror is at $z = L$

The irradiance β_+ at these two locations is related to the parameters α_0, γ_0 and L through the following transcendental equation obtained from (3.53, 55)

$$\gamma_0 L - \ln\left(\frac{\beta_1}{\beta_2}\right) = \frac{\alpha_0}{\sqrt{(\alpha_0-\gamma_0)^2 - (2\gamma_0\beta_0)^2}}\ln\left(\frac{F(\beta_1)}{F(\beta_2)}\right) \, . \tag{3.56}$$

For mirror reflectivities $r_1 = 1$ and $z = 0$ (perfectly reflecting) and $r_2 < 1$ at

$z = L$, we have

$$\beta_0 = \beta_1 = \beta_2\sqrt{r_2} \ . \tag{3.57}$$

Replacing β_0 and β_1 by $\beta_2\sqrt{r_2}$ in (3.54, 56), one obtains a transcendental equation relating β_2 and r_2 with α_0, γ_0, and L. It can be shown [3.34] that $|2\gamma_0\beta_0/(\alpha_0 - \gamma_0)| \leq 1$. Therefore it is convenient to substitute

$$\left(\frac{2\gamma_0}{\alpha_0 - \gamma_0}\right)\beta_0 = \sin 2\lambda \ . \tag{3.58}$$

Equation (3.54) yields then for β_+ equal β_1 or β_2 the form

$$F(\beta_{1,2}) = \frac{\beta_0 \cos \lambda - \beta_{1,2} \sin \lambda}{\beta_{1,2} \cos \lambda - \beta_0 \sin \lambda} \text{cotg} \, \lambda \ . \tag{3.59}$$

By virtue of the condition in (3.57), one obtains explicitly

$$F(\beta_1) = \text{cotg} \, \lambda \tag{3.60a}$$

$$\frac{F(\beta_1)}{F(\beta_2)} = \frac{\cos \lambda - \sqrt{r_2} \sin \lambda}{\sqrt{r_2} \cos \lambda - \sin \lambda} = \frac{1 - \sqrt{r_2} \tan \lambda}{\sqrt{r_2} - \tan \lambda} \ . \tag{3.60b}$$

The transcendental equation (3.56) assumes now the more convenient form

$$\gamma_0 L - \ln \sqrt{r_2} = \frac{\alpha_0}{\alpha_0 - \gamma_0} \frac{1}{\cos 2\lambda} \ln \left(\frac{1 - \sqrt{r_2} \tan \lambda}{\sqrt{r_2} - \tan \lambda}\right) \ . \tag{3.61}$$

The maximum available irradiance from the medium can be calculated by considering the same active medium as an amplifier with no losses. The maximum power is then obtained by using (3.37) with $I/I_0 \rightarrow \infty$ or

$$\frac{dI}{dz} = \alpha_0 I_0 \tag{3.62}$$

so that the maximum available irradiance of the medium becomes

$$I_{\text{max}} = I_0 \alpha_0 L \quad \text{or} \tag{3.63a}$$
$$\beta_{\text{max}} = \alpha_0 L \ . \tag{3.63b}$$

Returning to our laser system, the extracted power is equal to $(1 - r_2)\beta_2 I_0$ (assuming no absorption loss of the output mirror). Using (3.63a) one obtains for the extraction efficiency

$$\eta = \beta_2 \left(\frac{1 - r_2}{\alpha_0 L}\right) \ . $$

With $\beta_2 = \beta_0/\sqrt{r_2}$ and β_0 in terms of λ according to (3.58) one finds

$$\eta = \left(\frac{\alpha_0 - \gamma_0}{2\alpha_0 g_0 L}\right)\left(\frac{1 - r_2}{\sqrt{r_2}}\right)\sin 2\lambda \ . \tag{3.64}$$

Equations (3.61,64) serve to determine mirror reflectance $r_2 = r_2(\lambda)$ and extraction efficiency $\eta = \eta(\lambda)$ as a function of the parameter λ. The constants α_0, γ_0 and L enter into the explicit form of the parameter representations $r_2(\lambda)$ and $\eta(\lambda)$. Using the abbreviations $\gamma_0 L = \gamma$, $\sqrt{r_2} = x$, and $\alpha_0/(\alpha_0 - \gamma_0) = y$ eqs. (3.61,64) can finally be written as

$$(\gamma - \ln x)\cos 2\lambda = y \ln\left(\frac{1 - x\tan\lambda}{x - \tan\lambda}\right) \ , \tag{3.65a}$$

$$\eta = \frac{1}{2\gamma y}\left(\frac{1 - x^2}{x}\right)\sin 2\lambda \ . \tag{3.65b}$$

The optimum extraction efficiency is obtained for a particular value λ_0 of the parameter λ that makes $\eta(\lambda_0)$ maximal compared with its value at any other parameter point. The optimum output coupling is then given by

$$T_{\text{opt}} = 1 - r_{2\text{opt}} = 1 - x^2(\lambda_0) \tag{3.66a}$$

while the maximum extraction efficiency is

$$\eta_{\text{max}} = \eta(\lambda_0) = \frac{1}{2\gamma y}\left(\frac{1 - x^2(\lambda_0)}{x(\lambda_0)}\right)\sin 2\lambda_0 \ . \tag{3.66b}$$

Optimum output irradiance is then determined from

$$\beta_{\text{opt}} = \alpha_0 L \eta_{\text{max}} = \frac{1}{2(y - 1)}\left(\frac{1 - x^2(\lambda_0)}{x(\lambda_0)}\right)\sin 2\lambda_0 \ . \tag{3.66c}$$

The value for η_{max} has to be determined numerically. Since η must be positive, it follows that λ is confined to $0 < \lambda < \pi/4$. The best procedure is to subdivide the relatively short parameter interval $0 < \lambda < \pi/4$ by sufficiently many points λ_i, determine the corresponding roots $x_i = x(\lambda_i)$ from (3.65a), and insert these values into (3.65b) to obtain $\eta_i = \eta(\lambda_i)$. Sorting the discrete values η_i in ascending order yields the maximum η_{max} as well as the corresponding value λ_0, from which, in turn, $x(\lambda_0)$ and the optimum output coupling is obtained. In this way η_{max} and T_{opt} can be calculated as a function of $\alpha_0 L$ and $\gamma_0 L$. These values for η_{max} and T_{opt} are plotted along the axes of a Cartesian coordinate system (Fig. 3.8) with the lines $\alpha_0 L = \text{const}$ and $\gamma_0 L = \text{const}$ as parameters. For example, if $\alpha_0 L = 2$ and $\gamma_0 L = 0.1$ maximum extraction efficiency and optimum output coupling are read off as $\eta_{\text{max}} = 0.6$ and $T_{\text{opt}} = 0.5$. The optimum output irradiance in this case is $\beta_{\text{opt}} = 1.2$.

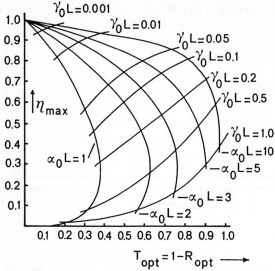

Fig. 3.8. Maximum extraction efficiency versus optimum output coupling for constant $\alpha_0 L$ and constant $\gamma_0 L$ [3.34]

For a continuous discharge CO_2 laser the tube diameter is limited because of efficient cooling of the gas. This means that the losses are mainly caused by diffraction that can be estimated from the beam waist (ω) and the tube diameter. For a single-mode system the value of $\gamma_0 L$ is typically between 0.01 and 0.05 and the gain $\alpha_0 L$ between 0.5 and 2. For pulsed systems the gain $\alpha_0 L$ can be even larger than 4. For these high small-signal gain factors the optimized output is not very critical with respect to outcoupling reflectivity. For such high-gain systems the use of a plane parallel NaCl window yields nearly the optimized output. The choice of an uncoated material is usually made because of its high damage threshold. For continuous systems the outcoupling mirror is often a plane parallel flat made of Ge or ZnSe. However, these mirrors have internal absorption and consequently the reflectivity is power dependent. The absorption and reflectivity depend on the temperature so that cooling of the mirror in the case of continuous systems is required. For output powers up to 100 W air-cooled Ge windows can be used. The reflectivity of a plane parallel flat is about 78%. At higher output powers up to 5 kW the use of air-cooled ZnSe is recommended.

4. Continuous Discharge Lasers

The continuous CO_2 laser with its high efficiency of above 10 % may be considered as the most practical laser. Its construction and operation are relatively simple. The positive column of the electrical discharge in a cylindrical tube plays a central role in the production of a high gain medium in a CO_2 laser gas mixture. The production rate of the inverted medium depends on the electrical discharge parameters, the cooling properties of the medium, and the partial pressures of the gas components. The success of continuous laser operation has become possible because of the understanding of the molecular glow discharge behavior in a cylindrical tube. Since the laser process is between low lying vibrational levels, the temperature profile and the cooling of the inverted medium is also very important.

In this chapter we start with an analysis of a glow discharge in a cylindrical tube and its impact on the temperature distribution. After that the optical aspects and their relation to the discharge parameters are treated. Knowing the medium parameters, special performances are described like single-mode, sequence and hot-band operation. Finally construction details of a high-performance tunable system are presented.

4.1 The Behavior of the Discharge

A continuous CO_2 laser uses a glow discharge in a cylindrical tube. A typical example is a discharge length of 130 cm and a diameter of 10 mm filled with 2.5 torr CO_2, 3.5 torr N_2, 12 torr He, 0.6 torr Xe, and 0.2 torr H_2O. The output is about 70 W for a discharge voltage and current of about 17 kV and 30 mA, respectively.

Because of its negative dynamic resistance the laser tube is connected in series with a ballast resistor to the power supply. In Fig. 4.1 the operating current and voltage in the discharge tube are found by the intersection of the voltage-current characteristic with the line $U = E_a - I R_b$, where E_a and R_b are the voltage of the power supply and the ballast resistor, respectively. In order to get a stable discharge current the condition has to be fulfilled that the absolute value of the dynamic resistance at the operating current is smaller than the ballast resistor or

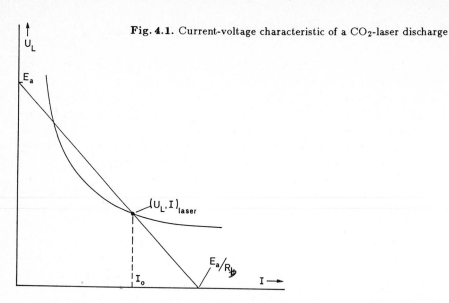

Fig. 4.1. Current-voltage characteristic of a CO_2-laser discharge

$$R_b > \left| \frac{dU_L}{dI} \right|_{I=I_0} . \tag{4.1}$$

The absolute value of the dynamic resistance is much smaller than the static resistance.

The laser power is usually regulated by the current. Since for low power a small current is required, the ballast resistor must be quite large. In practice, R_b is, depending on the adjustable power range, about 20 to 50 % of the static resistance at optimum performance. This has the disadvantage that for operation at larger currents the heat dissipation of the ballast is large. To circumvent this problem vacuum high-voltage triodes are used instead of linear resistors. Such triodes can operate with a small static and a large dynamic resistance. Electronic regulation and back coupling to the grid allow a large dynamic range with low heat dissipation and high current stabilization.

The discharge itself is characterized by the cathode and anode falls, in relatively small regions close to respectively the cathode and anode, and the positive column. Although the extension of the cathode fall is very small in practical systems, its voltage drop depending on the cathode material is of the order of 500 V. The voltage gradient along the positive column is of the order of 100 V/cm, so that for small systems the cathode energy loss can be substantial. The anode fall is negligible for usual systems. The positive column is the largest part of the laser discharge.

4.2 Elementary Theory of the Positive Column

For laser action only the positive column is important. As discussed in Chap. 3, the laser power depends on the pumping rate of the gas and the various temperatures. The pumping rate and the vibrational temperatures are mainly determined by the electron density and energy in the discharge. Therefore, in order to get some understanding of the experimental behavior we shall present an elementary theory of the positive column. This theory [4.1] can give a good quantitative description of the relations between axial and radial electric fields, the pressure, and the tube radius.

From experimental observations it is known that the axial gradient dU_L/dx is constant. This means that

$$\frac{d^2 U_L}{dx^2} = 0 \qquad (4.2)$$

so that the total charge is zero or the number of positive ions n_p and electrons n_e per unit length of column is equal: $n = n_p = n_e$. The number of electrons crossing unit area per second in one direction (random current) is almost equal to the number passing in the opposite direction. The relatively small difference between these currents represents the drift velocity. The drift current of the electrons can be separated into a current in the direction of the axis and a current normal to the axis, which ultimately goes to the wall. Since the ion current along the axis is small, due to the large impact with the neutral particles, the electron current along the axis is practically the tube current. The tube itself being an insulator, the electron current normal to the axis must be equal to the positive ion current in that direction. Since the diffusion coefficient and the mobility are much smaller for the ions than for the electrons, this bipolar current is only possible if the potential of the wall is negative with respect to the axis of the tube. In other words, the equipotential surfaces are shells curved convex with respect to the cathode (Fig. 4.2). At the wall the ions and electrons can recombine. Volume recombination can be neglected because of a very low three-body collision frequency at normal laser gas pressures. This means that the current of each species going to the wall must be equal to the number of ionizations per second. Using this condition the dependence of concentration of electrons n on the radial distance r can be derived. Thus we have

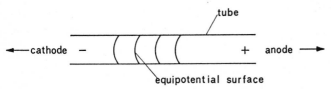

Fig. 4.2. Curved equipotential surfaces along the positive column of a discharge

$$j_e = j_p = j \ , \tag{4.3}$$

where j_e and j_p are the electron and ion current density perpendicular to the axis, respectively. This current is driven by diffusion and the radial field E_r. The diffusion of ions is negligible as compared to that of electrons. Thus we may write

$$j = \mu_p n E_r = -D\frac{dn}{dr} - \mu_e n E_r \ , \tag{4.4}$$

where μ_p and μ_e are the mobilities of ions and electrons, respectively, and D is the diffusion coefficient of the electrons.

Eliminating E_r in (4.4) and using the fact that $\mu_p \ll \mu_e$ we get

$$j = -\frac{\mu_p}{\mu_e}D\frac{dn}{dr} \ . \tag{4.5}$$

The increase of j with r is caused by ionization. If we assume that the ionization per unit time and per unit volume is equal to νn, ν being the ionization frequency per electron, we obtain

$$\frac{d(2\pi r j)}{dr} = 2\pi r \nu n \ . \tag{4.6}$$

Eliminating of j from (4.5,6) gives the Bessel differential equations

$$\frac{d^2 n}{dr^2} + \frac{1}{r}\frac{dn}{dr} + \frac{\mu_e \nu}{\mu_p D}n = 0 \ . \tag{4.7}$$

The quantity $\mu_p D/\mu_e = D_a$ is known as the ambipolar diffusion coefficient. The solution of (4.7) with real argument is

$$n = n_0 J_0\left(r\sqrt{\frac{\nu}{D_a}}\right) \tag{4.8}$$

which is an oscillating function with a variable period (Fig. 4.3). Since the concentration is a positive quantity and $n = n_0$ at $r = 0$, the first part of

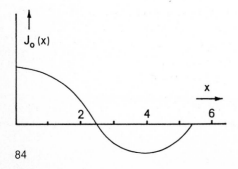

Fig. 4.3. Zero-order Bessel function

the curve has to be applied here. As the concentration near the wall $(r = R)$ is very small in comparison to the concentration at the axis, we write

$$J_0\left(R\sqrt{\frac{\nu}{D_a}}\right) = 0 \ . \tag{4.9}$$

Hence, from (4.9)

$$R\sqrt{\frac{\nu}{D_a}} = 2.405 \ . \tag{4.10}$$

Substituting (4.10) into (4.8) we obtain the distribution

$$n = n_0 J_0\left(2.4\frac{r}{R}\right) \ . \tag{4.11}$$

The concentration of charges in the discharge tube thus varies with r in a nearly parabolic manner. This has been confirmed experimentally by measuring the current distribution flowing to an anode that consisted of several concentric electrodes. Since both the axial field and μ_p do not depend on r, the current distribution over these anodes is proportional to $n(r)$.

The potential distribution across a cross section, $U(r)$, can be found from (4.4), as the drift current to the wall is small with respect to the random current, i.e.,

$$D\frac{dn}{dr} = -\mu_e E_r n = \mu_e n \frac{dU_L}{dr} \quad \text{or}$$

$$n = n_0 \exp\left\{\frac{\mu_e}{D}[U_L(r) - U_L(0)]\right\} \ . \tag{4.12}$$

4.3 The Similarity Rules

The number of ionizations of an electron per second, ν, depends on the probability $W(\varepsilon)$ that an electron with energy ε will ionize a particle, on the number of collisions per second which itself is proportional to the gas density ϱ [equal to $\varrho f(\varepsilon)$], and on the energy distribution function $n(\varepsilon)$. Thus we may write

$$\nu = \varrho \int_{\varepsilon_i}^{\infty} W(\varepsilon)f(\varepsilon)n(\varepsilon)d\varepsilon = \varrho F(\varepsilon_i) \ , \tag{4.13}$$

where ε_i is the ionization energy. The integration of $n(\varepsilon)$ over all energies gives unity. The ambipolar diffusion coefficient is inversely proportional to ϱ, or

$$D_a = \overline{D}_a / \varrho \tag{4.14}$$

where $\overline{D}_a$ is the coefficient for unit density. Substitution of (4.13, 14) into (4.10) gives

$$R\varrho = 2.4 \sqrt{\frac{\overline{D}_a}{F(\varepsilon_i)}} \ . \tag{4.15}$$

The distribution function $n(\varepsilon)$ in (4.13) depends on the axial field E_a. This field in the positive column is given by balancing the energy the electrons gain from the electric field and the energy they lose by collisions. An electron loses in one collision an energy $\Delta(\varepsilon)$. This can be transferred to translation, rotation, excitation or ionization of the gas particles. The energy gained per second is given by the electric force times the drift velocity v_d. Equating loss and gain we have

$$eE_a v_d = \varrho \int_0^\infty \Delta(\varepsilon) f(\varepsilon) n(\varepsilon) d\varepsilon \ . \tag{4.16}$$

The drift velocity is equal to the mobility times the axial field. Furthermore the mobility is inversely proportional to the gas density. Thus we can write

$$v_d = \frac{\overline{\mu}_e}{\varrho} E_a \ , \tag{4.17}$$

where $\overline{\mu}_e$ is the mobility for unit density.

Considering (4.15–17) we find that if the quantities $R\varrho$ and E_a/ϱ of one discharge are equal to the respective quantities of another discharge with the same gas composition, the energy distributions of the electrons and the excitation processes of both discharges are similar. Thus the optimized energy distribution in the positive column with respect to the excitation rate of the inversion process follows the similarity rules

$$\varrho D = \text{const} \ , \tag{4.18}$$

$$E_a/\varrho = \text{const} \ . \tag{4.19}$$

Combining (4.18, 19) we also have

$$E_a D = \text{const} \ , \tag{4.20}$$

where D is the diameter of the tube.

The E_a/ϱ value near maximum output does not depend on the current in the usual molecular gas mixtures, as is observed experimentally. The energy distribution of the electrons is, in practice, only a function of the product ϱD. The excitation rate of the inversion density being proportional to the electron density it is also proportional to the current since the drift velocity, proportional to E_a, does not depend on the current.

Applying the similarity rules to different discharges, a complication arises from the fact that ϱ is not uniquely defined in a discharge. The gas pressure must be constant everywhere in a continuous discharge but the density varies inversely with the temperature. The temperature distribution depends on the heat production and conductivity. Therefore, in order to apply the similarity rules we also have to compare the temperature profiles. Moreover, the temperature distribution in the discharge is very important for determining the depopulation rates and inversion density of the laser levels, as we have seen in Chap. 3, and also for its effect on the optical properties of the medium.

4.4 Thermal Effects and Similarity

In the following we shall analyse the temperature distribution of a discharge. From this we shall obtain the thermal similarity between discharges in relation with the previous section. Then the temperature distribution and the refractive index will be calculated.

The thermal effects in a gas discharge are mainly caused by the direct transfer of the thermal energy of the electrons to the translation and rotation energies of the gas, which are very strongly coupled. Another contribution to the thermal energy of the gas comes from the deexcitation of the lower laser level. The temperature distribution in the stationary state results from this heat production. In long lasers having axial symmetry the problem can be treated as a cylindrically symmetric one of infinite length. The temperature distribution T can then be described by the following second-order differential equation

$$ \kappa \frac{d^2 T}{dr^2} + \frac{\kappa}{r}\frac{dT}{dr} + A + \frac{d\kappa}{dT}\left(\frac{dT}{dr}\right)^2 = 0 \; , \tag{4.21} $$

where A is the heat production per unit volume, and κ the heat conduction coefficient. This is a differential equation of the Riccati type for the gradient dT/dr. The solution for the gradient is

$$ \frac{dT}{dr} = \frac{1}{\kappa r}\int\limits_0^r -r' A \, dr' \; , \tag{4.22} $$

where the constant of integration is zero because $dT/dr = 0$ at $r = 0$. Before we are able to solve the temperature distribution from (4.22) we have to find the heat production A as a function of r.

In a CO_2-laser discharge the energy of the spontaneous and laser emission is relatively small as compared with the energy transfer to translation and rotation. For this reason we assume that the discharge energy is all converted into heat. The heat dissipation per electron is given by (4.16). Multiplying this by the electron density given by (4.11) we obtain for this dissipation per unit volume

$$A(r) = eE_a v_d n_0 J_0\left(2.4\frac{r}{R}\right) . \tag{4.23}$$

The electron density n_0 can be expressed in terms of the current I, because I is given by

$$I = 2\pi e \int\limits_0^R v_d n r \, dr . \tag{4.24}$$

Using (4.17) and (4.11) we get

$$I = 2\pi\bar{\mu}_e E_a e n_0 R^2 \int\limits_0^1 \frac{J_0(2.4x)}{\varrho(x)} x \, dx \tag{4.25}$$

where we have substituted $x = r/R$. Combining (4.23, 25 and 17) we obtain

$$A(x) = E_a I J_0(2.4x)\left[2\pi\varrho(x)R^2 \int\limits_0^1 \frac{J_0(2.4x')}{\varrho(x')} x' \, dx'\right]^{-1} . \tag{4.26}$$

Since the pressure is constant in the tube, the density $\varrho(x)$ is inversely proportional to $T(x)$. Substituting (4.26) into (4.22) and replacing $\varrho(x)$ by $T(x)$ we obtain the following implicit equation for the temperature

$$\frac{dT}{dx} = \frac{E_a I}{2\pi\kappa x}\left[\int\limits_0^1 J_0(2.4x')T(x')x' \, dx'\right]^{-1} \int\limits_0^x -x' J_0(2.4x')T(x')dx' . \tag{4.27}$$

It is seen from (4.27) that the temperature profiles of two discharges are similar if the values $E_a I$ of both discharges are equal. Combining this result with (4.20) we find the similarity rule with respect to currents, i.e.,

$$I/D = \text{const} . \tag{4.28}$$

It is clear from the similarity rules that for optimized systems the input energies per unit length do not depend on the tube diameters. Further, because of equal electron energies and gas temperature profiles it is also expected that the stimulated emissions per unit length are equal. Both conclusions with respect to input and output powers are indeed in agreement with experimental observations. Narrow tubes that allow only single-mode operation are as powerful as multi-mode systems. The efficiency and the output power per unit length of a single-mode system are, in principle, independent of the total cavity length whereas the tube diameter is proportional to the square root of the cavity length, as will be discussed in Sect. 4.5.

Next we calculate the variation of the refractive index, which is proportional to the gas density according to

$$n(r) = 1 + \gamma \varrho(r) \; , \tag{4.29}$$

where γ is the Gladstone-Dale constant depending on the gas composition. For a radial symmetric distribution an expansion of $n(r)$ contains only even powers of r. When we write

$$n(r) = n_0 + \tfrac{1}{2} n_2 r^2 + \tfrac{1}{24} n_4 r^4 + \ldots \tag{4.30a}$$

we find

$$n_2 = \left(\frac{d^2 n}{dr^2} \right)_{r=0} = \gamma \left(\frac{d^2 \varrho}{dr^2} \right)_{r=0} \; . \tag{4.30b}$$

If ϱ_0 is the initial constant density in the absence of a discharge, we obtain from the conservation of mass that

$$\varrho_0 = (2/R^2) \int\limits_0^R \varrho(r) r \, dr \; .$$

The density is related to the temperature by

$$\varrho(r) = \frac{\varrho_0 R^2}{T(r) \int\limits_0^R \frac{2r \, dr}{T(r)}} \; . \tag{4.31}$$

For the quadratic term of the refractive index it then follows

$$n_2 = -\frac{\gamma \varrho_0}{S_1 T_0^2} \left(\frac{d^2 T}{dr^2} \right)_{r=0} \; , \tag{4.32}$$

where T_0 is the temperature at the axis of the tube and

$$S_1 = \int_0^1 \frac{2x\,dx}{T(x)} \quad . \tag{4.33}$$

From (4.27) we calculate that

$$\left(\frac{d^2T}{dr^2}\right)_{r=0} = -\frac{E_aIT_0}{4\pi\kappa_0R^2S_2} \quad , \quad \text{where} \tag{4.34}$$

$$S_2 = \int_0^1 J_0(2.4x)T(x)x\,dx \tag{4.35}$$

so that

$$n_2 = \frac{\gamma\varrho_0 E_a I}{4\pi\kappa_0 R^2 T_0^2 S_1 S_2} \quad , \tag{4.36}$$

where κ_0 is the heat conductivity at the temperature T_0. Equation (4.36) together with (4.33, 35) determine the quadratic term in the refractive index profile. However, the numerical value of n_2 can only be obtained if we know the temperature distribution from (4.27). The thermal conductivity in (4.27) is not constant, but is strongly dependent on temperature. Looking at tabulated values of κ for the usual gas mixtures in cw CO_2 lasers, where the thermal conductivity is mainly determined by He, we find that κ can be fairly well approximated by

$$\kappa = \kappa_w + \kappa_1(T - T_w) \quad , \tag{4.37}$$

where κ_w is the thermal conductivity at the wall temperature T_w, and κ_1 is given by

$$\kappa_1 = \frac{\kappa_0 - \kappa_w}{T_0 - T_w} \quad . \tag{4.38}$$

According to the literature we can deduce for a mixture containing 2 torr CO_2, 4 torr N_2, and 12 torr He that $\kappa_w = 0.075\,[\mathrm{Wm^{-1}\,K^{-1}}]$ at $T_w = 290\,\mathrm{K}$ and $\kappa_1 = 0.00021\,[\mathrm{Wm^{-1}\,K^{-2}}]$. Equation (4.27) can be solved by iteration. As a start we consider $T(x')$ in the two integrals of (4.27) as constant. The approximation obtained in this way allows the solution of (4.27) by separation of variables

$$\kappa_w(T_x - T_w) + \tfrac{1}{2}\kappa_1(T_x - T_w)^2 = \frac{-E_0I}{2\pi}[\int_0^1 J_0(2.4x)x\,dx]^{-1}$$

$$\int_1^x \frac{1}{x'}[\int_0^{x'} J_0(2.4x'')x''\,dx'']dx' = F(x) \tag{4.39}$$

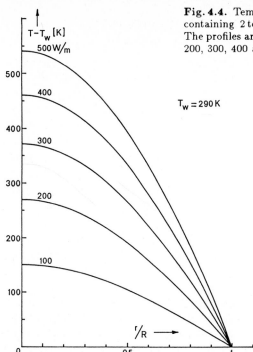

$T_w = 290$ K

or

$$T_x - T_w = -\frac{\kappa_w}{\kappa_1} + \left(\frac{\kappa_w^2}{\kappa_1^2} + \frac{2F(x)}{\kappa_1} \right)^{1/2} . \qquad (4.40)$$

The next step is the substitution of T_x from (4.40) into the integrals on the right-hand side of (4.27) and solving again for $T_x - T_w$. This iteration process can be repeated until sufficient accuracy has been obtained.

The temperature profiles according to the procedure described are plotted in Fig. 4.4 for several values $E_0 I$ of a gas mixture containing 2 torr CO_2, 4 torr N_2, and 12 torr He. The corresponding values of $n_2 R^2$ are presented in Table 4.1.

Table 4.1.

$E_0 I$ [W/m]	100	200	300	400
$n_2 R^2$	2.24×10^{-6}	2.68×10^{-6}	2.79×10^{-6}	2.81×10^{-6}

4.5 Optical Aspects of Single Mode Operation

As discussed previously, the optimized laser power per unit length does not depend on the tube diameter. In general, Gaussian-mode operation is highly desirable for most applications because of its spectral purity and the minimum spot size when the beam is focused. In designing such a laser the radiation distribution in the laser tube that can be expected, has to be known in order to optimize the cooling of the gas by minimizing the tube diameter. In some cases, especially for small laser systems, say less than 50 cm length, the problem can be simplified by considering only the geometry of the cavity and not taking into account the interaction with the active medium. This is done in the well-known theory of *Kogelnik* and *Li* [4.2]. This is justified because in small cavities the influence of the active medium is negligible and the observed radiation distribution does not show any perceptable deviation from that predicted on the basis of the empty-space resonator theory. However, for increasing output power the Gaussian mode deviates considerably from predictions of the empty-space theory.

Several physical processes within the active medium can be studied in relation to wave propagation in the optical resonator. The most important one that influences the beam width is the heat dissipation within the gas discharge. Next, the inhomogeneity of the gain and the dispersion of the active medium may effect the radiation distribution in the laser tube. These interactions with the medium are axially symmetric and can be described in terms containing only even powers of r. The variation of the refractive index as caused by the heat production in the discharge can, to a good approximation, be expressed by the quadratic term alone. Higher-order terms of the heating effect are negligible. This is in general not possible for the gain variations. On the other hand, the effect of gain variations on the beam size is relatively small.

Since the mode structure is mainly influenced by the heat production we shall analyze the problem by considering only the quadratic terms in the variations of refractive index and gain [4.3]. This simplifies the problem considerably because it has been shown [4.4] that if the propagating constant varies only quadratically with distance from the axis, a Gaussian beam remains Gaussian while propagating through such a medium. In fact, the appearance of high-power Gaussian laser beams supports the assumption that a quadratic description is adequate.

The derivation of the propagation laws for Gaussian beams in quite general lens-like media (only quadratic terms) starts with the scalar wave equation

$$\nabla E + k^2 E = 0 \ , \tag{4.41}$$

where the quadratic dependence of k^2 near the optic axis is given by

$$k^2(r, z) = k_0^2(z) - k_0(z)k_2 r^2 \ . \tag{4.42}$$

The constants of the medium are generally complex to allow for loss or gain. k_0 is the part of the propagation constant that does not depend on r, whereas k_2 describes the variation of the propagation constant with r and includes thermal effects, gain variations, and also the associated dispersive effects of the gain profile.

In solving (4.41), the main interest is in plane waves that propagate primarily in the z direction [4.4]. Therefore we substitute into (4.41)

$$E = E_1 \psi(r, z) \exp(-\mathrm{i} k_0 z) \ . \tag{4.43}$$

Substituting (4.43) into (4.41) and assuming k_0 to be constant we obtain

$$\frac{\partial^2 \psi}{\partial r^2} + \frac{1}{r}\frac{\partial \psi}{\partial r} - 2\mathrm{i}k_0\frac{\partial \psi}{\partial z} - k_0 k_2 r^2 \psi = 0 \ , \tag{4.44}$$

where the term $\partial^2 \psi/\partial z^2$ is neglected with respect to $k_0(\partial \psi/\partial z)$.

Next, we substitute for ψ the following expression

$$\psi = \exp\{ -\mathrm{i}[T(z) + \tfrac{1}{2}Q(z)r^2]\} \tag{4.45}$$

and obtain

$$-Q^2 r^2 - 2\mathrm{i}Q - k_0 r^2\frac{dQ}{dz} - 2k_0\frac{dT}{dz} - k_0 k_2 r^2 = 0 \ . \tag{4.46}$$

Since (4.46) holds for all values of r we separate it into

$$Q^2 + k_0\frac{dQ}{dz} + k_0 k_2 = 0 \ , \tag{4.47a}$$

$$\frac{dT}{dz} = -\frac{\mathrm{i}Q}{k_0} \ . \tag{4.47b}$$

For solving (4.47) we make the substitution

$$Q = \frac{k_0}{S}\frac{dS}{dz} \tag{4.48}$$

and obtain for (4.47a)

$$\frac{d^2 S}{dz^2} + \frac{k_2}{k_0}S = 0 \ . \tag{4.49}$$

Equation (4.49) has the solution

$$S(z) = a \sin \left(z \sqrt{k_2/k_0} \right) + b \cos \left(z \sqrt{k_2/k_0} \right) , \tag{4.50a}$$

$$\frac{dS}{dz} = a \sqrt{k_2/k_0} \cos \left(z \sqrt{k_2/k_0} \right) - b \sqrt{k_2/k_0} \sin \left(z \sqrt{k_2/k_0} \right) . \tag{4.50b}$$

Next we define a beam parameter q by

$$\frac{Q(z)}{k_0} = \frac{1}{q(z)} = \frac{1}{R(z)} - \frac{i\lambda}{\pi \omega^2(z)} . \tag{4.51}$$

The physical meaning of the beam parameter q is obtained by substituting (4.51 and 45) into (4.43) so that

$$E = E_1 \exp \left\{ -i[k_0 z + T(z)] - r^2 \left(\frac{1}{\omega^2(z)} + \frac{ik_0}{2R(z)} \right) \right\} . \tag{4.52}$$

It is now seen that $\omega(z)$ is the beam radius for which the field amplitude falls to $1/e$ times that on the axis, and $R(z)$ is the radius of curvature of the phase front.

The solution of q is obtained by substituting (4.50) into (4.48).

$$q(z) = \frac{q_1 \cos \left(z \sqrt{k_2/k_0} \right) + \sqrt{k_0/k_2} \sin \left(z \sqrt{k_2/k_0} \right)}{-q_1 \sqrt{k_2/k_0} \sin \left(z \sqrt{k_2/k_0} \right) + \cos \left(z \sqrt{k_2/k_0} \right)} , \tag{4.53}$$

where $q = q_1$ for $z = 0$.

The last expression describes the beam parameter q through the lens-like media. It is called the ABCD law [4.2], because the beam parameter is directly obtained from the (complex) matrix array that describes the paraxial propagation through lens-like media. In the present case this matrix is given by

$$\begin{vmatrix} A & B \\ C & D \end{vmatrix} = \begin{vmatrix} \cos \left(z \sqrt{k_2/k_0} \right) & \sqrt{k_0/k_2} \sin \left(z \sqrt{k_2/k_0} \right) \\ -\sqrt{k_2/k_0} \sin \left(z \sqrt{k_2/k_0} \right) & \cos \left(z \sqrt{k_2/k_0} \right) \end{vmatrix} . \tag{4.54}$$

4.5.1 Gain of a Gaussian Beam

We consider the gain near the optical axis. The propagation constants can be separated in terms of real and imaginary parts

$$k_0 = \beta_0 + i\alpha_0 , \tag{4.55a}$$

$$k_2 = -\beta_2 - i\alpha_2 . \tag{4.55b}$$

By differentiating (4.43) with respect to z and using (4.45, 47b) we find E on the axis ($r = 0$)

$$\frac{1}{E}\frac{dE}{dz} = i\left(k_0 - \frac{i}{q}\right) . \tag{4.56}$$

By adding the complex conjugates on both sides (4.56) and using (4.51) the incremental gain along the axis is found to be

$$\frac{1}{I}\frac{dI}{dz} = 2\left(\alpha_0 - \frac{1}{R(z)}\right) . \tag{4.57}$$

By substituting (4.51,55) into (4.47a) and taking the imaginary part, we get

$$\frac{2}{R(z)} = \frac{d\ln\omega^2}{dz} - \frac{1}{2}\alpha_2\omega^2 , \tag{4.58}$$

where we have assumed $\beta_0 \gg |\alpha_0|, |\beta_2|$.

By substituting (4.58) into (4.57) we finally obtain for the incremental gain on the optical axis

$$\frac{1}{I}\frac{dI}{dz} = 2\alpha_0 - \frac{d\ln\omega^2}{dz} + \frac{1}{2}\alpha_2\omega^2 . \tag{4.59}$$

The first term on the right-hand side of (4.59) describes the local conversion of inverted states into stimulated emission. This is the same term as found for a one-dimensional treatment with plane waves. The second term is related to the continuity of propagating electromagnetic energy, and describes focusing or defocusing of the beam. The last term accounts for the contribution of the radial radiation transport to the incremental gain related to a quadratic gain profile. In the absence of gain or loss the product $(\omega^2 I)$ of the beam cross section and intensity at the optical axis of a Gaussian beam is constant along the axis. The energy flux can therefore also be written as

$$\frac{d\ln(I\omega^2)}{dz} = 2\alpha_0 + \frac{1}{2}\alpha_2\omega^2 . \tag{4.60}$$

For homogeneous line broadening, the saturated amplitude gain coefficient α is related, according to Sect. 3.5, to the unsaturated or small-signal gain coefficient G, which depends on the line shape of the transition by

$$\alpha(r) = \frac{\frac{1}{2}G(r)}{1 + I(r)/I_0} , \tag{4.61}$$

where I_0 is the saturation parameter. Since we are interested in the constant and quadratic terms of the gain profile, we substitute into (4.61) for the unsaturated gain profile

$$G(r) = G(1 + \tfrac{1}{2}\gamma r^2) \tag{4.62}$$

and for the intensity

$$I(r) = I \exp(-2r^2/\omega^2) \ . \tag{4.63}$$

For the constant term of the amplitude gain we obtain

$$\alpha_0 = \frac{G/2}{1 + I/I_0} \ . \tag{4.64}$$

For the quadratic term we find

$$\alpha_2 = \frac{G/2}{1 + I/I_0} \left(\gamma + \frac{4}{\omega^2} \frac{I/I_0}{1 + I/I_0} \right) \ . \tag{4.65}$$

By substituting (4.64,65) into (4.60), we find for the incremental gain

$$\frac{d\ln(I\omega^2)}{dz} = \frac{G}{1 + I/I_0} \left(1 + \frac{1}{4}\gamma\omega^2 + \frac{I/I_0}{1 + I/I_0} \right) \ . \tag{4.66}$$

It is seen that the contribution of radial radiation transport to the incremental gain comes from both the small-signal gain profile and the gain-induced saturation. In general, the small-signal gain decreases with the distance from the axis (γ is negative). Such a profile broadens the wave front, and radiation energy is radially transported out of the center. This means that the threshold condition to start oscillations requires *higher* inversion densities along the axis than that predicted on the basis of a one-dimensional plane-wave interaction.

This conclusion can also be reached more simply by stating that the integrated gain over the mode volume must be equal to the losses. Then one finds that the inversion along the axis is much higher than the average inversion needed to reach threshold. It is assumed, however, that the geometry of the beam is unchanged by the gain profile. In other words, it is assumed that if the effective diameter of the gain region became smaller, the effective beam diameter would not become smaller. In Sect. 4.5.2 we shall show that the beam width is, indeed, practically unchanged by the gain profile.

In order to estimate the threshold condition, let us consider a small-signal gain profile proportional to the electron density of the discharge. In dealing with a positive column discharge, this gain profile can then, according to Sect. 4.2, be desribed by a zero-order Bessel function. The quadratic term of such a gain profile is

$$\gamma = -\frac{2.88}{R^2} \ , \tag{4.67}$$

where R is the radius of the tube diameter.

The second term of (4.66) becomes $-0.72\omega^2/R^2$. In many single-mode oscillators the value of ω^2/R^2 is about $\frac{1}{2}$. Substituting this value into (4.66) makes the incremental gain for starting oscillations about 36 % smaller than in the case where the small-signal gain profile is ignored.

The last term on the right-hand side of (4.66) describes the part of the radial radiation transport that comes from the change of the saturated gain profile by the Gaussian intensity distribution. For optimized laser systems, when $I/I_0 \gg 1$ the last term of (4.66) approaches one. Thus at high intensities the intensity profile itself leads to a radial radiation transport that contributes as much to the incremental gain as the stimulated emission from the local medium. It is interesting to note that the latter conclusion is independent of the beam width. Thus the amplifying medium near the beam center is about twice as efficiently depleted as expected on the basis of a plane-wae approximation. This has a large bearing on the determination of the saturation parameter, as will be pointed out in Sect. 4.5.3.

4.5.2 Width of a Gaussian Beam in an Oscillator

In the following we apply the ABCD law discussed in Sect. 4.5 to calculate a complete round-trip through a resonator of length L and with mirrors with radii of curvature R_1 and R_2. However, before doing this, we introduce dimensionless quantities for the matrix elements as well as for the beam parameters. The results are, of course, not different but the advantage of this, as we shall see, will be that the beam waist and radius of curvature are explicitly obtained as a function of one or two dimensionless parameters that contain the nonlinearity of the medium, the wavelength, and the cavity length. The derivation thus is general without specifying explicitly all quantities like cavity length, mirror curvatures, wavelength, or quadratic terms of the active medium. The complex dimensionless beam parameter $P(z)$ is

$$P(z) = L/q(z) = U(z) - iQ(z) \ , \tag{4.68}$$

where according to (4.51)

$$U(z) = L/R(z) \quad \text{and} \tag{4.69}$$

$$Q(z) = \lambda L/\pi\omega^2(z) \ . \tag{4.70}$$

Using the dimensionless parameter $P(z)$ and applying the ABCD law the matrix elements now also become dimensionless and will appear according to (4.53) in the following form.

$$\begin{pmatrix} e & f \\ g & h \end{pmatrix} = \begin{pmatrix} \cos\left(L\sqrt{k_2/k_0}\right) & \frac{1}{L}\sqrt{k_0/k_2}\sin\left(L\sqrt{k_2/k_0}\right) \\ -L\sqrt{k_2/k_0}\sin\left(L\sqrt{k_2/k_0}\right) & \cos\left(L\sqrt{k_2/k_0}\right) \end{pmatrix} \tag{4.71}$$

where k_0 and k_2 are given by (4.55).

The dimensionless matrix corresponding to a complete round trip starting at $z = 0$ through a resonator defined by two mirrors with radius of curvature R_1 and R_2 at $z = 0$ and $z = L$, respectively, is then given by

$$
\begin{pmatrix} a_0 & b_0 \\ c_0 & d_0 \end{pmatrix} = \begin{pmatrix} 1 & 0 \\ -2U_1 & 1 \end{pmatrix} \begin{pmatrix} e & f \\ g & h \end{pmatrix} \begin{pmatrix} 1 & 0 \\ -2U_2 & 1 \end{pmatrix} \begin{pmatrix} e & f \\ g & h \end{pmatrix}
$$
$$
= \begin{pmatrix} e^2 - 2efU_2 + fg & ef - 2f^2U_2 + fh \\ -2e^2U_1 + eg + (h - 2fU_1)(g - 2eU_2) & -2efU_1 + fg + (h - 2fU_1)(h - 2fU_2) \end{pmatrix} ,
$$
$$(4.72)$$

where $U_1 = L/R_1$ and $U_2 = L/R_2$. The values of U_1 or U_2 are positive if the reflected beam is made convergent; otherwise, they are negative. Starting at $z = 0$ with P_0, the P parameter after a full round trip becomes

$$
P_2 = \frac{c_0 + d_0 P_0}{a_0 + b_0 P_0} . \tag{4.73}
$$

In order to obtain the steady-state beam parameters we apply the condition $P_2 = P_0$. Substituting (4.71,72) into (4.73) we find

$$
P_0^2 + 2U_1 P_0 - \frac{(-U_1 - U_2)\cos^2\eta + U_1 \sin^2\eta + \left(\dfrac{2U_1U_2}{\eta} - \eta\right)\sin\eta\cos\eta}{\dfrac{1}{\eta}\sin\eta\left(\cos\eta - \dfrac{U_2}{\eta}\sin\eta\right)} = 0
$$
$$(4.74)$$

in which

$$
\eta = L\sqrt{\frac{k_2}{k_0}} . \tag{4.75}
$$

Since η is complex, (4.74) is not directly suitable for calculating the beam parameters U and Q. Therefore we substitute (4.55) into (4.75) and assume $\beta_0 \gg |\alpha_0|$. We obtain

$$
\eta = \left[-\frac{\lambda}{2\pi}(\beta_2 L^2 + i\alpha_2 L^2) \right]^{1/2} . \tag{4.76}
$$

We define further real parameters β and α

$$
\beta = \frac{\lambda}{2\pi}\beta_2 L^2 , \tag{4.77}
$$

$$
\alpha = \frac{\lambda}{2\pi}\alpha_2 L^2 , \tag{4.78}
$$

which are direct dimensionless measures for the quadratic index and gain profiles.

It can be shown [4.3] that the solution for the imaginary part Q of P as obtained from (4.74) depends only on the absolute value of α, whereas the solution for U shows that the difference between the reciprocal values of the radius of curvature of the wave front and the mirror curvature changes sign when the sign of the gain variation is reversed. Although the above equations describe the beam parameters for laser cavities with arbitrary mirror curvatures we give numerical data only for configurations with one plane mirror. This type of cavity is often used in practice because of the desired low divergence of the outcoming beam or because of the use of a plane parallel uncoated germanium outcoupling plate, having a reflectivity of 78 %. In Figs. 4.5 and 4.6 the Q values at both mirrors are plotted as a function of β for different values of U_1 and zero values of α.

It is seen that for a certain β values, stable oscillations (i.e., Q larger than zero) can only be obtained for U_1 values between a lower and upper limit. Further, the U_1 value for minimum spot size depends on β. It should be noted that in the absence of gain variations the curvature of the wave front near a mirror is equal to that of the mirror. Similarly, in Figs. 4.7 and 4.8 we plot the Q values at the two mirrors as a function of $|\alpha|$ for several values of U_1 but now for $\beta = 0$. It is seen that the Q value increases (spot size decreases) with increasing $|\alpha|$ and that the asymptotic value tends to be independent of U_1. It is seen from Figs. 4.7 and 4.8 that small values of α, say less than 0.2, have negligible effect on the beam width near the mirrors. The effect of β is much more pronounced. It is also seen that the smallest beam width is obtained for the semi-confocal configuration, i.e. $L/R = 0.5$.

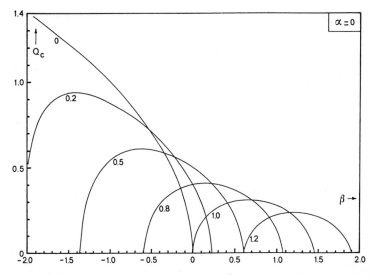

Fig. 4.5. The beam parameter $Q_c = \lambda L/\pi\omega^2$ near the curved mirror is calculated for systems with one flat mirror having only refractive-index variations and no gain variations. This is done for the indicated values of $U_1 = L/R_1$

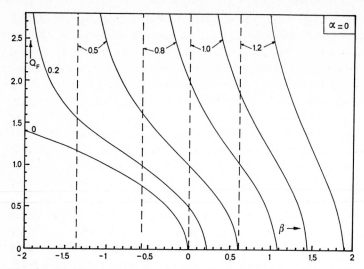

Fig. 4.6. The beam parameter Q_F near the flat mirror is calculated for systems having only refractive-index variations. Different values of U_1 are indicated. The dotted lines are the asymptotes

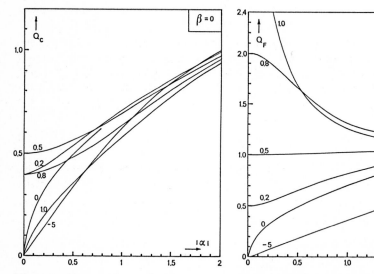

Fig. 4.7. The beam parameter Q_c is calculated for systems with one flat mirror having gain variations but no refractive-index variations. The results depend only on the absolute value of α. Values of U_1 are indicated

Fig. 4.8. The beam parameter Q_F near the flat mirror is plotted as a function of α for $\beta = 0$. The results again depend only on the absolute value of α. Values of U_1 are indicated

For this configuration the Q value near the curved mirror which determines the diameter of the laser tube, changes by 20 % for $\beta = 0.2$. The value of β, related to the refractive index by

$$\beta = n_2 L^2 \ , \tag{4.79}$$

is considerably larger for high-power systems with long discharges. Thus, although in the absence of medium distortions the beam width for given U_1 and U_2 according to (4.70) is proportional to $\sqrt{L}$, the effective medium parameters with respect to a further change of the beam width are proportional to L^2.

In designing the geometry of the laser cavity we have to set an upper limit to the diffraction losses. Looking at Fig. 3.8 and aiming at an extraction efficiency close to 90 % we find that the diffraction losses $\gamma_0 L$ may not be larger than about 2 %. For an optimized discharge with an input energy of 350 W/m we calculate n_2 according to Sect. 4.4. With this result β is calculated according to (4.79). For a discharge length of 130 cm, a diameter of 10 mm, and a cavity length of 150 cm we find $\beta = 0.25$. Constructing a semi-confocal system ($R_1 = 300$ cm and $R_2 = \infty$) the Q value found from Fig. 4.5 is about 0.4. From this we obtain by using (4.70) $w^2 = 12.6 \, \mathrm{mm}^2$. The diffraction losses are due to the limited tube diameter.

The total beam energy of the Gaussian mode is given by

$$W = I_0 \int_0^\infty \exp(-2r^2/w^2) 2\pi r \, dr = \frac{I_0}{2} \pi w^2 \ , \tag{4.80}$$

where I_0 is the intensity on the axis.

The diffraction loss can be considered as what is excluded from the limited tube diameter D. The loss occurs continuously while the beam propagates along the axis. The fractional loss may then be estimated as

$$\gamma_0 L = \exp(-D^2/2w^2) \ . \tag{4.81}$$

Taking a tube diameter of 10 mm we find $\gamma_0 L = 0.018$ which is acceptable. If beam distortions by the medium were not taken into account a tube diameter of about 9 mm would be found. Such a geometry would reduce the output power by 10–20 % as was also found experimentally.

4.5.3 Saturation Parameter Measurements

In Sect. 3.11 we have seen that the maximum available power density per unit length of a system is the product of the small-signal gain and the saturation intensity. Accurate determination of these parameters is necessary to evaluate the performances of various systems. In principle, the small-signal gain and the saturation parameter can be obtained by following the gain versus input intensity of an amplifying medium. For low intensities the amplified signal yields the small-signal gain whereas the derivative of the amplified intensity with respect to input intensity gives us the saturation intensity. In analyzing the measurements with the assumption that the center part of a Gaussian beam near the axis can be approximated by a plane wave, the deduced values of the parameters are questionable. As we pointed out in Sect. 4.5.1, the gain depends also on the gain profile which is ignored in a plane-wave approximation. The same holds for the saturation behavior of a Gaussian beam, as will be discussed below.

The growth of the intensity along the axis for a Gaussian beam will be obtained by solving (4.47a) and (4.60) numerically. In fact, the solution of (4.47a) is identical with the ABCD law if k_2 is constant along the optical axis. This, however, is not the case in a saturating medium with an increasing intensity of the propagating beam with distance. This is also indicated by (4.65) with the α_2 dependence of I. Therefore the numerical procedure will be performed by splitting the amplifier into a large number of parts of equal length. Then the α_0, α_2 and β_2 values of any part can be calculated from the beam properties at the end of the former part. With these values of α_0, α_2 and β_2 taken as constant over the amplifier part the increase in intensity over that part can be calculated by means of (4.60) and the beam parameter q by means of (4.53). In each small section of the amplifier α_0 and α_2 are given by (4.64) and (4.65), respectively. In this way the effect of saturation on the beam properties can be taken into account. A numerical example is displayed in Fig. 4.9 for the case $\beta_2 = 0$ and $\gamma = 0$ [4.5]. The intensity increase ΔI is plotted versus the intensity near the entrance normalized to the saturation intensity. Since the results of the calculations depend on the beam width and radius of curvature of the incoming beam at the entrance of the amplifier, these parameters have to be specified. In the

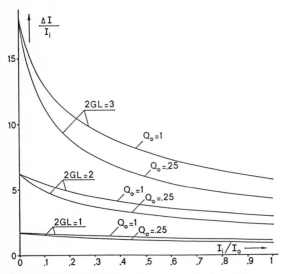

Fig. 4.9. The intensity gain on the optical axis versus the normalized input intensity with the small signal gain $2GL$ as parameter

present example these parameters have been chosen in such a way that in empty space the beam waist ($R = \infty$) would be half-way along the amplifier and has a value ω_0. The parameter $Q_0 = \lambda L/\pi\omega_0^2$ in the figure characterizes this situation. The total length of the amplifier has been chosen as $2L$ so that the small-signal gain of the amplifier is $2GL$.

From a theoretical point of view, the determination of the saturation parameter from the intensity-gain measurements is preferable to a determination from power-gain measurements because influences due to terms higher than the constant and quadratic ones of the gain mainly affect the power gain, whereas the intensity gain at the axis of the tube will only be slightly altered, especially for Q_0 values smaller than 1. However, in practice, one often has to determine the power gain as a function of the incident power, simply because measurements of the intensity gain are not accurate enough. Moreover by taking the derivative near zero input intensity the nonlinear terms are very small. For that reason we continue to calculate the power gain as a function of the input intensity. Again the beam waist would be in the middle of the amplifier in the case of empty space. We choose $2GL = 0.41$ because this value can be used to evaluate experimental data published in the literature [4.6]. It turns out that the results for Q_0 values in the range between 1 and 2.5 can not be distinguished at this value of $2GL$. The results are plotted in Fig. 4.10. We also plotted the numerical results in the case of plane-wave amplification; i.e. the lower curve in Fig. 4.10.

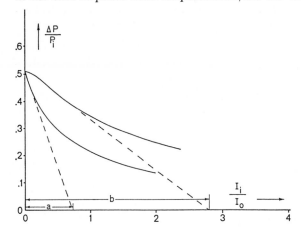

Fig. 4.10. The power gain versus the normalized input intensity. The small-signal gain $2GL = 0.41$. The lower curve shows the result for a plane-wave interaction

Considering a plane wave the intensity gain is given by

$$\frac{1}{I}\frac{dI}{dx} = \frac{G}{1 + I/I_0} \tag{4.82}$$

with the solution

$$\ln\left(\frac{I_u}{I_i}\right) + \frac{1}{I_0}(I_u - I_i) = 2GL \tag{4.83}$$

103

where I_i and I_u are the incoming and outgoing intensity, respectively. Taking the derivative of I_u/I_i with respect to I_i near $I_i = 0$ we obtain the saturation intensity I_0. It is readily found that

$$\left(-\frac{\partial I_u/I_0}{\partial I_i}\right)_{I_i=0} = \frac{e^{2GL}(e^{2GL}-1)}{I_0} . \tag{4.84}$$

The left-hand side of the last expression is the tangent to the intensity or power gain curve at $I_i = 0$. This tangent intersects the abscissa at the intensity I_m. The ordinate of the tangent is equal to $\exp(2GL) - 1$, so that we have the relation

$$\frac{e^{2GL}-1}{I_m} = \frac{e^{2GL}(e^{2GL}-1)}{I_0} \quad \text{or} \tag{4.85}$$

$$I_0 = I_m e^{2GL} . \tag{4.86}$$

In Fig. 4.10 it is seen that the tangent in the case of a Gaussian beam is much lower than in the case of a plane wave. This means that if the above procedure of approximating a Gaussian beam by a plane wave is chosen to deduce the saturation parameter from the observed power-gain curve its value will be too large. The correction will then be obtained by multiplying the measured tangent by b/a, where a and b are determined by the intersections of the two tangents with the line $\Delta P/P_i$, as indicated in Fig. 4.10. In our example $b/a = 3.57$.

Using the plane-wave approximation the saturation intensity is deduced as $18 \, \text{W/cm}^2$ [4.6]. With the above correction one finds a/b. $18 = 5 \, \text{W/cm}^2$ [4.5] which is in agreement with a value that is obtained from the life time of the excited upper laser state, as pointed out in Sect. 3.5.

4.6 Sealed-off CO_2 Lasers

Since CO_2 always dissociates in an electrical discharge, the lifetime of sealed-off CO_2 lasers is most probably limited by the irreversible chemistry of the dissociation products like CO, and O, and C and the gas clean up at the cathode by the sputtered film or on the surface of the discharge tube. If no special precautions are taken, a gradual decrease of output power is observed. To circumvent this problem of CO accumulation and the possible formation of complex molecules with surface materials a continuous flow of gas is often applied. Experiments have shown that a rate of about $20 \, \text{cm}^3/\text{min}$ is sufficient to maintain the initial output [4.7]. With flowing systems a maximum output power of about $60 \, \text{W/m}$ can be obtained.

From the invention of the CO_2 laser onwards, research has been carried out to decrease the dissociation of the laser gas and its interactions with surface materials in order to find appropriate conditions for a sealed-off laser. The advantages of operating a sealed-off system are that the stability of the output spectrum and power are much better, it is more compact and less complicated to operate, and finally a reduction in operating costs is connected with the lower gas consumption.

Long-term operation has been seen as a complicated problem depending on the types of molecules that are produced in the discharge and on the materials used for the discharge tube and electrodes. We found that a great deal of the decrease in radiation output during the first hour was due to the kind of glass used [4.8], for instance Pyrex, so that we supposed that, with some glasses, either gas components are absorbed or impurities were released from the wall during the discharge. We also found the type of electrode material used to be very critical. Ordinary cathode materials like tungsten, nickel, aluminium, stainless steel or iridium have been found to be unsatisfactory. Most materials are known to form carbonyls in a discharge containing CO_2. This problem may be eliminated by heating the cathode, as has been tried with nickel [4.9]. We found that the problem of the construction materials can be solved satisfactorily by the use of fused silica for the discharge tube and an inert metal like gold or platinum for the electrodes [4.10]. The advantage of fused silica is not only its clean surface and the absence of impurities that are otherwise released from the wall but also that fused silica absorbs less water in comparison to glasses with a high alkali content [4.11]. The use of inert metals as cathode material has the disadvantage of high sputtering rate compared with most other metals so that low-pressure gas component removal by a sputtered film may be a problem in the long term. This problem can be reduced by the construction of a hollow cylinder for the cathode in such a way that the electrons are emitted only from the inner cylindrical surface fitted over a glass sleeve. The sputtering due to collisions of ions with sharp edges of the cylinder is then minimized.

As far as the dissociation problem is concerned we found that the catalytic process which has to regenerate the carbon dioxide from the dissociation products during the discharge can be promoted by small amounts of water vapor (or hydrogen) added appropriately to the laser gas mixture [4.10]. Figure 4.11 shows the original lifetimes and output powers obtained with a 1.5 m version of sealed-off systems with platinum electrodes during continuous operation. The internal diameter is 20 mm and the gas mixture is 1 torr CO_2, 2.5 torr N_2, 11 torr He, 0.2 torr H_2, and 0.1 torr O_2. The outcoupling is by means of a Ge etalon. The result is compared with an identical laser as far as construction, cleaning and the gas composition of CO_2, N_2 and He are concerned. The only difference in gas content was the addition of H_2 and O_2. We have also plotted the discharge currents for maximum

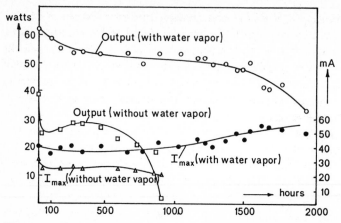

Fig. 4.11. The maximum radiation output, and the corresponding current, I_{max}, as a function of the operation time with a gas mixture of 1 torr CO_2, 2.5 torr N_2, and 11 torr He, and with a gas mixture that contains also 0.2 torr H_2 and 0.1 torr O_2

radiation production. The maximum lifetime of about 1500 h is limited by the disappearance of H_2 or H_2O as we investigated experimentally [4.12].

For this purpose the gas composition of a system was studied with a single focusing mass spectrometer as a function of operation time. The laser system had the same construction parameters and gas composition as described above except that several gas-sample bottles for mass spectro-scopic analysis were connected to the tube. At given time intervals during

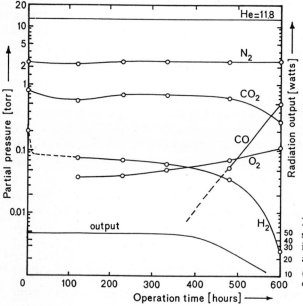

Fig. 4.12. The partial pressures of the gas components in the CO_2 laser *(left scale)* and the radiation output *(right scale)* as a function of operation time

the continuous operation of this system the gas bottles were sealed-off and analysed. In Fig. 4.12 the gas composition and radiation output as a function of operation time are plotted. Water vapor could not be detected with the mass spectrometer used. However, since the hydroxyl spectrum is detected from the side-light emission and since we did not find the initial hydrogen concentration, we assume that the mixture also contained water vapor.

According to the detected hydrogen concentration of 0.07 torr after 118 h, it follows that at that time about 0.13 torr water vapor was also contained in the laser mixture. We supposed that this amount of water vapor was formed shortly after the operation of the laser started. Therefore the hydrogen content before the first gas sample was taken off, is indicated in Fig. 4.12 by the dotted line.

It can be seen from this figure that the quantities of CO_2, N_2 and He remain more or less constant during the first 475 h whereas the hydrogen concentration decreases slowly. At a relatively low concentration of hydrogen we were able to observe for the first time the presence of CO. Its concentration increases strongly when the hydrogen disappears below the detection limit. During this period of increased production of CO the concentration of CO_2 and the radiation output decreases fast. These observations together with the spectroscopic measurements of the side-light emission show clearly that the presence of hydrogen and probably also water vapor coincide with a negligibly low concentration of CO, whereas for a decreasing concentration of H_2 the CO concentration increases at the expense of the CO_2 concentration. The results support the suggestion that a recombination of (vibrationally excited) CO with OH to $CO_2^* + H$ occurs. It is interesting to note that this reaction has also been proposed to explain the absence of absorption bands of CO molecules in the Venus atmosphere [4.13]. This suggestion is also supported by the results obtained from spectrographic analysis of side-light emission of the discharge tube. In the absence of water vapor and hydrogen, the emission spectrum (between 240 and 800 nm) has been firmly established as second positive bands of N_2, third positive bands of CO, and NO bands from the γ-system. No emission from CN or C_2 was observed. On the other hand, if 0.2 torr hydrogen is added to the same laser tube, emission originating from CO or NO molecules is no longer observed, even if the exposure time is increased by a factor two. The emission spectrum shows only the same nitrogen bands with pratically the same intensity, and hydroxyl bands from the A system.

We must conclude from the above described observations that for mixtures containing water vapor (or hydrogen) the CO concentration during the discharge, if present at all, must be very low.

For longer systems the output per unit length usually increases somewhat because of relative smaller losses and of the more effective use of the discharge length. In Table 4.2 we show the output powers and efficiencies of three sealed-off CO_2 lasers having the same gas compositions and inter-

Table 4.2. The output powers and efficiencies of three sealed-off CO_2 lasers which are distinguished only by the lengths of the discharge tubes; η being the overall efficiency including anode and cathode fall

Length [m]	Output [W]	W/m	η
1.5	60	40	11 %
2.4	120	50	12 %
3	190	63	15 %

nal diameters as described above but which are distinguished only by their length.

In conclusion we mention that properly designed sealed systems yield the same output per unit length as flowing systems.

4.7 Single Mode CO_2 Lasers

As we discussed in Sects. 4.4 and 4.5, the output power does not depend on the diameter so that a geometry containing only the zero-order or Gaussian mode may yield as much radiation as a multimode system. For obtaining high output with only the zero-order mode it is desirable to construct the tube diameter not too narrow to avoid too high diffraction losses. This approach may yield space for the next higher-order mode. However, the oscillations of these modes can be suppressed by making use of the competition phenomena between the modes. Due to the homogeneous line broadening and the large spatial overlap of these neighboring modes the interactions between the modes are very strong. Single-mode operation is then obtained, if the ratio of the excitation rate and the loss factor of one mode is much larger than that of the others [4.14].

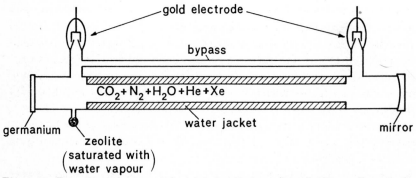

Fig. 4.13. Experimental setup of closed molecular laser (length 150 cm, diameter 1 cm). The gas mixture contains 0.3 torr H_2O, 2 torr CO_2, 4 torr N_2 and 12 torr He, 0.6 torr Xe. The maximum output power of 70 W is found in single-mode operation at a discharge current of 30 mA

Experimentally we observed that for a cavity length of 150 cm this was so in the case of diameters less than about 10 mm.

A sealed-off single-mode (zero-order) laser construction is shown in Fig. 4.13. The cathode and anode are here identical. Since the anode does not sputter and the energy dissipation is relatively small, it can be constructed as a simple pin also made of gold or platinum. The laser is a water-cooled device with a cavity length of 150 cm. It was made of quartz and equipped with gold electrodes. The totally reflecting gold-coated mirror has a radius of curvature of about 2500 cm. The transmitting mirror is a plane parallel germanium plate about 2 mm thick. The gas mixture contains 2 torr CO_2, 4 torr N_2, 12 torr He, 0.6 torr Xe, and 0.3 torr H_2O. The maximum output power is 70 W in single-mode operation at a discharge current of about 30 mA. We observed that the life of some lasers was reduced by the absorption of hydrogen from the water vapor at the sputtered products of the platinum or gold cathode, so that after some hundreds of hours the water content was too low and the CO_2 gas started to dissociate. Furthermore we observed that during the dc discharge, the gas composition was not homogeneous along the length of the tube. The water became partially segregated from the CO_2. This might be caused by cataphoresis.

In order to overcome these drawbacks, we constructed our laser with a water vapor replenisher, consisting of about 20 mg of zeolite saturated with water vapor at the desired pressure. Furthermore, our design included a bypass, as indicated in Fig. 4.13, so that diffusion through this bypass would reduce the gas segregation in the discharge tube.

Both the face and back of the germanium plate contribute to the reflection. The reflected beam is the result of the interference between these two surfaces, which means that the reflection depends on the frequency of the radiation and varies between 0 and 78 %. The frequencies for maximum reflection are about 2×10^{10} Hz apart. The frequency interval between two laser transitions with successive rotational quantum numbers in the center of P branch is about 5.5×10^{10} Hz. Therefore, the reflection in the case of radiation originating in one rotational transition will differ considerably from that originating in other rotational transitions. This means that, in general, the ratio of excitation rate and loss factor of modes originating in one transition is much higher than those of other modes. This will lead to the survival of only one lasing rotational transition. Thus the germanium outcoupling plate virtually discriminates among the available rotational transitions by selecting only one. Experimentally it was verified with a spectrometer that the laser beam, indeed, only originated in one rotational-vibrational transition.

As mentioned above the tube diameter of 10 mm allows under appropriate conditions the oscillations of higher order modes. This is demonstrated with the following experiment [4.15]. The discharge current was varied between 0 and 35 mA. The gas content was initially varied within the ranges

0.2 to 0.4 torr H_2, 1.5 to 2.5 torr CO_2, 2.5 to 5 torr N_2, and 4 to 10 torr He. This was done with a laser containing no zeolite and having no by-pass. With a large number of gas compositions, we observed the following phenomena. In the case of low partial pressure of He (4 torr), we observed, depending on the discharge current, the TEM_{00} and TEM_{03} modes where the two indices are, respectively, the radial and azimuthal mode number. For discharge currents above 25 mA, the TEM_{00} mode changes gradually to the TEM_{03} mode. This might be related to the fact that higher discharge currents increase the gas temperature of the center part of the tube, thus, exerting an unfavorable influence on the gain conditions of the TEM_{00} mode.

The addition of more helium results in the operation of lowest-order modes only, with competition between the TEM_{00} and TEM_{01} modes. The radiation output and mode pattern found for a gas mixture of 0.4 torr H_2, 2 torr CO_1, 4 torr N_2, and 8 torr He is plotted in Fig. 4.14 as a function of the discharge current. For discharge currents up to 25 mA, we observed only the TEM_{00} mode, whereas at 30 mA both TEM_{00} and TEM_{01} appeared and at 35 mA only the TEM_{01} mode could be observed.

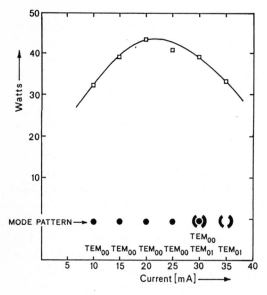

Fig. 4.14. Mode pattern and output power as a function of discharge current for a gas mixture of 0.4 torr H_2, 2 torr CO_2, 4 torr N_2, and 8 torr He

It is interesting to note that the competition phenomena are accompanied by hysteresis. If the discharge current is changed rapidly from 20 to 35 mA, or vice versa, it takes about 15 s before the new stable operation is completed. During the transition time one sees both modes operating.

4.8 The Sequence- and Hot-Band Lasers

In Sect. 3.9 we derived for the small-signal gain ratio of the corresponding lines in the sequence and regular bands the relation

$$\frac{\alpha_s}{\alpha_r} = 2\exp\left(-\frac{h\nu_3}{kT_3}\right) . \tag{3.49}$$

When T_3 is about $1800\,\mathrm{K}$ as found in the absence of laser action (Fig. 3.6) we calculate that α_s/α_r is about 0.44. For a system lasing in a regular band, T_3 is about $1000\,\mathrm{K}$ (Table 3.2) and the gain ratio is only 0.13. Thus considerable gain in the sequence band can be obtained if laser action in the regular bands is suppressed. The sequence lines were initially seen in a laser cavity with good wavelength discrimination [4.16]. In fact, it is not only a matter of sufficient gain in the sequence band but also of the highest gain. Since all transitions are collisionally coupled there is strong competition between lines; not only within a band but also of different bands. In a stationary process the line with highest gain will survive at the expense of weaker lines.

An effective way to favor the sequence lines is the addition of a cell containing hot CO_2 within the laser cavity [4.17]. The in-cavity hot cell enables a conventional CO_2 laser to be converted into one which lases only on the sequence lines. Since the sequence lines lie so close in frequency to the stronger regular lines, it is very difficult to select them with only an adjustable outcoupling. However, the absorption properties of hot CO_2 can be chosen in such a way that the regular bands are suppressed and the sequence bands are allowed. This can be understood from the fact that the lower levels of the regular bands lie only $1300\,\mathrm{cm}^{-1}$ above the ground state, whereas the lower levels of the sequence bands lie about $3200\,\mathrm{cm}^{-1}$ above the ground state. The densities of the absorbing particles are $N_{CO_2}Q^{-1}\exp(-h\nu_1/kT)$ for the $10.4\,\mu\mathrm{m}$ regular band and $N_{CO_2}Q^{-1}\exp[-(h\nu_1 + h\nu_3)/kT]$ for the $10.4\,\mu\mathrm{m}$ sequence band, where Q is the partition function given by (3.41) and T the temperature of the hot cell. It is clear that the lower levels of the sequence bands lie so high above the ground state that for a large temperature range above room temperature their thermal populations are negligible. This means that the hot CO_2 cell in the cavity introduces significant losses at all the regular line frequencies, but negligible loss at the sequence frequencies. A successful configuration contains a $40\,\mathrm{cm}$ in-cavity cell filled with $50\,\mathrm{torr}$ CO_2 at a temperature of about $400°\mathrm{C}$. Using an adjustable plane grating of 150 lines per mm on one side and a concave end-reflector on a piezoelectric transducer on the other side of a $140\,\mathrm{cm}$ long CO_2 laser as much as 80 sequence lines were observed by *Reid* and *Siemsen* [4.17]. While turning the cavity over the first sequence band they also observed many additional lines which were identified as vibrational-rotational transitions in higher sequence bands.

Identifying the sequence lines and distinguishing them from regular lines which may lie close in frequency one may use a low pressure absorbing CO_2 cell at room temperature. The regular lines passing this cell are absorbed and strong fluorescence at $4.3\,\mu m$ can be detected as side-light from molecules which are pumped to the 00^01 level and then radiate to the ground level. As the sequence lines are not absorbed, they show no fluorescence. A strong signal can be observed from as little as 0.1 W of a regular line; no fluorescence was detected from 10 W of sequence line [4.17].

Since the gain of the sequence lines is smaller than that of the regular lines, the cavity losses including outcoupling must be much smaller than that of a conventional laser. The inversion density of the lasing transition is, in a steady-state situation, proportional to the losses. Decreasing the losses of a sequence laser system in comparison to a regular system the inversion density during laser action may become the same as that of a regular system. Since all transitions of the CO_2 molecule are coupled by molecular collisions it is expected that for a well-designed sequence laser system the output power and efficiency will closely approach those of a regular system of the same discharge length.

So far we discussed the performance of a sequence laser system. The main argument that a hot cell discriminates against the regular bands is also applicable for laser action on the hot-band lines. The density of the absorbing molecules of the hot band is equal to $N_{CO_2}Q^{-1}\exp[-(h\nu_1 + h\nu_2)/kT]$ which is a factor $\exp(-h\nu_2/kT)$ smaller than that of the regular band. Since $h\nu_2/k$ is about 975 K and $h\nu_3/k$ about 3400 K the densities of the absorbers of hot lines are much more sensitive to the temperature change than in the case of sequence lines. This means that the hot-band operation requires a much lower temperature of the hot cell. Among all potential lasing lines there is strong competition. The line with the highest gain-to-loss ratio will oscillate at the expense of all other lines. It must be clear that this competition effect can be strongly influenced by the hot-cell temperature.

4.9 Transition Selection with Adjustable Outcoupling

For line tuning of the CO_2 laser it is desirable to have a narrow-band reflector which can be continuously scanned over a large frequency region. A possible solution to this problem is the use of a grating in the Littrow arrangement as one end mirror. This technique does not work for the low-gain transitions with wavelengths away from the blazed wavelength of the grating because of the reduced reflectivity. Furthermore, this technique is inadequate for high-power where due to heating effects the quality of the grating is lost.

We shall discuss the possibilities and realization of an outcoupling device that simultaneously offers high effective reflectivity even for high-power systems together with narrow bandwidth and easy tunability. It consists, in principle, of an optical cavity with one broad-band reflector and a rotatable-wavelength selective grating reflecting in zero and first order [4.18]. It will be shown that this construction will not only lead to considerably higher reflectivity than the Littrow arrangement but, more important, it avoids the exposure of the grating to high power.

These transitions of a CO_2 laser are highly competitive in the sense that each of them competes to oscillate not only on its own inversion but on the inversion of the vibrational levels and their rotational sublevels. As a result, only one or two transitions with the largest gain-to-loss ratio will oscillate [4.19]. Using a wavelength discriminating device with sufficient reflectivity, it should be possible to obtain also laser action on any low-gain transition irrespective of output power.

Since the output power and mode pattern are influenced by any slight disturbance, the experimental setup, as indicated in Fig. 4.15, is mounted on a 10-cm thick 2-m by 1-m cast-iron table. The cavity has a length of 150 cm and an internal diameter of 10 mm. There is a totally reflecting gold-coated mirror with a radius of curvature of 2500 mm on one side and a 36 % reflecting germanium flat AR coated on the other side. The length of the discharge region is 130 cm. The cavity is strongly over-coupled and would not lase with the usual dc discharge current of 30 mA. The device is water cooled.

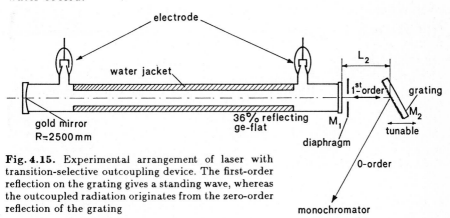

Fig. 4.15. Experimental arrangement of laser with transition-selective outcoupling device. The first-order reflection on the grating gives a standing wave, whereas the outcoupled radiation originates from the zero-order reflection of the grating

The gold-coated diffraction grating has 1800 lines per 2.54 cm and is blazed for 10.6 μm. Together with the 36 % reflecting germanium flat it forms an additional outcoupling cavity with a length of about 10 cm. The grating is mounted on a tunable piezoelectric translator for position scanning. The grooves of the grating are perpendicular to the optical axis of the system. The translator, with grating, is mounted on an angular orientation device with angular resolution of about 0.1 arcsec. In a region of interest, for instance between 9.1 and 9.7 μm, the zero- and first-order reflectivities of the grating are, respectively, 26 and 46 % for incident intensities of about 15 W. For low-power densities the first-order reflectivity may be considerably higher. In order to suppress higher-order modes a 6 mm diaphragm is placed between the grating and the germanium flat. This outcoupling cavity is effectively a Fabry-Perot (FP), where one reflecting surface is the 36 % reflecting germanium flat and the other one the wavelength selecting diffraction grating. The first-order reflection on the grating is used for intracavity reflection, whereas the zero-order reflection is used for outcoupling. Only radiation for which the first-order reflection is parallel to the incident beam can build up a standing wave. This type of FP offers a continuously adjustable narrow-band reflection. In this way any of the available transitions may oscillate if the inversion is sufficient to overcome the comparatively small losses of this outcoupling device. Without the grating the reflectivity of this germanium flat would be too low to obtain oscillations on any transition.

4.9.1 Analysis of Three-Mirror Configuration

For a plane wave propagating along the axis of the cavity, assuming the first-order reflection is parallel to the incident beam, one obtains the following expression for the complex amplitude reflectivity

$$\sqrt{R_{\text{tot}}}\, e^{\,i\phi} = -\sqrt{R_1} + \frac{(1 - R_1)\sqrt{R_{21}}}{e^{-2i\gamma} - \sqrt{R_1 R_{21}}} \ . \tag{4.87}$$

Here, R_1 is the intensity reflectivity of the germanium flat and R_{21} is the first-order intensity reflectivity of the grating; $\gamma = \omega L_2/c$ is the optical phase of the outcoupling cavity; ω, L_2, and c are the angular radiation frequency, the distance between M_1 and M_2, and the velocity of light, respectively. According to (4.87), the phase shift $\Delta\phi$ for the amplitude reflection depends on the distance L_2. This in turn produces a frequency shift of the standing wave in the cavity. The frequency shift is

$$\Delta\nu = -\frac{c}{2L_1}\frac{\Delta\phi}{2\pi} \ , \tag{4.88}$$

where $L_1 = 150$ cm is the length of the cavity containing the discharge. This frequency shift in turn influences the reflectivity according to (4.87). However, since L_1 is much larger than L_2, the latter shift can be neglected. Hence, the reflectivity and frequency change depends only on L_2, as described by (4.87,88). The frequency shift is about 25 MHz as the outcoupling cavity is scanned over the high-reflection region. This is well within the Doppler width of any relevant transition. Comparing these values with experimental observations, it turns out that for many transitions the frequency shifts are not more than about 10 MHz.

In the following we assume that absorption and scattering losses on the grating are considerably higher than those on the gold-coated end mirror and transmitting mirror M_1. Therefore we consider only the losses on the grating. For our laser cavity with one perfectly reflecting end mirror and a CO_2 active medium with homogeneous line broadening, the incident radiation I_1 on the outcoupling mirror M_1 is given by [4.20]

$$\frac{I_1}{I_0} = \frac{g_0 L_1 + \frac{1}{2} \ln R_{\text{tot}}}{1 - R_{\text{tot}}} \quad , \tag{4.89}$$

where I_0 and g_0 are the intensity saturation parameter and the small-signal gain, respectively. From (4.89) the maximum radiation output is obtained for the maximum value of R_{tot}. The output power I_{out}, leaving the grating, is then

$$\frac{I_{\text{out}}}{I_0} = \frac{(1 - R_{21} - \alpha)}{(1 - R_{21})} \left[g_0 L_1 + \ln \left(\frac{\sqrt{R_1} + \sqrt{R_{21}}}{1 + \sqrt{R_1}\sqrt{R_{21}}} \right) \right] \quad , \tag{4.90}$$

where we used the relation $1 = R_{21} + R_{20} + \alpha$. The parameter α represents the fractional losses on the grating and R_{20} is the zero-order reflectivtiy of the grating.

In Fig. 4.16 the calculated radiation intensity according to (4.90) is plotted versus the first-order reflectivity R_{21} of the grating for several values of the parameters $g_0 L_1$ and α. The value of R_1 is kept constant and equal to 0.36. From Fig. 4.16 the maximum value of I_{out} is obtained if the value of R_{21} decreases with increasing values of $g_0 L_1$. This indeed has been observed with our experimental arrangement for strong lines by rotating the grating $180°$ about its normal. Then, for the same wavelength, R_{21} becomes less and the output was found to increase considerably. In Fig. 4.17 we have plotted the maximum output as a function of R_1 according to (4.90) for $g_0 L_1 = 0.2$ and several values of the parameter α. For each value R_1 the calculated radiation is maximized with respect to R_{21} and the corresponding values of R_{21} are also indicated. Next we calculate the ratio of the power I_1, incident on M_1, to the power I_{grat} impinging the grating. The derivation is straightforward and results in

$$\frac{I_{\text{grat}}}{I_1} = \frac{1 - R_1}{(1 + \sqrt{R_1}\sqrt{R_{21}})^2} \quad . \tag{4.91}$$

The calculated results in Fig. 4.18 show how much the exposure of the grating to high power is reduced as a consequence of the inserted mirror M_1. It is interesting to see that a small reflection on the inserted mirror already leads to a considerable reduction of the power impinging on the grating.

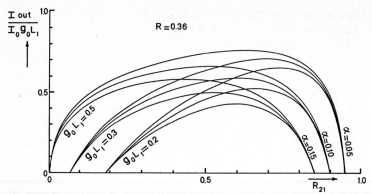

Fig. 4.16. Calculated outcoupled radiation intensity versus first-order reflectivity R_{21}, of the grating for $R_1 = 0.36$ and several values of the parameters $g_0 L_1$ and α

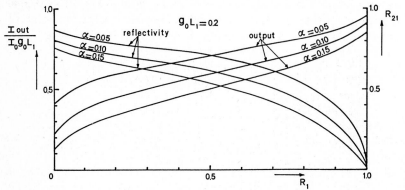

Fig. 4.17. Maximum output power with corresponding values of R_{21} as a function of R_1 for $g_0 L_1 = 0.2$

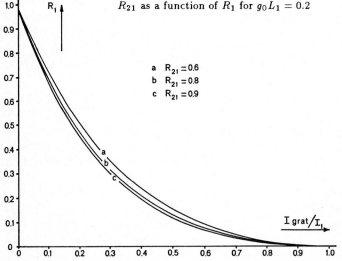

Fig. 4.18. Fractional change of radiation power impinging on the grating as a function of the reflectivity of the inserted mirror

4.9.2 Experiments with a Tunable Outcoupling

With the experimental setup described, we were able to tune continuously over all 65 transitions between P(2) and P(66) and between R(0) and R(62) of the (00^01)-(II) vibrational mode, all 66 transitions between P(2) and P(68) and between R(0) and R(62) of the (00^01)-(I) vibrational mode, and furthermore, 13 unidentified transitions between 11.0 and 11.3 μm wavelength presumably belonging to the hot band. It should be noted that for low-power transitions R_{tot} may be far above 90 % in the absence of thermal deformations of the grating. This may explain why between 9.08 and 11.30 μm wavelength we found a total of 144 transitions with maximum spacing of 0.02 μm. Due to the poor power-handling capabilities of the grating and its increased reflectivity close to the blazed wavelength, it was not possible with this technique to obtain stable oscillations for the high-gain transitions of the (00^01)-(I) mode. We shall come back to this problem in Sect. 4.9.3 and propose another technique for the high-gain band. All other transitions were oscillating stably with maximum output powers up to 15 W in a large central region of the (00^01)-(II) mode and a maximum of about 1 W for the hot band. For the R branch of the (00^01)-(II) mode there is a region where the spacing between two transitions is not more than 0.008 μm. There, at one angular position, it was observed that two neighboring transitions may oscillate depending on the translational position scanning. They did not oscillate simultaneously, but apparently each one requires an R_{tot} value above its minimum for oscillation. The latter depends of course on the gain of the transition and the cavity losses.

Figures 4.19–21 show the R(10) and R(12) transitions of the (00^01)-(II) vibrational mode as a function of the position scanning of the piezoelectric translator. After a scan of a half-wavelength the output is repeated. The upper part shows the ramp voltage applied to the translator and the lower part the output both as a function of time. The oscillograms of Figs. 4.19 and 20 were obtained from the monochromator output. The oscillogram of Fig. 4.21 is a direct observation of the laser output beam. It is seen that, depending on the position of the translator, two transitions are observed but not at the same time. The traces were obtained by scanning the outcoupling cavity very slowly and the whole recorded graph was made in about 30 s. The intensities were measured by means of a thermopile having a time constant of about 20 ms. The asymmetry of the stronger line of Fig. 4.19 is related to a change of mode pattern from Gaussian to a low-order off-axial mode as the reflectivity decreases. This change in the mode pattern is easily demonstrated by illuminating a piece of asbestos with the laser beam. The strength of the relatively weaker line of Fig. 4.20 can be increased by changing the angular orientation, which was set to maximize the stronger line of Fig. 4.19.

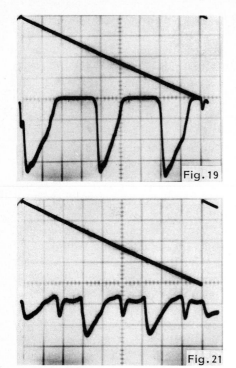

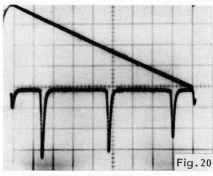

Fig. 19

Fig. 20

Fig. 21

Fig. 4.19. Oscillogram of R(10) transition obtained from the monochromator over a range of M_2 travel. The upper part shows ramp voltage applied to the translator

Fig. 4.20. Oscillogram of R(12) transition obtained from the monochromator over a range of M_2 travel

Fig. 4.21. Oscillogram of directly observed laser output over a range of M_2 travel

4.9.3 Performance at High Stability

The maximum reflectivity calculated according to (4.87) is limited by the condition that the reflectivity R_1 must be just below its minimum value for spontaneous oscillation of the strongest transition. This maximum value of R_{tot} is, however, not limited by absorption and scattering losses on the grating and can be increased if, for any system, the value of R_1 can be chosen higher. Thus, this technique can give high reflectivity (high Q values) at any selected wavelength and reduces considerably the power imginging on the grating. In the case of high gain and high power, as for the central part of the (00^01)-(I) vibrational band, the effective reflectivity for maximum output should be considerably below 100 %. In such a case it is advantageous to use the present outcoupling device as a wavelength-selective end mirror with maximum reflectivity. It is seen from Fig. 4.18 that, for instance, a value of 90 % for R_1 or more results in very small values of I_{grat} and therefore in negligibly small fractional losses. The effective reflectivity becomes 99 %. Then the outcoupling may be achieved on the other side of the cavity by replacing the gold-coated end mirror of Fig. 4.15 with a partly transmitting dielectric mirror.

118

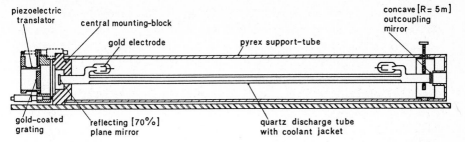

Fig. 4.22. Schematic drawing of the cross section of a stable tunable CO_2 laser. Note that the pyrex support-tube is only on one side supported

For applying such a system there are some practical requirements. The first one is that the laser should give a very stable output power for each of the selected laser lines. The second requirement is to build up a system that can be easily transported from one place to another without any need for re-adjustments of either the mirror or the grating. These requirements can be fulfilled with the construction shown in Fig. 4.22. Inside a pyrex support tube the quartz discharge tube has been mounted. Both the pyrex support tube and the quartz laser tube are on one side glued in the central mounting block. In this solid block also the grating is mounted, with the possibility to turn it very precisely around a vertical axis perpendicular to the optical axis. We used an 1800 l/2.54 cm gold-coated grating, blazed at 10.6 μm. The grating is mounted on a piezoelectric translator, in order to tune the length of the Fabry-Perot at the oscillating wavelength. The other side of the quartz discharge tube near the outcoupling mirror is connected to the pyrex support tube by means of a slightly adjustable support mechanism. It should be noticed that the pyrex support tube is fixed only on one side to the central mounting block. This block, in turn, is mounted on a thick, rigid, iron profile. The laser tube has a discharge length of about 90 cm and an internal diameter of 9 mm. The electrode material is gold. Near the grating a flat mirror with 90 % reflectivity is used. The outcoupling mirror has a curvature of 2 m and a reflectivity of 80 %. The stabilizatioin of this device on a single line can be accomplished by means of the opto-voltaic effect, as will be described in Sect. 4.10.

4.10 Frequency and Output Stabilization by the Opto-Voltaic Effect

The phenomenon of static resistance fluctuation in a cw laser discharge due to the intra-cavity coherent radiation is well-known as opto-voltaic or opto-galvanic effect [4.21]. It can be used successfully for stabilizing a single-line CO_2 laser with high accuracy [4.21, 22, 23], the alignment of a laser cavity [4.24], and the detection of radiation [4.25] in CO_2 lasers.

In general, the resistance fluctuations of the discharge due to stimulated emission in a current stabilized or ballast resistor regime result in fluctuations of the power that is dissipated in the discharge, and hence in temperature variations of the plasma and the discharge tube. The relative changes of static resistance of the laser tube due to changes of stimulated emission inside the laser cavity can be expressed in terms of relative changes of the tube voltage U_L (or longitudinal electric field) and the discharge current I

$$\frac{\Delta R_s}{R_s} = \frac{\Delta U_L}{U_L} - \frac{\Delta I}{I} \ . \tag{4.92}$$

We shall consider the discharge laser tube as a nonlinear resistance with parametrical change of its voltage-current characteristic by the coherent radiation in the discharge tube. The operating current and voltage on the characteristic are determined by the voltage E_a of the power supply and the serial balast resistor R_b, the value of which must be larger than the absolute value of the negative dynamic resistance of discharge tube (Fig. 4.23).

The operating values of U_{L0} and I_0 change with the radiation power along the working-line determined by balast resistor R_b and supply voltage E_a

$$U_{L0} = E_a - R_b I_0 \ . \tag{4.93}$$

It is seen from Fig. 4.23 that the displacement along this line involves the change of discharge voltage, ΔU_L, and simultaneously the change of discharge current, ΔI. The impedance of the discharge tube, however, increases

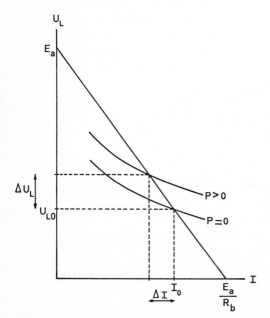

Fig. 4.23. Illustration of voltage-current characteristic of a laser tube parametrically changed by the power of radiation inside the cavity

with the radiation produced. This means that the variation of the discharge voltage, ΔU_L, called the opto-voltaic effect, is in phase with the variation of the radiation power P, and simultaneously the variation ΔI, called the opto-galvanic effect, has the opposite phase. The opto-galvanic signal is usually measured over a probe resistor connected in series between cathode and earth.

A first-order approximation to the static current-voltage characteristic of the discharge can be obtained by an expansion in power series in the neighborhood of the point: $I = I_0$, $P = 0$. We obtain

$$U_L(I, P) = U_L(I_0, P = 0) + \frac{\partial U_L(I, P)}{\partial I}(I - I_0) + \frac{\partial U_L(I, P)}{\partial P} P$$

or

$$U_L(I, P) = E_b - R_d I + \alpha P \quad \text{where} \tag{4.94}$$

$$R_d = -\frac{\partial U_L}{\partial I} \quad \text{for} \quad P = 0 \quad \text{and} \quad I = I_0 \quad \text{is the dynamic resistance},$$

$$E_b = U_{L0} + R_d I_0,$$

$$\alpha = \frac{\partial U_L}{\partial P} \quad \text{for} \quad P = 0 \quad \text{and} \quad I = I_0.$$

The opto-galvanic and opto-voltaic signals can be found from (4.93, 94) as a function of power

$$\Delta U_L = \frac{\alpha R_b \Delta P}{R_b - R_d}, \tag{4.95}$$

$$\Delta I = \frac{-\alpha \Delta P}{R_b - R_d}. \tag{4.96}$$

The relative change of the impedance becomes

$$\frac{\Delta R}{R} = \left(\frac{\alpha R_b}{E_b R_b - E_a R_d} + \frac{\alpha}{E_a - E_b} \right) \Delta P. \tag{4.97}$$

In the case of current stabilization only the opto-voltaic effect is present. This kind of operation is useful for determining the static voltage-current characteristics of the laser discharge. Voltage stabilization of the discharge is impossible due to the negative dynamic resistance.

It is interesting to make an analysis of voltage drop $U(x)$ along the laser tube. Fig. 4.24 shows the simple models for two typical cases:

a) the laser tube is connected only to the supply voltage through the ballast resistor R_b (Fig. 4.24a), and

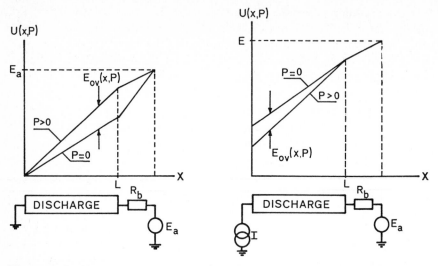

Fig. 4.24a,b. Voltage drop $U_L(x, P)$ along the laser tube for two typical cases of operation, (a) discharge tube with balast resistor, (b) additional application of current stabilizer

b)   when it is connected from cathode side to a current stabilizer (Fig. 4.24b). In the second case the current stabilizer can be treated as a current-controlled source.

In both cases the change of voltage as a function of distance along the discharge axis due to internal radiation power, $E_{ov}(x, P)$, is given by

$$E_{ov}(x, P) = U(x, P = 0) - U(x, P) \ . \tag{4.98}$$

In the set-up of Fig. 4.24a it is seen that the largest opto-voltaic effect is obtained near the ballast resistor R_b of the laser tube. In the case of Fig. 4.24b the opto-voltaic effect is largest near the current stabilizer. It is clear that these two techniques give different values for the opto-voltaic signal in both amplitude and phase.

In the following we consider the modulation of radiation. This can be done by either periodically tuning the line selecting cavity or by internal chopping of the radiation. The signal $E_{ov}(x, P)$ is then a modulated part of plasma potential and it is linearly distributed along the tube (Fig. 4.24b). For current stabilization the total ballast resistance is $R_b + R_s$, R_s is the dynamic resistance of the stabilizer which is much larger then R_d. With this condition (4.95) becomes

$$\Delta U_L = \alpha \Delta P$$

and thus

$$E_{ov}(x, P) = \alpha P \frac{L - x}{L} . \tag{4.99}$$

The extra power Q that is delivered to the discharge, the so-called opto-voltaic input due to laser action, is then given by

$$Q = E_{ov}(x = 0, P)I . \tag{4.100}$$

As we discussed before, the discharge tube is surrounded by a water jacket. In this way the plasma is capacitively coupled to the water jacket. The initial linear distribution of the polarization charges due to $E_{ov}(x, P)$ will equilibrate over the water jacket because of its conductivity. For a normal quartz tube laser construction the capacity over the quartz wall can be estimated as $6\,pF/cm$ and the resistance through the cooling water as $1.5\,k\Omega/cm$. For a one meter tube the corresponding equilibrium time τ is about $9.10^{-5}\,s$. With this time constant the potential of the water jacket reaches a maximum amplitude $\overline{E}_{ov} = \frac{1}{2}E_{ov}(x = 0, P)$. This value can be observed experimentally and turns out to be appropriate in a range of typical modulation frequencies up to $1\,kHz$.

The current of a positive column of a glow discharge tube with radius R can be expressed according to (4.24) as

$$I = 2\pi e \int_0^R v_d(r) n(r) r \, dr , \tag{4.101}$$

where $n(r)$ is the electron density and $v_d(r)$ the drift velocity as a function of the radial distance from the axis. The drift velocity is given by

$$v_d = E_a \mu_a , \tag{4.102}$$

where E_a is the field strength in axial direction and μ_e is the electron mobility. The impedance of the discharge tube is then

$$z = \frac{L}{2\pi e \int_0^R \mu_e n r \, dr} . \tag{4.103}$$

In the positive volumn there is a balance between the energy that electrons gain from the electric field and the energy they lose in collisions. Both the mobility and electron density increase with higher electron energies. In a molecular discharge there is also a balancing of energy between the molecular excitation and electron cooling. The energy transfer flow to the molecules depends on the molecular transition losses such as spontaneous decay, molecular relaxations and stimulated emission.

In the presence of laser action there is thus a larger cooling effect on the electrons and this on its turn decreases both μ_e and n. In the case of current stabilization stimulated emission results to a larger U_L and thus to opto-voltaic input energy.

Experimentally this can be demonstrated as follows: The laser tube is connected in series with a current stabilizer with a high internal dynamic resistance ($200\,\mathrm{M\Omega}$) and a wide frequency range of stabilization ($6\,\mathrm{kHz}$). The modulation occurs by means of mechanical chopping with a frequency of $100\,\mathrm{Hz}$. The changes of the cathode voltage were measured with an impedance probe. We measure the opto-voltaic signal $E_{\mathrm{ov}}(x = 0, P)$ at the cathode and obtain the opto-voltaic input power Q in accordance with (4.100). The optical output is simultaneously measured. In Fig. 4.25 the opto-voltaic input and optical output powers are shown as a function of discharge current with the reflectivity of the output mirror as a parameter.

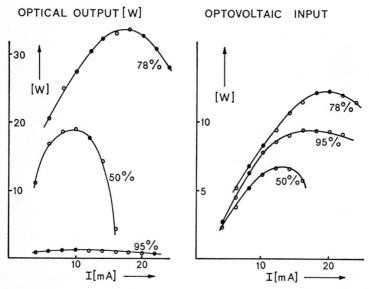

Fig. 4.25. Optical output and opto-voltaic input as a function of discharge current for several reflectivities of the outcoupling mirror. (Notice the difference of the power scales)

From these figures one can observe that for a given reflectivity, optical output P and opto-voltaic input Q are correlated. It is seen that the calculated ratio Q/P does not vary significantly except for the high reflectivity of 95 %. One might expect that the large ratio of Q and P at 95 % reflectivity is due to large internal radiation losses. At such low outcoupling the internal losses are relatively large. The detected radiation is then only a small part of the stimulated emission.

4.10.1 Opto-Voltaic Signal in the Water-Cooling Jacket

We detected the opto-voltaic signal in the water-jacket by putting a needle-shaped probe in the electrically nonconducting water tube [4.23]. The signals measured on both ends of the water cooling tube (i.e., from input and output of water stream) are equal in agreement with the analysis that the potential of the water jacket balances with the time constant τ. The signal measured directly in the water-jacket is the response to a nearly rectangular opto-voltaic signal in the differential circuit RC_1, where R is the resistance of about $1\,M\Omega$ of the scope and C_1 is the capacity between the discharge and the water jacket. For the reproduction of the shape of the opto-voltaic signal we use a capacitor $C_2 = 2.2\,\mu F$ in series with the needle. In this way we effectively construct a capacitive voltage divider (because the resistance of the water between the needle-probe and earth was much higher than the reactance of the C_2 capacitor) consisting of the capacitance C_1 between the plasma and the water jacket and C_2 between the water jacket and earth.

In Fig. 4.26 we plot the signals S_1 measured on the electrode ($x = 0$), S_2 measured on the needle with a resistance of $1\,M\Omega$, and S_3 measured with

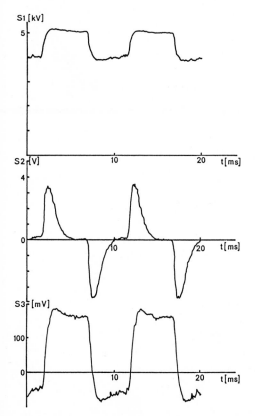

Fig. 4.26. Oscillograms of the opto-voltaic signals due to intra-cavity chopping obtained at the electrodes (S_1), directly on the needle (S_2), and by means of a voltage divider with a capacity of $2.2\,\mu F$ in series with the needle (S_3)

125

the voltage divider between C_1 and C_2. We observed that S_1 and S_3 have the same shape and the relative amplitudes are in agreement with the ratio of the capacitive divider.

Usually opto-voltaic detection is done by means of a high impedance probe on the cathode or anode depending on the supply regime. Here we show that the opto-voltaic signal can also be obtained capacitively from the water jacket with a simple needle probe. The advantage of this method is its high sensitivity and the separation from the high voltage circuit. Because of its sensitivity it can be successfully used for frequency stabilization of the three mirror system. Applying a small oscillating voltage to the piezo-electric translator of the construction shown in Fig. 4.22 the capacitively coupled voltage signal (S_3) of the water jacket is phase sensitively detected and used to optimize the output power by means of the d.c. voltage component of the piezo-electric element. In other words the oscillation frequency is stabilized near the line center frequency. It is observed that the minimum amplitude of this translator is about one nanometer which corresponds to a frequency variation of about 40 kHz.

Further the opto-voltaic input is correlated with the optical outpout and the ratio does not vary substantially with cavity parameters and discharge current, except the case of low outcoupling with relatively high internal losses. This offers practical possibilities for output detection.

Finally we remark from a comparison of output and opto-voltaic input power that stimulated emission cools the plasma.

5. Fast Flow Systems

Continuous output powers of the order of many kilowatts can be obtained with fast flowing systems. The so-called gasdynamic laser where the radiation is produced by rapidly cooling a pre-heated gas during the acceleration of the gas to supersonic velocities will not be treated here and can be found elsewhere [5.1]. In the present chapter we shall treat the fast flowing systems that are, in principle, discharge lasers with a fast renewal of the laser gas. In the case of an axial flow the system is closely related to the non-flowing system described in the previous chapter. The addition of a fast flow has a large cooling effect on the medium. This means that the input power, i.e. the current, can be much higher. Since the current density is directly related to the excitation rates of the molecules a higher production rate of the inverted medium occurs.

The maximum axial gas flow is limited by the speed of sound. A larger flow can then be obtained by a so-called transverse flow perpendicular to the optical axis. Since such a gas flow has a large effect on the discharge behavior special excitation techniques have been developed.

5.1 Convection-Cooled Laser

The results obtained with sealed-off tubes have shown that an output power of 70 W per meter length of discharge or approximately 0.5 W/cm^3 is an upper limit of cw power available from electric-discharge CO_2-lasers. The reason for this limitation is the low cooling capacity of the medium that occurs by diffusion. Heating of the gas causes degradation of the inversion by thermal population of the lower laser level. For these diffusion cooled lasers the cooling power is independent on the gas density because it is proportional to the product of the density and the mean free path of the particles being inversely proportional to the gas density. Furthermore, as we have discussed in the previous chapter, the available power per unit volume decreases with increasing tube diameter at the same rate at which the volume itself is increasing. Thus the power per unit length is independent on both gas density and diameter. The only alternative means for increasing the output power in diffusion cooled lasers is to increase the length. Although kilowatt systems have been realized in this way the technique is very impractical and difficult because of its length and alignment of the optical cavity with small diameter-to-length ratio. Also the negative lens effect of the heated gas in such long discharge tubes is detrimental to the optical beam quality.

The above-mentioned limitations to power and laser performance can be overcome by a fast flow of the discharge medium. If the gas flow transit time is much faster than the characteristic time for diffusion to the wall, the excess heat absorbed by the gas can be removed by the flowing gas. In this manner the laser medium is cooled convectively rather than by diffusion to the wall. The electrical excitation density can then be much larger than for a diffusion-cooled system. This fast flow technique allows an increase of obtainable power per unit length of discharge by several orders of magnitude. Multikilowatt electrically excited lasers in reasonable sizes can be built. A number of different fast-flow configurations have been described in literature [5.3–10].

In principle, the convective cooling can be obtained by fast axial flow parallel or by a transverse flow perpendicular to the optical beam as indicated by Fig. 5.1a, b, respectively. Since the gas temperature rise limits the maximum output and the output power is a given fraction, say 15 % for an optimized system, it follows that the maximum output is proportional to the heat transport by the convectively flowing gas. If the extracted laser power from the volume is P_L and the heat power by convectively flowing the gas P_H we have the relation

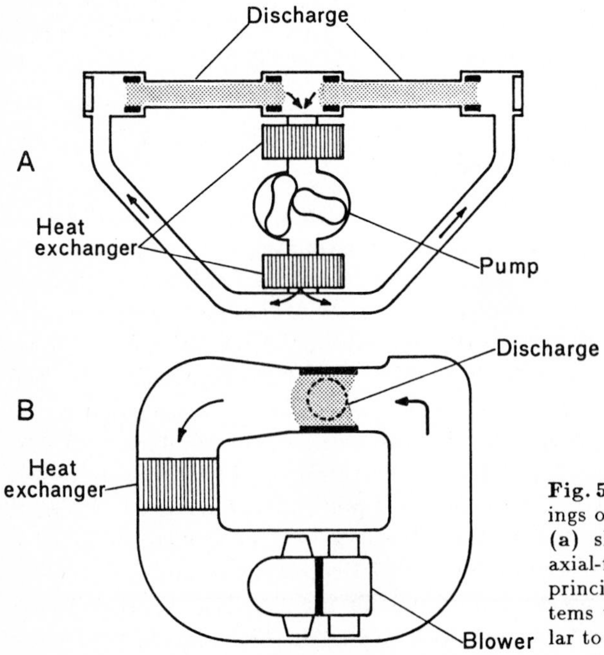

Fig. 5.1a,b. Schematic drawings of convection cooled lasers; (a) shows the principle of an axial-flow system, (b) shows the principle of transverse-flow systems with the flow perpendicular to the optical axis

$$P_L = \left(\frac{\eta}{1-\eta}\right)P_H \ , \tag{5.1}$$

where η is the fraction of extracted input energy.

The temperature rise in the gas by passing the volume is given by

$$\Delta T = \frac{P_H}{\dot{m}C_p} \tag{5.2}$$

where C_p is the specific heat and $\dot{m}$ is the mass flow. Using (5.1) we obtain

$$P_L = \left(\frac{\eta}{1-\eta}\right)\dot{m}C_p\Delta T \ . \tag{5.3}$$

For maximum output power the gas temperature rise must be limited to about 250 K. For typical laser gas mixtures with an efficiency of $\eta = 0.15$ the specific output power $P_L/\dot{m}$ is about 100 J per gram or an input power of about 600 J per gram gas flow. This is the limitation caused by the temperature rise of all convection-cooled CO_2 lasers. Usually the gas is cycling in a closed system with sufficient cooling capacity to return the gas to the initial temperature in one period. Thus, the laser power in all sufficiently convection-cooled systems scales simply with the mass flow through the laser volume. The discharge parameters are typically those of cw diffusion-cooled systems discussed in the previous chapter. The gas pressure turns out to be somewhat higher. This is because the power is not only a function of the discharge parameters such as electron energy distribution, but also of the mass flow. The effect of mass flow shifts the optimized gas pressure to somewhat higher values. Typical values are: 2 torr CO_2, 8 torr N_2, and 20 torr He.

5.2 Principles of Laser Design

The concepts of axial- and transverse-flow systems are shown schematically in Fig. 5.1. Both these principles of mass flow through the volume have their own merits and demerits. The axial flow, especially for narrow tubes, has a relatively simple discharge behavior. In fact, this system is developed from the diffusion stabilized positive column discharge. Further, because of the discharge symmetry with respect to the axis, high performance of the optical beam and high discharge efficiency can be obtained. The conditions for a stable resonator with only the Gauss-mode (TEM$_{00}$) distribution that overlaps practically all active volume can be reached because this discharge matches the cylindrical configuration of Gaussian-mode resonators.

The disadvantages of the axial-flow type are low specific output power per laser volume and high flow speed with a relatively large pressure gradient in axial flow direction. For instance a 50 cm glass tube with an internal diameter of 16 mm may have due to wall friction a pressure drop of 25 mbar for an average flow speed of 300 m/s. Since the discharge properties of small-diameter tubes are to a great extent similar to those of the wall-stabilized discharges, which as we have seen in Chap. 4 give the best laser performance at total partial pressures of CO_2 and N_2 below 10 torr, such a large pressure rise along the laser tube diminishes the specific output power. Nevertheless powers of up to 5 kW and discharge efficiencies above 20 % in multimode laser beams have been realized in this way [5.10]. For larger tube diameters, even above 10 cm, the discharge can also be stabilized by aerodynamic turbulence in the gas flow. This requires a well-designed gas flow in the discharge. For instance, the gas mixture is injected by means of many small nozzles evenly distributed around a circular section of the upstream portion of the discharge tube [5.11]. The injectors are used both as anode and expansion nozzles for the turbulent flow. For these large diameters the energy is usually extracted by means of unstable resonators. Stable output and good beam quality is obtained with axial flow in large diameters having output powers up to 20 kW [5.12].

However, cw powers above 5 kW can be much more practically obtained with transverse systems. Comparing axial- and transverse-flow systems of roughly the same laser volumes and output powers the respective flow velocities are proportional to the length and the width of the laser volume. This means that the flow velocity in the transverse system is an order of magnitude smaller. How far this property of transverse systems actually leads to much higher specific laser powers depends greatly on the discharge configuration and the beam optics. The main problem of the discharge in a transverse flow is the fact that the unstabilized plasma column is bowed strongly under the force of the gas flow (Fig. 5.1b). In order to achieve efficient laser operation, it is neccessary to straighten and align the axis of the discharge collinear with the optical axis of the laser resonator. Under pulsed operation the bowing of the discharge presents no problems and the fast flow allows for high repetition rates. Using a supersonic transverse flow a pulse repetition frequency of 17 kHz has been reported [5.13].

Different discharge techniques have been proved to be successful. The basic principles of various configurations are shown in Fig. 5.2.

a) A normal transverse dc discharge by means of many small cathodes coupled in parallel with ballast resistors to ensure a uniform current distribution. The resistors are roughly proportional to the discharge width and for usual mixtures are about 10 kΩ/cm.

b) A magnetic cross-field stabilization of a glow discharge perpendicular to the drawing. The bowing of the discharge by the gas flow is prevented

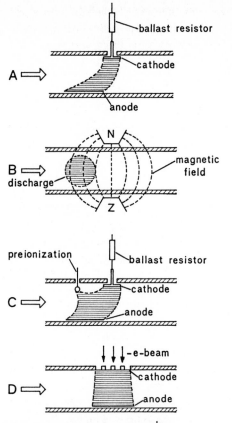

Fig. 5.2a–e. Typical discharge techniques developed for transverse flow systems. (a) normal dc discharge with many cathodes that are separately ballasted with a resistor, (b) stabilization and confinement of the discharge region by an external magnetic field, (c) normal dc discharge with rf preionization, (d) e-beam sustained discharge, and (e) a capacitively coupled rf preionization for stabilizing a normal discharge with several ballasted cathodes

in this configuration by the application of a magnetic field perpendicular to both the gas flow and the collinear optical and discharge axis [5.14]. The laser discharge is also confined by the magnetic field so that a short transit time through the plasma discharge is obtained. The discharge geometry is more or less cylindrical in shape and therefore easily matched to Gaussian modes. The stabilization conditions can be obtained as follows. The magnetic force acting on an electron is given by

$$F_{\mathrm{m}} = e v_{\mathrm{d}} \times B \; , \tag{5.4}$$

where v_d is the drift velocity under the influence of the electric field along the direction of the discharge, and B is the magnetic flux. The magnetic force causes a drift velocity v_m in the direction of this force and relative to the flow velocity given by

$$v_m = \mu_e v_d \times B \ . \tag{5.5}$$

The drift velocity v_d is given by $\mu_e E$, E being the electric field. Similarly the drift velocity for the ions in the magnetic field can be derived. Further, since the charged particles drift upstream collisional forces from the flowing neutral particles interact with the charged particles in the direction of the flow. These frictional forces are much larger for ions than for electrons so that the charges are separated and a space-charge field E_s is created. The field E_s, in turn, influences also the drift velocity upstream. The total transverse drift velocity of the electrons in the case where the magnetic field is oriented perpendicular to both gas flow and electric discharge field is given by

$$v_e = \mu_e^2 EB - \mu_e E_s \ ; \tag{5.6}$$

and similarly for the ions

$$v_i = \mu_i^2 EB + \mu_i E_s \ . \tag{5.7}$$

For stationary conditions we must have the condition that the absolute values of both v_e and v_i are equal to the transverse flow velocity. Then by eliminating E_s from (5.6, 7) the transverse flow velocity is

$$v = \mu_e \mu_i EB \ . \tag{5.8}$$

Thus for a given flow velocity the required magnetic field can be found by using (5.8). The magnetic field is strongest in the center between the pole pieces. The conditions must be chosen such that this center field creates an upstream drift velocity larger than the flow velocity. Then the discharge area moves upstream where the field is weaker and stabilizes. The measured stabilizing magnetic field is indeed observed to be proportional to the gas flow. Figure 5.3 is a plot of B versus v for a gas mixture $CO_2 : N_2 : He = 1 : 1.2 : 4$ at a total pressure of 18.7 torr [5.14]. The E/p value is 18.2 V/cm torr. The low magnetic fields of the order of 10 G/m/s can be obtained easily with both electromagnets and permanent magnets.

   c) A dc discharge similar to configuration a) but with the addition of high-frequency preionization by means of pins. The discharge geometry is considerably improved and a better homogeneity across the laser beam is obtained [5.15].

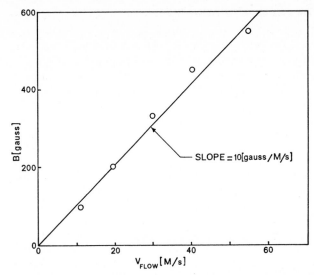

Fig. 5.3. The dependence of the stabilizing magnetic field upon the gas flow velocity. $E/p = 18.2\,\mathrm{V/cm\,torr}$; $CO_2 : N_2 : He = 1 : 1.2 : 4$; pressure $18.7\,\mathrm{torr}$ [5.14]

d) This configuration is a combination of a non-self-sustained discharge and electron beam ionization. The construction is costly but can be optimized with respect to the E/N value. High efficiencies are obtained.

e) A stable and homogeneous discharge is obtained by a combination of dc and rf excitation. The dc discharge is maintained in the direction of the flow between an array of many uniformly spaced pin cathodes and cylindrical anodes positioned as shown in the figure. Each cathode is again separately ballasted. The rf field is capacitively coupled into the gas transverse to the flow direction by metal plates outside the flow channel [5.9].

High performance of all these transverse configurations requires in particular the realization of a homogeneous active medium and also a full optical interaction with this medium. Good homogeneity in the axial direction of the optical beam can be realized, but is more complicated in the direction of the gas flow. Except for the magnetic stabilization they have bad conditions for cylindrical symmetry and hence, unless only a fraction of the medium is used, no favorable conditions for single-mode operation are present. Efficient transverse systems are multi-mode and often of poor optical quality as compared with axial flow systems.

5.3 Gain and Saturation Parameter

For laser systems the relation between inversion density and field intensity for given excitation and cavity parameters is described by the rate equations for the radiation field and the upper and lower level densities. Let us consider a small volume with a flowing medium in the x direction. The density N_2 of the upper level can be described by the equation

$$\frac{\partial N_2}{\partial t} = P_2 - \frac{N_2}{\tau_2} - (N_2 - N_1)\frac{\sigma I}{h\nu} - v\frac{\partial N_2}{\partial x} \ , \tag{5.9}$$

where P_2 is the upper level pumping rate per unit volume, τ_2 is the upper level lifetime which is in fact the collisional relaxation time, N_1 is the density of the lower level, I is the beam intensity, σ is the cross section for stimulated emission, $h\nu$ the photon emission of the laser process, and v is the flow velocity. The last term accounts for local change of N_2 by the flow in the x direction.

Let us assume that the gas enters the laser volume with the densities N_{20} and N_{10} for the upper and lower laser levels, respectively. During the passage through the volume the gas is excited and the inversion $N_2(x,t) - N_1(x,t)$ increases with x (or t). This increase saturates due to stimulated emission. In the case where the maximum excitation rate is matched to the flow speed, i.e. P_L is maximized with respect to the mass flow according to (5.3), the excitation rate near the end of the passage will be equal to the stimulated emission rate. The gas leaves the laser volume with the maximum inversion density.

In order to calculate the gain and saturation we average the population densities over the length L of the laser volume in the direction of the flow. The field intensity I in the cavity is the sum of the intensities of the running waves to the left and the right, respectively. It can be shown that to a good approximation this sum is nearly constant in the direction of the optical axis but not perpendicular to it. However, in the case of transverse flow we consider the averaging over the distance L where the strong field is more or less constant. We may make the approximation $\partial \overline{N}_2/\partial x \sim (\overline{N}_2 - N_{20})/L$. Equation (5.9) then becomes for the average value

$$\frac{\partial \overline{N}_2}{dt} = P_2 - \frac{\overline{N}_2}{\tau_2} - (\overline{N}_2 - \overline{N}_1)\frac{\sigma I}{h\nu} - \frac{\overline{N}_2 - N_{20}}{\tau_g} \tag{5.10}$$

where $\tau_g = v/L$ is the transit time of the gas in the laser volume. Similarly, we can derive for the density of the lower level

$$\frac{d\overline{N}_1}{dt} = P_1 - \frac{\overline{N}_1}{\tau_1} + (\overline{N}_2 - \overline{N}_1)\frac{\sigma I}{h\nu} - \frac{\overline{N}_1 - N_{10}}{\tau_g} \ . \tag{5.11}$$

Under equilibrium conditions the average rates of change, $d\overline{N}_2/dt$ and

$d\overline{N}_1/dt$, are zero and the average gain $\overline{\alpha}$ can be obtained from (5.10, 11)

$$\overline{\alpha} = \sigma(\overline{N}_2 - \overline{N}_1) = \frac{\sigma P_2' \frac{\tau_2 \tau_g}{\tau_2 + \tau_g} - \sigma P_1' \frac{\tau_1 \tau_g}{\tau_1 + \tau_g}}{1 + \frac{\sigma I}{h\nu}\left(\frac{\tau_2 \tau_g}{\tau_2 + \tau_g} + \frac{\tau_1 \tau_g}{\tau_1 + \tau_g}\right)} \quad , \tag{5.12}$$

where $P_2' = P_2 + (N_{20}/\tau_g)$, and $P_1' = P_1 + (N_{10}/\tau_g)$. The densities N_{20} and N_{10} are negligible because it is essential for high-power flow systems that the entering gas is cooled. The small signal gain α_0 is thus given by

$$\alpha_0 = \sigma\left(\frac{P_2 \tau_2 \tau_g}{\tau_2 + \tau_g} - \frac{P_1 \tau_1 \tau_g}{\tau_1 + \tau_g}\right) . \tag{5.13}$$

For small flow velocities, $\tau_g \gg \tau_1, \tau_2$ we obtain for the gain

$$\alpha_0 = \sigma(P_2 \tau_2 - P_1 \tau_1) . \tag{5.14}$$

For fast flow, $\tau_g \ll \tau_1, \tau_2$ we find

$$\alpha_0 = \sigma \tau_g (P_2 - P_1) . \tag{5.15}$$

The excitatioin rate for an optimized system, $P_2 - P_1$, is according to (5.3) proportional to $\dot{m}$ and hence proportional to $1/\tau_g$. This means that, in principle, the small-signal gain of an optimized system is independent on the flow speed. However, in practice, this can be somewhat different because the volume of the active medium or its discharge parameters may change with the flow velocity [5.8]. For instance, for axial flow there is an increasing density gradient with flow speed.

The saturation intensity defined as the intensity I_s for which the gain is half the small signal value, can be deduced from (5.12) and is given by

$$I_s = \frac{h\nu}{\sigma} \frac{1}{\frac{\tau_2 \tau_g}{\tau_2 + \tau_g} + \frac{\tau_1 \tau_g}{\tau_1 + \tau_g}} . \tag{5.16}$$

For small flow velocities, $\tau_g > \tau_1, \tau_2$, the saturation becomes

$$I_s = \frac{h\nu}{\sigma \tau_2}, \tag{5.17}$$

where we have neglected τ_1 with respect to τ_2, because $\tau_2 \gg \tau_1$, as we have seen in Chap. 4.

For large flow velocities, $\tau_g \ll \tau_1, \tau_2$, we find

$$I_s = \frac{h\nu}{2\sigma \tau_g} . \tag{5.18}$$

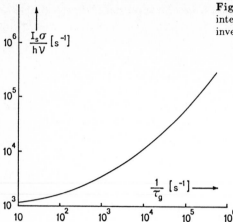

Fig. 5.4. The variation of the saturation intensity divided by $h\nu/\sigma$ as a function of the inverse of the transit time τ_g

Figure 5.4 illustrates the variation of the saturation intensity with the inverse of the transit time. It is seen that for low speed, i.e. large τ_g-values, the saturation is constant and at minimum. For fast flow the saturation intensity increases in proportion to $1/\tau_g$ or to the mass flow.

Since the radiation production by stimulated emission per unit volume is equal to αI we deduce from (5.12) that the maximum power production of the system per unit volume is given by $\alpha_0 I_s$. The saturation intensity being proportional to v the total output is then also proportional to v in agreement with (5.3). The experimentally observed dependence of output power on flow velocity of a 1 m long transverse laser is shown in Fig. 5.5.

The small signal gain, as we mentioned above, does not depend on the flow velocity provided the excitation rate is optimized. This means that the optimized reflectivity of outcoupling mirror is in principle independent of

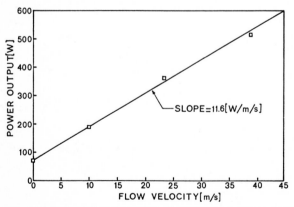

Fig. 5.5. Laser output power as a function of gas flow velocity. ($CO_2 : N_2 : He = 1 : 1.3 : 5.7$; pressure: 23.0 torr; output coupling: 20 %; active discharge length: 1 m) [5.8]

the gas flow and its value depends only on the total gain. It can be calculated according to the theory of Sect. 3.11. The difference with non-flowing systems is the much higher power to be transmitted. Ge windows which have a power dependent absorption coefficient are not suitable. In general, one uses ZnSe of good optical quality. These mirrors have been successfully used for power densities of about $5\,\mathrm{kW/cm^2}$.

6. Pulsed Systems

The major problem in increasing the output energy of a continuous CO_2 laser system is the temperature rise with increased input energy. As discussed in Chaps. 3 and 4, the population density of the lower laser level and its depopulation rate to translational and rotational energies depend strongly on the gas temperature. The gas temperature, in turn, results from the balance between the input and the heat conduction energies. Since the heat conductivity is independent of the gas density, the heat transport by conduction is fixed for a given maximum gas temperature and hence is the limitation on the input energy. Higher output energies of cw systems can then only be obtained by also applying heat convection for cooling such as by rapidly moving the gas through the electric discharge region. For efficient operation of these systems, the discharge power input per unit mass flow rate does not exceed $600\,\mathrm{J/g}$, as we have seen in Chap. 4.

The third approach to this heat handling problem is the use of the heat capacity of the medium itself in pulsed operation. This will be treated in the present chapter. Because the heat capacity is proportional to the density, the maximum output energy per pulse will be proportional to the density.

The pulse duration is determined, to a large extent, by the relaxation times of the excited molecules. Since these times become shorter at higher pressures, the pulse length also depends on the gas pressure. As a result, the peak power is proportional to the square of the pressure. Thus larger energy pulses with higher peak powers and shorter pulse lengths are obtained when the laser gas is at a higher pressure. At low gas pressures stable glow discharges giving uniform excitation are normal. At elevated pressures the stable discharge is not simply achieved. As the pressure increases, the conventional low-pressure glow discharge changes at about 50 torr; the glow constricts to an arc. Arc discharges are unstable, not uniform, and not suitable for efficient excitation of molecular gas lasers.

Since the electromagnetic field in a discharge is given by balancing the energy the electrons gain from the electric field and the energy they lose by collisions, the field is inversely proportional to the electron free path, i.e. the field is proportional to the density. For an axial system ultrahigh voltages have to be used. This is not very practical. However, for transverse excitation across the laser beam the discharge voltage is much lower. Furthermore, the conditions for obtaining stable glow discharges at high pressure are more easily obtained for transversal excitation, too.

6.1 Basic Principles of Laser Operation

The first transversely excited atmospheric (TEA) CO_2 gas lasers were reported by French and Canadian scientists [6.1,2]. These two groups independently devised techniques which would allow maintenance of reasonably large volume glow discharges in atmospheric-pressure CO_2-N_2-He mixtures. Following these early developments of the TEA CO_2 laser a continuous series of advances in excitation technique have been made, progressing from the primitive method of discharge current limiting through arrays of pins or a large set of trigger electrodes to the more sophisticated uv-preionized discharges and e(electron)-beam techniques. These improved methods have resulted in significant progress in both power density and total output power in large volumes.

These developments have also led to advancements in the understanding of the discharge plasma mechanism at atmospheric and multi-atmospheric pressure with respect to laser excitation and maintenance of a stable glow discharge. A fundamental parameter that governs the plasma behavior, in the quasi-steady-state regime, is the ratio of the electric field E to the gas density N. This E/N-parameter is different for uv-preionized and e-beam systems. The basic difference between the two techniques is the degree of ionization produced by the ionizing source. In the uv-preionized system the discharge has initially a low electron density ($10^4 < n < 10^8 \, cm^{-3}$). The applied electric field accelerates the initial electrons to an energy level that is sufficient to further ionize the discharge. In other words, the E/N value is sufficiently large to promote avalanche ionization in the discharge until a quasi-steady-state electron density is reached. This type of discharge is therefore called "self-sustained". In the case of the e-beam technique the ionization necessary to maintain the discharge is produced by an external electron beam. The technique involves the use of high-energy electrons (100–200 keV). The applied electric field, or the E/N-value, is relatively small and causes practically no further ionization; the discharge is "non-self-sustained". The great advantage of this technique is its flexibility in the choice of electric field and gas mixture so that efficient vibrational excitation by the electrons may be obtained. Also, large systems operating at multi-atmosphere gas pressure can be realized. The disadvantages of the e-beam technique is mainly the cost and complexity, as compared with uv systems. The uv preionization is in particular very suitable for small systems.

6.2 Electron-Energy Transfer and $I - V$ Characteristics of Laser-Gas Mixtures

We shall first treat the physical mechanisms involved in the production of excited vibrational states, the transport coefficients, and the operating E/N value for self-sustained glow discharges as a function of current density for various laser-gas mixtures. As we have discussed in Chap. 3, the energy given to the CO_2 upper laser level is due to inelastic collisions with electrons and by resonant energy transfer from vibrationally excited N_2, which, in turn, is also excited by inelastic collisions with electrons. In order to determine the distribution of the discharge input energy transferred to the various vibrational modes, excited electronic states, dissociation, ionization, and also transport coefficients, detailed knowledge of the electron density n and its distribution function $f(u)$ is required. The most important transport coefficients determining macroscopic properties of the discharge are the electron drift velocity, ionization, and attachment coefficients which can all be calculated as a function of E/N.

In Sect. 4.2 the continuity equation for the electrons is controlled by the ionization rate and the ambipolar diffusion. For typical high-power laser discharges the operating gas pressure is one atmosphere or higher. Further, a stable transverse excitation between the electrodes produces a more or less uniform excitation in a medium with relatively large aperture as compared with a low-pressure discharge tube. Under these conditions, the electron density profile instead of being controlled by diffusion is determined by the electron-ion recombination in three-body collisions or by attachment. The continuity equation for the electrons is then given by

$$D_a \nabla^2 n + \alpha v_d n - a v_d n - \gamma n^2 = 0 \;, \tag{6.1}$$

where D_a is the ambipolar diffusion coefficient, v_d the drift velocity, n the electron density, γ the recombination coefficient, and α and a are the ionization and attachment coefficient, respectively. Comparing (6.1) with (4.7) it is seen that $\alpha v_d = \nu$ is the ionization frequency per electron. The ion and electron loss by diffusion is, for TEA lasers, much less than by recombination and attachment. Therefore, (6.1) reduces to

$$\alpha v_d n - a v_d n - \gamma n^2 = 0 \;. \tag{6.2}$$

From (6.2), $n = (\alpha - a)v_d/\gamma$ and using the current density $j = nev_d$ we obtain

$$\frac{j}{N} = \frac{e v_d^2 (\alpha - a)}{\gamma N} \;. \tag{6.3}$$

This equation expresses j/N as a function of E/N through the dependences

140

of the transport coefficients γ, v_d, α/N, and a/N on E/N, as will be derived below.

Lowke et al. [6.3] have numerically calculated the electron distribution function for a variety of experimentally interesting electric discharge conditions using the measured cross sections for elastic and inelastic collisions. We shall follow their analysis in which it is assumed that spatial uniformity and steady-state calculations are applicable because of the slow change of the discharge parameters compared to the electron-neutral inelastic collision rate.

For average electron energies in high-pressure laser discharges of less than 5 eV, the dominant collision process associated with He is the elastic scattering process. In He all inelastic processes have very high threshold energies above 20 eV and small cross sections and, hence, result in negligible power loss. This, however, is not the case for N_2 and CO_2.

6.3 Derivation of Boltzmann's Equation

The Boltzmann equation is one of the most practical tools for describing transport phenomena in gas discharges. It is used to calculate the electron distribution function in a plasma in the presence of field parameters and allows for the impact of elastic and inelastic collisions of the electrons. By solving this equation, the energy flow to excited states of the gas can be calculated.

Let us define the one-particle distribution function $f(v,t)$ of a spatially uniform medium so that $n_0 f dv$ is the number of particles in the velocity-space volume element dv at the time t. The electron density is n_0 so that f is normalized to unity or

$$\int f dv = 1 \ . \tag{6.4}$$

Consider the number of particles $dn = n_0 f(v,t) dx\, dv$ in the phase-space volume element $dx\, dv$. In the time δt, the particle coordinates, in the absence of interparticle interactions, change to

$$x' = x + v\delta t \ , \tag{6.5a}$$

$$v' = v + \left(\frac{F}{m}\right)\delta t \ , \tag{6.5b}$$

where F is any external force field which may be applied to the discharge. If there were no interparticle interactions, dn would be equal to the number of particles $dn' = n_0 f(v', t + \delta t) dx'\, dv'$. If F is constant in time, the Jacobian $\partial(x',v')/\partial(x,v) = 1$, and $dx'\, dv'$ can be replaced by $dx\, dv$. Next we consider that δt is much larger than the duration of a collision and that only binary collisions occur. Further, we assume

$$\left|\frac{f}{\partial f/\partial t}\right| \gg \delta t \ ,$$

and that several collisions will occur during δt, some of which scatter particles into the velocity interval $d\boldsymbol{v}$ and some out of $d\boldsymbol{v}$.

Writing the rate of change of f due to electron-molecule collisions as $(\delta f/\delta t)_c$ we equate the change in the number $dn' - dn$ to the effect of these collisions and get

$$\left(\frac{\partial f}{\partial t} + \frac{\boldsymbol{F}}{m}\cdot\frac{\partial f}{\partial \boldsymbol{v}}\right)\delta t = \left(\frac{\delta f}{\delta t}\right)_c \delta t \ . \tag{6.6}$$

The left-hand side of (6.6) is the time derivative of f computed along the trajectory given by (6.5). In a laser discharge the acceleration of an electron is caused by the applied electric field $\boldsymbol{E}$ and is equal to $-e\boldsymbol{E}/m$, so that (6.6) becomes

$$\frac{\partial f}{\partial t} - \frac{e\boldsymbol{E}}{m}\cdot\frac{\partial f}{\partial \boldsymbol{v}} = \left(\frac{\delta f}{\delta t}\right)_c . \tag{6.7}$$

The electron distribution function f can now be calculated from the Boltzmann equation (6.6) if we know the collision term $(\partial f/\partial t)_c$. Before calculating this term we shall introduce a perturbation term that describes the deviation from an isotropic distribution. This facilitates the further evaluation of the Boltzmann equation.

6.3.1 Near-Isotropic Expansion of Boltzmann's Equation

In the CO_2 laser discharge the average kinetic energy of the electrons is much larger than that of the molecules. This means that the electrons are moving, relatively, very fast and the electrons see the molecules as being nearly at rest. For an elastic collision of an electron with a molecule there is, due to the large mass difference, only a relatively small amount of energy transferred compared to the electron energy, and a large deflection of the electron trajectory. Provided the electron mean free path is small compared to the volume occupied by the gas, the large deflection results in an electron distribution that is nearly independent of the direction of $\boldsymbol{v}$. This enables $f(\boldsymbol{v})$ to be written as

$$f(\boldsymbol{v}) \simeq f_0(\boldsymbol{v}) + \frac{\boldsymbol{v}}{v}\cdot\boldsymbol{f}_1(\boldsymbol{v}) \ , \tag{6.8}$$

where f_0 and $\boldsymbol{f}_1$ are functions of the magnitude of $\boldsymbol{v}$ only, and $|f_1| \ll f_0$. This expansion will now be substituted into (6.7). We first consider in the second term of the left-hand side of (6.7) $E{\cdot}(\partial f/\partial v)$. Using (6.8) one obtains

$$E{\cdot}\frac{\partial f}{\partial v} = \frac{1}{v}\frac{\partial f_0}{\partial v}(E{\cdot}v) + \frac{E{\cdot}f_1}{v} - \frac{1}{v^3}(E{\cdot}v)(v{\cdot}f_1) + \frac{1}{v^2}(E{\cdot}v)\left(v{\cdot}\frac{\partial f_1}{\partial v}\right) ,$$

(6.9)

where we have used $\partial v/\partial v_x = v_x/v$, etc.

Working out the last two terms we replace the quadratic contributions of the velocity components by their averages or $v_x^2 \simeq \frac{1}{3}v^2$, $v_x v_y = 0$, etc. This is justified because we are dealing with a nearly isotropic distribution of v. This then leads to

$$E{\cdot}\frac{\partial f}{\partial v} \simeq \frac{1}{v}\frac{\partial f_0}{\partial v}(E{\cdot}v) + \frac{E{\cdot}f_1}{v} - \frac{1}{3}\frac{E{\cdot}f_1}{v} + \frac{1}{3}E{\cdot}\frac{\partial f_1}{\partial v}$$

or

$$E{\cdot}\frac{\partial f}{\partial v} = \frac{E{\cdot}v}{v}\frac{\partial f_0}{\partial v} + \frac{1}{v^2}\frac{\partial}{\partial v}\left(\frac{v^2}{3}E{\cdot}f_1\right) .$$

(6.10)

Using the last result we obtain by substituting (6.8) into (6.7)

$$\frac{\partial f_0}{\partial t} + \frac{v}{v}{\cdot}\frac{\partial f_1}{\partial t} - \frac{e}{m}\left[\frac{E{\cdot}v}{v}\frac{\partial f_0}{\partial v} + \frac{1}{v^2}\frac{\partial}{\partial v}\left(\frac{v^2}{3}E{\cdot}f_1\right)\right]$$
$$= \left(\frac{\delta f_0}{\delta t}\right)_c + \frac{v}{v}{\cdot}\left(\frac{\delta f_1}{\delta t}\right)_c .$$

(6.11)

Since (6.11) must hold for all directions of v we can equate the parts with and without v, or

$$\frac{\partial f_1}{\partial t} - \frac{eE}{m}\frac{\partial f_0}{\partial v} = \left(\frac{\delta f_1}{\delta t}\right)_c ,$$

(6.12a)

$$\frac{\partial f_0}{\partial t} - \frac{e}{3mv^2}\frac{\partial}{\partial v}(v^2 E{\cdot}f_1) = \left(\frac{\delta f_0}{\delta t}\right)_c .$$

(6.12b)

Equations (6.12a) can be further evaluated by considering the frequency of elastic colisions in $(\partial f_1/\delta t)_c$. This collision frequency $\nu_c(v)$ between an electron and the molecules depends strongly on the electron velocity. It can be expressed in terms of the momentum-transfer cross section for the scattering of an electron with velocity v, or

$$\nu_c(v) = NQ_m(v)v ,$$

(6.13)

where N is the neutral gas density, and $Q_m(v)$ the momentum-transfer cross section. The elastic collision frequency will be much higher than those for the non-elastic collisions. This prohibits the electrons to obtain a large velocity component along the E direction and thereby making approximation (6.8) acceptable. The elastic collisions will strongly reduce the asymmetry of the distribution function. It has been suggested [6.4] that this reduction is proportional to the collision frequency and may be written as

$$\left(\frac{\partial f_1}{\partial t}\right)_c = -\nu_c f_1 \ . \tag{6.14}$$

Substituting (6.13, 14) into (6.12) we obtain

$$-\frac{\partial f_1}{\partial t} + \frac{eE}{m}\frac{\partial f_0}{\partial v} = NvQ_m(v)f_1 \ , \tag{6.15a}$$

$$\frac{\partial f_0}{\partial t} - \frac{e}{3mv^2}\frac{\partial}{\partial v}(v^2 E \cdot f_1) = \left(\frac{\delta f_0}{\delta t}\right)_c \ , \tag{6.15b}$$

from which f_1 can be eliminated to give one equation for $f_0(v)$. Solving f_1 in (6.15a) we assume all quantities to be independent of t. We find

$$f_1 = \frac{eE(\partial f_0/\partial v)}{mNvQ_m} + \exp(-NvQ_m t) \ . \tag{6.16}$$

At atmospheric pressure $(N = 2.45 \times 10^{19}\,\mathrm{cm^{-3}})$ and an average electron energy of about $2\,\mathrm{eV}$ the cross section is of the order of $10^{-15}\,\mathrm{cm^2}$. The relaxation time $(NvQ_m)^{-1}$ is then about $10^{-12}\,\mathrm{s}$, so that for times larger than $10^{-12}\,\mathrm{s}$ we may omit the second term on the right-hand side in (6.16) to yield

$$f_1 = \frac{eE(\partial f_0/\partial v)}{mNvQ_m} \ . \tag{6.17}$$

6.3.2 Energy Transport by Elastic Collisions

For the binary elastic collisions we consider the relations for the conservation of energy and momentum:

$$\tfrac{1}{2}mv^2 + \tfrac{1}{2}MV^2 = \tfrac{1}{2}mv_1^2 + \tfrac{1}{2}MV_1^2 \ , \tag{6.18a}$$

$$mv + MV = mv_1 + MV_1 \ , \tag{6.18b}$$

where m and M are the electron and molecular mass, respectively, and v and V are their initial velocities. After the collision the velocities of electron and molecule are v_1 and V_1, respectively. It is convenient to consider the vector V in terms of a component along v and one perpendicular to it, i.e.,

$$V = av + w \; . \tag{6.19}$$

The energy ΔE lost by the electrons in an inelastic collision with a molecule is given by

$$\Delta E = \tfrac{1}{2}mv^2 - \tfrac{1}{2}mv_1^2 \; . \tag{6.20}$$

Eliminating V_1 in (6.18) we get

$$\Delta E = \frac{1}{2}\frac{m^2}{M}(v^2 + v_1^2 - 2v\cdot v_1) + ma(v^2 - v\cdot v_1) - mw\cdot v_1 \; . \tag{6.21}$$

Due to the large mass difference between the electron and the molecule the energy ΔE is much smaller than the initial electron energy $mv^2/2$, so that $v_1 \simeq v$. Equation (6.21) can now be written as

$$\Delta E = \left(\frac{m^2}{M}v^2 + mv\cdot V\right)(1 - \cos\theta) - mwv\,\cos\alpha \; , \tag{6.22}$$

where θ is the electron-scattering angle between v and v_1, and α the angle between w and v_1.

The power transfer W_e of the electrons by elastic collisions to the molecules can be calculated from ΔE multiplied by the differential cross section $I(\theta, |v - V|)$ for elastic collisions, the molecular density N, the electron number density n_0, the relative velocity between electron and molecule, the electron velocity distribution $f_0(v)$, and the molecular velocity distribution $f_M(V)$, all integrated over v and V, i.e.,

$$W_e = \iint \Delta E I N n_0 |v - V| f_0(v) f_M(V) dv\, dV \; . \tag{6.23}$$

Substituting (6.22) into (6.23) we obtain

$$W_e = \iint \left(\frac{m^2}{M}v^2 + mv\cdot V\right)(1 - \cos\theta) I N n_0 |v - V| f_0(v) f_M(V)$$
$$\times v^2 \sin\theta\, d\theta\, d\psi\, dv\, dV \; , \tag{6.24}$$

where the term $mwv\cos\alpha$ of (6.22) is zero in the above integral because for any value of $\cos\alpha$ there is also an opposite contribution by $\cos(\pi - \alpha)$ corresponding to the angle between $-w$ and v'.

Integrating (6.24) over θ and ψ yields the integral

$$\iint (1 - \cos\theta) I(\theta, |v - V|) \sin\theta\, d\theta\, d\psi \; ,$$

which is equal to the momentum transfer cross section $Q_m(|\boldsymbol{v} - \boldsymbol{V}|)$ in the direction of the incident colliding electron [6.5]. Equation (6.24) now becomes

$$W_e = \iint \left(\frac{m^2}{M}v^2 + m\boldsymbol{v}\cdot\boldsymbol{V} \right) Q_m(|\boldsymbol{v} - \boldsymbol{V}|) N n_0 |\boldsymbol{v} - \boldsymbol{V}| f_0(v) f_M(V) v^2 dv \, d\boldsymbol{V} \ .$$

(6.25)

Next we introduce the relative velocity $\boldsymbol{v}_r = \boldsymbol{v} - \boldsymbol{V}$. Keeping in mind that the electron velocity is much larger than the velocity of the molecule, i.e. $v \gg V$, we evaluate the following quantities:

$$v^2 \simeq v_r^2 + 2\boldsymbol{v}_r\cdot\boldsymbol{V} \ ,$$

(6.26)

$$v \simeq v_r + \frac{\boldsymbol{v}_r\cdot\boldsymbol{V}}{v_r} \ ,$$

(6.27)

$$v^2 + \frac{M}{m}\boldsymbol{v}\cdot\boldsymbol{V} \simeq v_r^2 + \frac{M}{m}\boldsymbol{v}_r\cdot\boldsymbol{V} \ ,$$

(6.28)

$$f_0(v) \simeq f_0(v_r) + \frac{\boldsymbol{v}_r\cdot\boldsymbol{V}}{v_r}\frac{\partial f_0(v_r)}{\partial v_r} \ .$$

(6.29)

Substituting (6.28, 29) into (6.25) we get upon dropping the subscript r

$$W_e = \iint \left(\frac{m^2}{M}v^2 + m\boldsymbol{v}\cdot\boldsymbol{V} \right)$$
$$\times Q_m(v) N n_0 f_M(V) \left[f_0(v) + \frac{\boldsymbol{v}\cdot\boldsymbol{V}}{v}\frac{\partial f_0}{\partial v} \right] v^3 dv \, d\boldsymbol{V} \ .$$

(6.30)

In the further evaluation of the integral the term constaining $\boldsymbol{v}\cdot\boldsymbol{V}$ becomes zero because the molecular distribution $f_M(V)$ will be considered Maxwellian so that for any value of $\boldsymbol{v}\cdot\boldsymbol{V}$ there is also a term containing $-\boldsymbol{v}\cdot\boldsymbol{V}$. We obtain

$$W_e = \iint \left[\frac{m^2}{M}v^5 f_0 + mv^2(\boldsymbol{v}\cdot\boldsymbol{V})^2 \frac{\partial f_0}{\partial v} \right] N n_0 Q_m f_M(V) dV dv \ .$$

(6.31)

The integration of $f_M(V)$ over the molecular velocity space gives unity. For the second term containing $(\boldsymbol{v}\cdot\boldsymbol{V})^2$ where the angle β between the two velocities is randomly distributed, we take the average value of $\cos^2 \beta$ equal to $1/3$ so that $(\boldsymbol{v}\cdot\boldsymbol{V})^2 = v^2 V^2/3$. The integration over $V^2 f_M(v)$ is related to the average kinetic energy of the molecules, viz.

$$\int \tfrac{1}{2}MV^2 f_M(V) dV = \tfrac{3}{2}kT \ .$$

(6.32)

We finally obtain for the power transfer by elastic collisions to the molecules per unit volume

$$W_e = \int \left(\frac{m^2}{M} v^5 f_0 + \frac{m}{M} kT v^4 \frac{\partial f_0}{\partial v} \right) N n_0 Q_m dv \ . \tag{6.33}$$

6.3.3 Energy Transfer by the Applied Field

Next we calculate the rate of change of electron energy by the field. The energy change per unit time for an electron with velocity v is given by

$$\frac{d\varepsilon}{dt} = m\dot{v}\cdot V = -eE\cdot v \ . \tag{6.34}$$

The total electron energy change per unit volume over all electron velocities due to the acceleration of the electrons by the applied field E is

$$W_t = - \int n_0 e E\cdot v \left[f_0(v) + \frac{v}{v}\cdot f_1(v) \right] dv \ , \tag{6.35}$$

where we have used (6.8) for the velocity distribution. Since $f_0(v)$ is an isotropic distribution function its integration with the term $E\cdot v$ becomes zero, we get

$$W_t = - \int n_0 e (E\cdot v) \left[\frac{v}{v}\cdot f_1(v) \right] dv \ . \tag{6.36}$$

Evaluating the integrand in (6.36) we are dealing with the terms v_x^2, v_y^2, $v_x v_y$, etc. The integration with the cross terms $v_x v_y$, $v_x v_z$, and $v_y v_z$ gives zero. The terms v_x^2, v_y^2, and v_z^2 will be averaged and set equal to $v^2/3$ because the direction of v with respect to f_1 and E is random. We now obtain

$$W_t = -\frac{1}{3} n_0 \int e E\cdot f_1 v^3 dv \ . \tag{6.37}$$

Since we consider the electron-molecule system to be in equilibrium the power transfer per unit volume to the electrons is equal to the power transfer from the electrons to the molecules by elastic and inelastic collisions. The power transfer by inelastic collisions is then equal to $W_t - W_e$. The power transfer by inelastic collisions of a single electron in the velocity space element $v^2 dv$ becomes

$$-\frac{1}{3} e v E\cdot f_1 - \frac{m^2}{M} f_0 N Q_m v^3 - \frac{m}{M} kT N Q_m \frac{\partial f_0}{\partial v} v^2 \ . \tag{6.38}$$

If we now replace the term $\frac{1}{3} e v E\cdot f_1$ in (6.15b) by the last expression then

147

the contribution to the right-hand side of (6.15b) due to elastic collisions will have been allowed for. Thus we obtain for (6.15b)

$$\frac{\partial f_0}{\partial t} - \frac{1}{mv^2}\frac{\partial}{\partial v}\left(\frac{ev^2}{3}\boldsymbol{E}\cdot\boldsymbol{f}_1 + \frac{m^2}{M}NQ_\mathrm{m}v^4 f_0 + \frac{mkT}{M}NQ_\mathrm{m}v^3\frac{\partial f_0}{\partial v}\right)$$
$$= \left(\frac{\delta f_0}{\delta t}\right)_{\mathrm{inel}}, \qquad (6.39)$$

where the right-hand side refers to inelastic collisions only. The quantity f_1 can be eliminated by means of (6.17) so that (6.39) becomes

$$\frac{\partial f_0}{\partial t} - \left(\frac{2e}{mu}\right)^{1/2}\frac{\partial}{\partial u}\left(\frac{E^2 u}{3NQ_\mathrm{m}}\frac{\partial f_0}{\partial u}\right.$$
$$\left. + \frac{2mNQ_\mathrm{m}u^2 f_0}{M} + \frac{2mkTNQ_\mathrm{m}u^2}{Me}\frac{\partial f_0}{\partial u}\right) = \left(\frac{\delta f_0}{\delta t}\right)_{\mathrm{inel}}, \qquad (6.40)$$

where we have expressed the kinetic energy of the electron in volts by the substitution

$$u = \frac{mv^2}{2e}, \quad \text{so that} \quad \frac{\partial}{\partial v} = \frac{mv}{e}\frac{\partial}{\partial u}.$$

6.3.4 Energy Transfer by Inelastic Collisions

When considering the inelastic collisions (electronic and molecular excitation, ionization) it is usual to consider the electron energy in units of electron volts. The kinetic energy in terms of u [V] is related by $eu = mv^2/2$. Similarly, the inelastic energy transfer with the molecule, E_j, is expressed in terms of u_j, with the relation $eu_j = E_j$. In the following we assume that the inelastic cross section for the excitation from the ground state to the jth internal state is independent of the molecular velocities, which are very small compared to the electron velocities. The cross section $Q_j(u^*)$ is then only dependent on the initial electron velocity v^* or energy u^*. After the collision the electron energy u is equal to $u^* - u_j$.

Due to the inelastic collisions there will be a production rate $(\delta f/\delta t)_{\mathrm{gain}}$ per unit volume and per unit velocity space at the energy u of an electron with the initial energy $u_j + u$. It is given by

$$\left(\frac{\delta f_0}{\delta t}\right)_{\mathrm{gain}} = \sum_j N_0 f_0(u + u_j)Q_j(u + u_j)v^*. \qquad (6.41)$$

The initial velocity v^* in (6.41) will be approximated in terms of u and u^* by

$$v^* = \left(\frac{2eu^*}{m}\right)^{1/2} \simeq u^* \left(\frac{2e}{mu}\right)^{1/2} , \qquad (6.42)$$

because substantial values of the cross section of an excitation process are found for $u \gg u_j$ in (6.41) so that $\sqrt{u^*} \simeq \sqrt{u}$. We now write (6.41) as

$$\left(\frac{\delta f_0}{\delta t}\right)_{\text{gain}} = \sum_j N f_0(u + u_j) Q_j(u + u_j)(u + u_j) \left(\frac{2e}{mu}\right)^{1/2} . \qquad (6.43a)$$

Similarly to (6.43a) electrons with energy u are lost due to inelastic processes. The loss term is then given by

$$\left(\frac{\partial f_0}{\partial t}\right)_{\text{loss}} = \sum_j N f_0(u) Q_j(u) u \left(\frac{2e}{mu}\right)^{1/2} . \qquad (6.43b)$$

Substracting (6.43b) from (6.43a) we get the rate of change of $f_0(u)$ due to inelastic processes, i.e.,

$$\left(\frac{\delta f_0}{\delta t}\right)_{\text{inel}} = \sum_j N [f_0(u + u_j) Q_j(u + u_j)(u + u_j)$$
$$- f_0(u) Q_j(u) u] \left(\frac{2e}{mu}\right)^{1/2} . \qquad (6.44)$$

So far we have considered the energy transfer from the electrons to the molecules. However, there are also collisions of the second kind in which the excitation energy goes back into the electron energy. The effect of these superelastic collisions to the distributon function is similar to (6.44) given by

$$\left(\frac{\delta f_0}{\delta t}\right)_{\text{super}} = \sum_j N_j [f_0(u - u_j) Q_{-j}(u - u_j)(u - u_j)$$
$$- f_0(u) Q_{-j}(u) u] \left(\frac{2e}{mu}\right)^{1/2} , \qquad (6.45)$$

where $Q_{-j}(u)$ is the cross section for a superelastic collision of an electron with energy u.

6.3.5 Stationary Boltzmann's Equation for the Electron Energy Distribution

For the further evaluation of the equation for the distribution function f_0 we substitute into (6.40) for the term $(\delta f_0/\delta t)$ the contributions from both the excitation processes and the collisions of the second kind as given by (6.44) and (6.45), respectively. In the stationary state, $\delta f_0 / \delta t = 0$, we then find

$$
\frac{1}{3}\left(\frac{E}{N}\right)^2 \frac{d}{du}\left(\frac{u}{Q_m}\frac{df}{du}\right) + \frac{2m}{M}\frac{d}{du}(u^2 Q_m f) + \frac{2mkT}{Me}\frac{d}{du}\left(Q_m u^2 \frac{df}{du}\right)
$$

$$
+ \sum_j [(u+u_j)f(u+u_j)Q_j(u+u_j) - uf(u)Q_j(u)] \tag{6.46}
$$

$$
+ \frac{1}{N}\sum_j N_j[(u-u_j)f(u-u_j)Q_{-j}(u-u_j) - uf(u)Q_{-j}(u)] = 0 ,
$$

where we have dropped the subscript of f. Equation (6.46) is the Boltzmann equation for the isotropic part of the nearly isotropic energy distribution function. The equation has been given in this form by *Frost* and *Phelps* [6.6]. It is seen that the distribution function f depends on the parameter E/N.

6.4 Solving Bolzmann's Equation for the CO_2-Laser Discharge

The parameter E/N plays an important role in the discharge behavior. It indicates the mean free path of an electron between two collisions. The larger the mean free path, the more energy the electron extracts from the field and consequently the larger the average electron energy. For values $E/N > 10^{-17}$ V cm^2 collisions of the second kind can be neglected in the Boltzmann equation as these terms are insignificant compared to the other terms in (6.46). In the further calculations we assume throughout that the population densities of the various inelastic levels of the gas are appropriate at room temperature.

So far, (6.46) has been derived for electrons in a gas of one species of molecules. For a gas mixture of different species the derivation of the Boltzmann equation is similar except that we have to sum over the contributions of the various species. The Boltzmann equation becomes

$$
\frac{1}{3}\left(\frac{E}{N}\right)^2 \frac{d}{du}\left(\frac{u}{\sum_k \delta_k Q_m^k}\frac{df}{du}\right) + 2m\frac{d}{du}\left(\sum_k \delta_k \frac{Q_m^k}{M_k}u^2 f\right)
$$

$$
+ \frac{2mkT}{e}\frac{d}{du}\left(\sum_k \delta_k \frac{Q_m^k}{M_k}u^2 \frac{df}{du}\right)
$$

$$
+ \sum_k \sum_j \delta_k [(u+u_{jk})f(u+u_{jk})Q_j^k(u+u_{jk}) - uf(u)Q_j^k(u)] \tag{6.47}
$$

$$
+ \sum_k \sum_j \delta_{jk}[(u-u_{jk})f(u-u_{jk})Q_{-j}^k(u-u_{jk}) - uf(u)Q_{-j}^k(u)] = 0 ,
$$

where $\delta_k = N_k/N$ is the fractional concentration of molecules of species k and Q_m^k the corresponding momentum transfer cross section, $\delta_{jk} = N_{jk}/N$ is the fractional concentration of excited molecules to the state j of the species k, and u_{jk} the corresponding energy.

Calculating the distribution function by means of eq. (6.47) the rotational excitation of nitrogen can be included by using the so-called continuous approximation [6.6]. The term $\sum_k \delta_k Q_m^k/M_k$ in eq.(6.47) is replaced by $\sum_k \delta_k Q_m^k/M_k + (C/mu)\delta_{N_2}$; the constant $C = (16/15)\pi B_0 q^2 a_0^2$, where B_0 is the rotational constant of N_2, $q = 1.04\,ea_0^2$ is the quadrupole moment of N_2, a_0 the Bohr radius, and δ_{N_2} is the fraction of N_2 in the mixture. The conditions for the validity of the continuous approximation are satisfied in the calculations since the mean electron energies are well above the mean energy of the gas. Rotational excitation of CO_2 is neglected because its influence is small in comparison with vibrational and electronic excitation of CO_2.

The input cross sections for elastic and inelastic collisions of CO_2, shown in Fig. 6.1 and listed in Table 6.1, are modifications of those given by *Hake* and *Phelps* [6.7]; the revisions being largely based on the work of *Andrick* et al. [6.8]. The cross sections of Fig. 6.1 were obtained to give consistency between predicted and measured transport coefficients. The detailed shape of the (010) cross section has been obtained from the Born

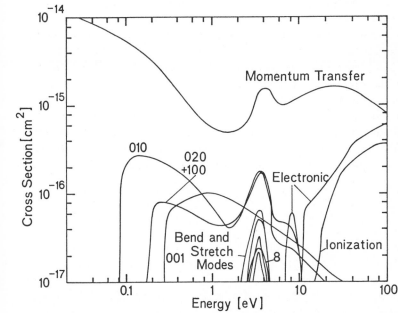

Fig. 6.1. Derived cross sections for momentum transfer and inelastic scattering for electrons in CO_2 [6.3]. The cross sections are adjusted to obtain D_e/μ and v_d consistent with the experiment, as in [6.7], but use the more recent cross-section data of [6.6–8]

Table 6.1. Tabulation of CO_2 inelastic cross section [6.3]

	Energy loss [eV]	Threshold [eV]	Process
1	0.083	0.083	$000 \rightarrow 010$
2	0.167	0.167	$000 \rightarrow 020 + 100$
3	0.291	0.291	$000 \rightarrow 001$
4	0.252	2.5	$000 \rightarrow 0n0 + n00$
5	0.339	1.5	$000 \rightarrow 0n0 + n00$
6	0.422	2.5	$000 \rightarrow 0n0 + n00$
7	0.505	2.5	$000 \rightarrow 0n0 + n00$
8	2.5	2.5	$000 \rightarrow 0n0 + n00$
9	3.85	3.85	$e + CO_2 \rightarrow CO + O^-$
10	7.0	7.0	electronic
11	10.5	10.5	electronic
12	13.3	13.3	$e + CO_2 \rightarrow CO_2^+ + 2e$

approximation. The steep rise and magnitude near threshold is necessary to obtain transport coefficients consistent with experiments at characteristic energies less than 0.02 eV. The near-threshold region for the (100) cross section is very arbitrary, except that the initial slope is chosen to be consistent with the relative threshold data of *Stamatovic* and *Schulz* [6.9]. The higher-energy portion is taken from *Andrick* et al. [6.8] and may include energy losses to the (020) level.

The cross section labeled 8 replaces the 3.1-eV energy loss process needed by *Hake* and *Phelps* [6.7] to fit transport coefficients. The magnitude and threshold of this 2.5-eV energy loss process is an attempt to account for the large number of energy-loss processes [6.10] which occur in connection with the 3.8-eV resonance. It is assumed that, as in the case for the vibrational excitation process observed by *Andrick* et al. [6.8], the resonance contributes only to the excitation of the bending and asymmetric stretch modes.

The inelastic cross section for nitrogen are those of *Engelhardt* et al. [6.11]. For helium the momentum-transfer cross section of *Frost* and *Phelps* [6.6] is used. The inelastic cross sections for ionization and for the electronic excitation of the 2^3S, 2^1S, and 2^1P levels are obtained from [6.12–15]. These approximate inelastic cross sections for helium are shown in Fig. 6.2. The dissociative attachment cross section for electrons in carbon dioxide was taken from *Schulz* [6.16].

Altogether 31 different cross sections have been used to represent the inelastic losses for vibrational and electronic excitation for the gas mixture.

Using all these cross-section data (6.47) was solved numerically for various gas mixtures and a wide range of E/N values. In Figs. 6.3–5 distribution functions are shown [6.3] for mixture ratios $CO_2 : N_2 : He$ of $1 : 1 : 8$, $1 : 2 : 3$, and $1 : 7 : 30$. The distribution function for these figures is given by $u^{1/2}f$, where f in this case is normalized by

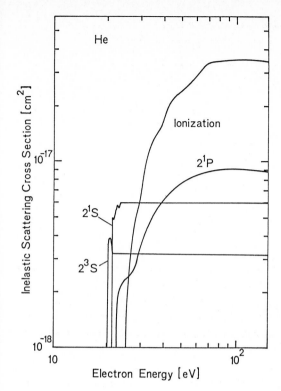

Fig. 6.2. Cross sections used for momentum transfer and inelastic scattering of electrons in He at high energy [6.3, 12–15]

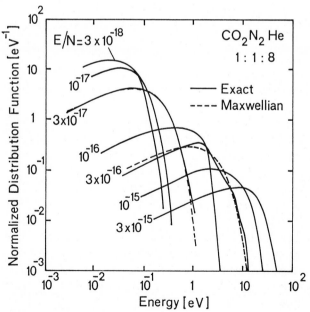

Fig. 6.3. Derived distribution functions electrons for $CO_2 : N_2 : He = 1 : 1 : 8$ for various values of E/N [V cm^2]. Broken curves indicate a Maxwellian distribution [6.3]

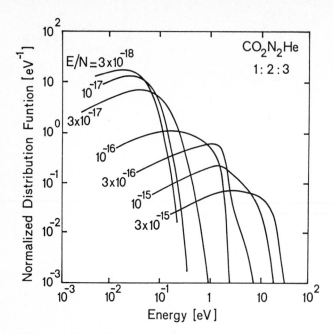

Fig. 6.4. Derived distribution functions of electrons for $CO_2 : N_2 : He = 1 : 2 : 3$ for various values of E/N [V cm^2] [6.3]

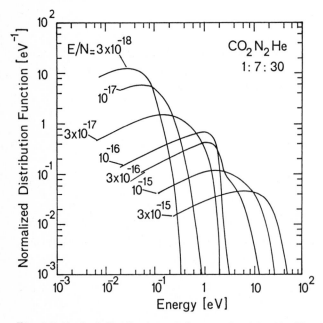

Fig. 6.5. Derived distributions of electrons for $CO_2 : N_2 : He = 1 : 7 : 30$ for various values of E/N [V cm^2] [6.3]

$$\int_0^\infty u^{1/2} f \, du = 1 \; . \tag{6.48}$$

In Fig. 6.3 a comparison is made with a Maxwellian distribution, which is indicated by the broken curve. The Maxwellian distribution is chosen to have the same average energy as the exact distribution functions, indicated by the solid curves.

6.4.1 Transport Coefficients

Once the distribution function f for a given E/N is obtained, the drift velocity v_d and characteristic energy D_e/μ_e are obtained, using the formula [6.17]

$$v_d = -\frac{E}{3N} \sqrt{\frac{2e}{m}} \int_0^\infty \frac{u}{\sum_k \delta_k Q_m^k(u)} \frac{df}{du} du \; , \tag{6.49}$$

where $\delta_k = N_k/N$ is the fractionial concentration of molecules of species k and Q_m^k is the corresponding momentum transfer cross section.

The diffusion coefficient D_e of the electrons is given by

$$D_e = \frac{1}{3N} \sqrt{\frac{2e}{m}} \int_0^\infty \frac{u f \, du}{\sum_k \delta_k Q_m^k(u)} \; . \tag{6.50}$$

The characteristic energy D_e/μ_e, where $\mu_e = v_d/E$ is the electron mobility, gives a measure of the average electron energy. In the case of a Maxwellian distribution of the electrons with temperature T_e one finds

$$\frac{D_e}{\mu_e} = \frac{kT_e}{e} \; .$$

The ionization coefficient α is given by

$$\frac{\alpha}{N} = \frac{1}{v_d} \sqrt{\frac{2e}{m}} \sum_k \int_0^\infty \delta_k Q_i^k(u) f u \, du \; , \tag{6.51}$$

where Q_i^k is the ionization cross section of species k. A similar relationship defines electron attachment, given by a/N. The power P_j^k delivered by each electron to the inelastic level j of species k is

$$P_j^k = u_j \delta_k N \sqrt{\frac{2e}{m}} \int_0^\infty u Q_j^k(u) f \, du \; , \tag{6.52}$$

where u_j is the energy loss to the jth level in eV. It should be noted that in the above-quoted expressions for v_d, D_e, α/N, and P_j^k the distribution function used is normalized according to (6.48).

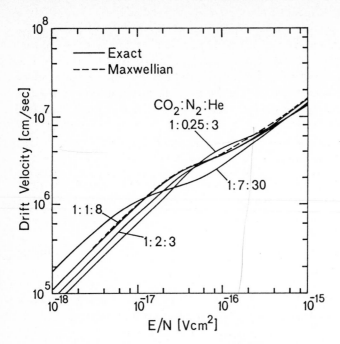

Fig. 6.6. Calculated drift velocities of electrons for various gas mixtures of CO_2, N_2 and He [6.3]

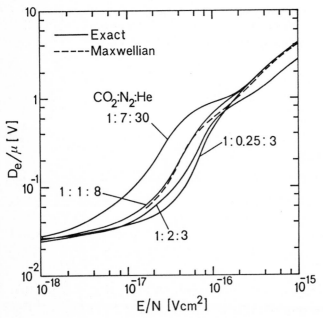

Fig. 6.7. Calculated values of D_e/μ_e for various gas mixtures of CO_2, N_2 and He [6.3]

The transport coefficients v_d and D_e/μ_e are shown in Figs. 6.6 and 7. The dashed curves indicate results for a Maxwellian distribution having the same average energy as derived from the exact calculations. It is seen that for transport coefficients such as v_d and D_e/μ_e, which depend on integrals over the entire distribution function, the assumption of a Maxwellian distribution does not introduce errors of more than 20 %.

Figure 6.8 indicates values of α/N as a function of E/N, where Penning ionization by metastable helium levels is included in α/N. Results obtained assuming a Maxwellian velocity distribution are again shown by broken curves. In Fig. 6.9 values of a/N are given for the dissociative attachment of electrons to CO_2, i.e. $CO_2 + e^- \rightarrow CO + O^-$.

6.4.2 Predictions of Optimum Laser Efficiency

From the calculated electron distribution functions the input power per electron delivered into each of the various inelastic processes can be calcu-

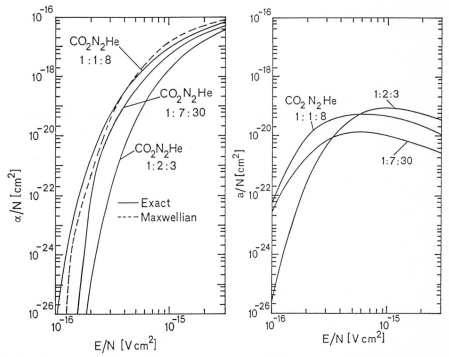

Fig. 6.8. Calculated values of α/N for various gas mixtures of CO_2, N_2, and He; α is the ionization coefficient [cm^{-1}]. The broken curve indicates values obtained assuming a Maxwellian distribution of the $1:1:8$ mixture [6.3]

Fig. 6.9. Calculated values of a/N for various gas mixtures of CO_2, N_2 and He, a is the attachment coefficient [cm^{-1}] [6.3]

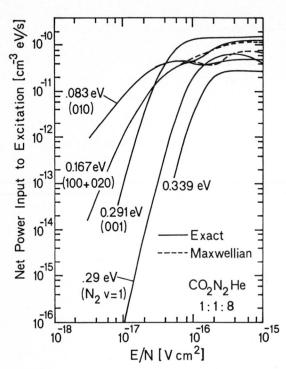

Fig. 6.10. Calculated net power input to various vibrational levels of CO_2 and N_2 for a mixture of $CO_2 : N_2 : He = 1 : 1 : 8$. The broken curve indicates values assuming a Maxwellian distribution [6.3]

lated. For example, in Fig. 6.10 the power to the first five vibrational levels is plotted for a laser mixture of $CO_2 : N_2 : He = 1 : 1 : 8$. The 0.083-eV process represents losses to the (010) CO_2 vibrational level, the 0.167-eV process gives the combined losses to (020) and (100) levels of CO_2, the 0.291-eV process corresponds to the (001) level of CO_2, which is the upper laser level, and the 0.29-eV process corresponds to the $v = 1$ nitrogen vibrational level. Calculations which assume a Maxwellian velocity distribution are also shown in Fig. 6.10 for $E/N > 10^{-17}\,\mathrm{V\,cm^2}$. Although differences with exact calculations are less than for values of α/N displayed in Fig. 6.8, differences of 30 % still occur.

The individual cross sections used for the results of Fig. 6.10 have, to a large extent, been obtained by fitting to get agreement between theory and experiment of the transport coefficients v_d and D_e/μ_e for the pure gases. As v_d and D_e/μ_e depend on averages over the entire distribution function, the question arises how accurate the detailed predictions of the power delivered to individual processes are. It is possible that experimental measurements of laser output may lead to improved individual cross sections, although the situation is complicated by the many kinetic processes which lead to the interchange of energy between the various levels.

It is seen from Fig. 6.10 that although 90 % of the input power is delivered to the lower bending mode level (010) at $E/N = 10^{-17}\,\mathrm{V\,cm^2}$, at

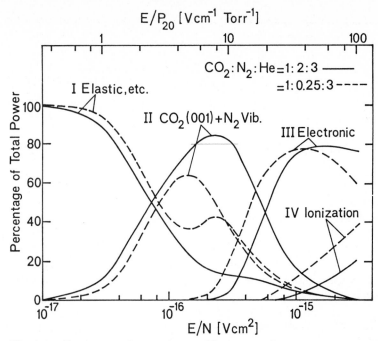

Fig. 6.11. Percentage of power lost to (I) elastic collisions, rotational excitation of N_2 and excitation of bend and stretch modes of CO_2, (II) CO_2 001 level and the first eight vibrational levels of N_2, (III) electronic excitation, and (IV) ionization. Increasing the ratio of N_2 to CO_2 increases the predicted efficiency given by (II) [6.3]

higher values of E/N direct excitation of the upper laser level (001) increases to over 40 %. Power fed to the various vibrational levels of nitrogen is closely coupled to the (001) level, and the predicted power available to the upper laser level is better shown by combining the various loss processes into four groups.

In Fig. 6.11 such a summation for different mixtures is shown. Curves marked by I indicate the sum of losses by elastic collisions with CO_2, N_2, and He, and also losses through the rotational excitation of nitrogen and all bending and stretch modes of CO_2 (Table 6.1). Curves marked II represent the sum of losses to the (001) level of CO_2, which is the upper laser level, and the sum of the losses to vibrational excitation of nitrogen. Because these vibrational levels are closely coupled to the (001) level of CO_2, curves denoted by II in Fig. 6.11 should indicate the excitation efficiency of laser performance. Here it is assumed that the energy extraction rate is low enough to allow energy transfer from N_2 molecules and fast enough so that collision relaxation of the 001 level of CO_2 is small.

Curves marked III represent losses due to the sum of eight processes of electronic excitation of CO_2 and N_2. Curves designated by IV represent losses due to the ionization of CO_2, N_2 and He, together with three processes for the electric excitation of He which can result in ionization due to Penning ionization.

The influence of the ratio of CO_2 to N_2 in a mixture is shown in Fig. 6.11 where comparison is made between results for a mixture $CO_2 : N_2 : He$ of $1 : 2 : 3$, shown by the full curves, and a mixture $CO_2 : N_2 : He$ of $1 : 0.25 : 3$, shown by the broken curves. The calculations indicate a higher maximum of curve II for the $1 : 2 : 3$ mixture, compared with the $1 : 0.25 : 3$ mixture, because less power is lost in the bending and stretch modes of CO_2 when the ratio of N_2 to CO_2 is increased. The slight maximum of curve I for the broken curve results from the excitation of bending and stretch modes of CO_2 represented by the process labeled 8 of Fig. 6.1 and Table 6.1.

Values of E/N for which curve II of Fig. 6.11 has a maximum are, of course, significant in the optimization of laser performance. Similar type-II curves for mixtures of $CO_2 : N_2 : He$ of $1 : 7 : 30$ and $1 : 7 : 0$ are given in Fig. 6.12, to be discussed later. For a mixture of $CO_2 : N_2 : He = 1 : 1 : 8$ curve II has a maximum of 74 % of the total power at $E/N = 1.2 \times 10^{-16}$ V cm^2. Curves of the total power delivered to the upper laser level, i.e., a value proportional to the percentage indicated by curve II, multiplied by $v_d E/N$, has a maximum at a higher value of E/N than for the maximum of curve II. For mixtures of $CO_2 : N_2 : He$ of $1 : 2 : 3$, $1 : 0.25 : 3$, $1 : 7 : 30$, and $1 : 1 : 8$ these maxima of total laser power occur for E/N values of 10^{-15}, 4×10^{-16}, 4×10^{-16}, and 4×10^{-16} V cm^2, respectively.

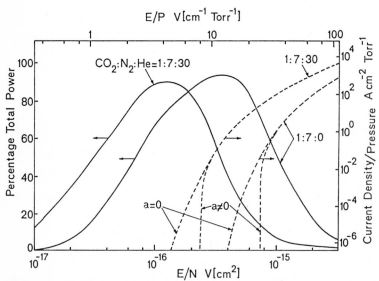

Fig. 6.12. Effect of increasing the proportion of helium, This decreases both the E/N for maximum efficiency and also the value of E/N at which the discharge operates for a given current density [6.3]

6.4.3 Operating E/N Values for Self-Sustained Glow Discharges

Calculating the current-voltage characteristics according to (6.3) we use the above results for v_d, α/N and a/N. The recombination rate γ depends on the gas composition and pressure. It depends only slightly on E/N [6.18]. Also an experimental analysis of the recombination rate by means of current voltage characteristics shows that $\gamma \sim 10^{-7}\,cm^3\,s^{-1}$ at one atmosphere in the range $2-4.5 \times 10^{-16}\,V\,cm^2$ [6.19]. The curves of Fig. 6.13 exhibit E/N versus j/N for various gas compositions. Also, the broken curves of Fig. 6.12 show the relation between current density/pressure and E/N. The second branch of these curves is drawn at low values of j/p appropriate to $a/N = 0$ as it is possible that O^- ions are destroyed when they collide with molecules of either CO, which may accumulate appreciably in steady-state lasers, or with hydrogen, which is sometimes added to laser discharges. In these cases electrons would be released back to the discharge, making a/N effectively zero. A lower operating value of E/N with a higher laser-excitation efficiency would be expected when the effect of attachment is suppressed.

It is interesting to note that at high current densities the discharge is always recombination controlled. At low current densities there is a large difference between attachment-controlled (broken curves of Fig. 6.13) and recombination-controlled. Although γ will be to some extent a function of gas mixture and pressure the $E - I$ characteristics of Fig. 6.13 predict, at least qualitatively, the operating E/N values for pulsed CO_2 systems.

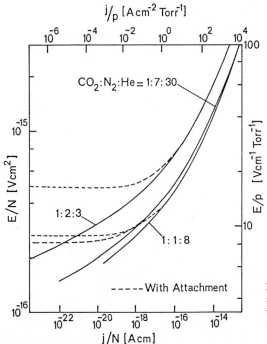

Fig. 6.13. Predicted E-I characteristics assuming there is a volume-controlled glow with $\gamma = 10^{-7}\,cm^3\,s^{-1}$ [6.3]

The effect of varying the proportion of helium in the laser mixture is also indicated in Fig. 6.12. For a given E/N, if the proportion of helium is increased, the average electron energy is also increased because helium has none of the inelastic vibrational losses that are present in N_2 and CO_2. Thus, electron energies sufficient to excite the upper laser levels occur at a lower E/N for the $1:7:30$ mixture than for the $1:7:0$ mixture where there is less helium. Thus, in Fig. 6.12 the maxima of the solid curves, which give the power delivered to the upper laser levels, occurs at $E/N = 1.2 \times 10^{-16}$ for the $1:7:30$ mixture and $4 \times 10^{-16}\,\mathrm{V\,cm^2}$ for the $1:7:0$ mixture.

6.5 Analysis of the Pulse-Forming Process

Let us for the moment start from a uniform gain medium produced by an electric discharge pulse. The shape of the optical pulse and its delay with respect to the electric pulse depend on the laser parameters such as the shape and energy content of the electric pulse, and the gas pressure and composition. The relations between the various physical processes are described by the energy-transfer processes. The relaxation processes can be fully described with the energy relaxation equations [6.20]. The densities of the various energy levels can then be expressed in terms of the vibrational and translational temperatures of the system, as has been done in Chap. 3.

Let us consider laser action on only the $P(20)$ line of the $(00^0 1)$-(I) vibrational band. The energies of the vibrational modes can be expressed as [6.21]

$$\frac{dE_1}{dt} = N_e(t)N_{CO_2}h\nu_1\chi_1 - \frac{E_1 - E_1(T_2)}{\tau_{12}(T_2)}$$
$$+ \left(\frac{\nu_1}{\nu_3}\right)\frac{E_3 - E_3(T, T_1, T_2)}{\tau_3(T, T_1, T_2)} + \left(\frac{\nu_1}{\nu}\right)\alpha I \ . \tag{6.53}$$

The first term on the right-hand side describes the excitation rate of the ν_1 vibration by the electron density in the discharge. The parameter χ_1 is the effective electron vibrational excitation rate. The second term is the relaxation rate of E_1 to the ν_2 vibration; $E_1(T_2)$ means the vibrational energy of the symmetric valence vibration at the temperature T_2. The third term is the part of the energy relaxation rate of the ν_3 vibration that goes to the ν_1 vibration. Each quantum $h\nu_3$ of E_3 is split up into one quantum $h\nu_1$ added to E_1 and one quantum $h\nu_2$ added to E_2, and the excess energy $h(\nu_3 - \nu_1 - \nu_2)$ goes into translation. The last term desribes the production of the lower level energy by the laser process; i.e. for each photon produced by stimulated emission a quantum $h\nu_1$ is added to E_1. The relaxation times τ_{12} and τ_3 are, respectively, given by [6.20]

$$\tau_{12} = \frac{1}{A_{12}(T)} \frac{\exp(h\nu_2/kT_2) - 1}{\exp(h\nu_2/kT_2) + 1} , \tag{6.54}$$

$$\tau_3 = \frac{1}{A_3(T)} \frac{[\exp(h\nu_1/kT_1) - 1][\exp(h\nu_2/kT_2) - 1]}{\{\exp[(h\nu_1/kT_1) + (h\nu_2/kT_2) + (h\nu_3 - h\nu_1 - h\nu_2)/kT] - 1\}} , \tag{6.55}$$

where $A_{12}(T)$ and $A_3(T)$ depend on the gas composition and the temperature T. The energy $E_3(T, T_1, T_2)$ is expressed by

$$E_3(T, T_1, T_2) = N_{CO_2} h\nu_3 \left[\exp\left(\frac{h\nu_1}{kT_1} + \frac{h\nu_2}{kT_2} + \frac{\Delta E_3}{kT} \right) - 1 \right]^{-1} . \tag{6.56}$$

The gain α is for a P transition according to (3.31) given by

$$\alpha = \left(n_{v'j'} - \frac{2j - 1}{2j + 1} n_{vj} \right) \sigma_{v'j' \to vj} , \tag{6.57}$$

where $\sigma_{v'j' \to vj}$ is given by (3.15).

It is shown in (6.53) that the ν_1 vibration relaxes to the ν_2 vibration. This process is, as we mentioned, very fast as compared with the other vibrational relaxation processes because of the near resonance between the (I) and (II) level. (There is practically no energy exchange with translation.) This relaxation process is the dominant path for draining of the lower laser level. The direct relaxation rate of this level to the translation can be neglected.

The relaxation of the ν_2 vibration is then

$$\frac{dE_2}{dt} = 2N_e(t)N_{CO_2}h\nu_2\chi_2 + \frac{E_1 - E_1(T_2)}{\tau_{12}(T_2)} - \frac{E_2 - E_2(T)}{\tau_2}$$
$$+ \frac{\nu_2}{\nu_3} \frac{E_3 - E_3(T, T_1, T_2)}{\tau_3(T, T_1, T_2)} . \tag{6.58}$$

The first term on the right-hand side is the excitation of the two degenerate bending modes by the electron density with the excitation rate χ_2. The third term is the relaxation of the ν_2 vibration into translational energy. The last one is the part of the E_3 energy that drains into the ν_2 vibration.

The energy density in the ν_3 vibration which contains the upper laser level is described by

$$\frac{dE_3}{dt} = N_e(t)N_{CO_2}h\nu_3\chi_3 - \frac{E_3 - E_3(T, T_1, T_2)}{\tau_3(T, T_1, T_2)}$$
$$+ \frac{E_4 - E_4(T_3)}{\tau_{43}(T)} - \left(\frac{\nu_3}{\nu} \right) \alpha I . \tag{6.59}$$

The first term on the right-hand side is again the excitation by the electrons. The third one represents the vibrational energy transfer from N_2 (Sect. 3.7). The last one gives the relation with the stimulated emission.

The energy stored in the N_2 vibration, E_4, decays dominantly by pumping the ν_3 vibration:

$$\frac{dE_4}{dt} = N_e(t)N_{N_2}h\nu_4\chi_4 - \frac{E_4 - E_4(T_3)}{\tau_{43}} \quad , \tag{6.60}$$

where the vibrational excitation rate for N_2 is given by χ_4. The rate equation for the radiation intensity I is given by the time evolution of the cavity-field intensity.

$$\frac{dI}{dt} = -\frac{I}{\tau_c} + \alpha c I + \left(\frac{dI}{dt}\right)_{\text{spont}} \quad , \tag{6.61}$$

where $\tau_c = -2L/c\ln R$ is the laser cavity lifetime.

The contribution to the spontaneous emission, $(dI/dt)_{\text{spont}}$, can be found as follows. The total number of modes in a volume V and the frequency interval between ν and $(\nu + d\nu)$ is

$$p(\nu)d\nu = \frac{8\pi\nu^2}{c^3}V\,d\nu \quad . \tag{6.62}$$

The radiation production W in a volume V originating from spontaneous emission of a P transition in the frequency interval between ν and $(\nu + d\nu)$ is given by

$$W = h\nu n_{v'j'}A_{v'j-1\rightarrow vj}V S\,d\nu \quad , \tag{6.63}$$

where $n_{v'j'}$ is the density of the upper laser level, $A_{v'j-1\rightarrow vj}$ is the spontaneous decay, and S is the line shape function. The spontaneous flux per mode, W_{mode}, is now obtained by dividing (6.63) by (6.62)

$$W_{\text{mode}} = n_{v'j'}ch\nu\sigma_{v'j-1\rightarrow vj} \quad , \tag{6.64}$$

where we have used (3.15) for the stimulated emission cross section. The radiation density production $d\varrho/dt$ in a mode with volume V originating from spontaneous emission becomes

$$\frac{d\varrho}{dt} = \frac{n_{v'j'}ch\nu\sigma_{v'j-1\rightarrow vj}}{V} \quad . \tag{6.65}$$

From this we find that the increase of radiation intensity by spontaneous emission is

$$\left(\frac{dI}{dt}\right)_{\text{spont}} = \frac{c^2 h\nu}{V} n_{v'j'} \sigma_{v'j-1 \to vj} \ . \tag{6.66}$$

Finally a complete set of rate equations is given by the addition of the time evolution of the translational and rotational temperatures. The coupling between rotation and translation is extremely fast so that on the time scale of the present time evolution these temperatures are equal. The vibrational energies as far as they are not converted into stimulated emission will be transferred to translation. During this process the translational temperature will increase toward an equilibrium which is equal to the equilibrium value of the vibrations. Thus for the translation together with the rotation we have

$$\frac{dE}{dt} = \frac{E_2 - E_2(T)}{\tau_2} + \frac{\nu_3 - \nu_1 - \nu_2}{\nu_3} \frac{E_3 - E_3(T, T_1, T_2)}{\tau_3(T, T_1, T_2)} \ , \tag{6.67}$$

where

$$E = \tfrac{3}{2} kT(N_{CO_2} + N_{N_2} + N_{He}) + kT(N_{CO_2} + N_{N_2}) \ .$$

Each of the five energy densities E, E_1, E_2, E_3 and E_4 are associated with a temperature T, T_1, T_2, T_3 and T_4, respectively, which represents the instantaneous temperature of the degree of freedom. Given enough time after the discharge pulse all temperatures will finally relax to the translational temperature T.

In order to calculate the optical pulse from several numerical constants must be specified. The time constants τ_2 and τ_3 can be obtained as a function of pressure and temperature from the relation

$$\tau_i = [p(\psi_{CO_2} k_{iCO_2} + \psi_{N_2} k_{iN_2} + \psi_{He} k_{iHe})]^{-1} \ . \tag{6.68}$$

The rate constants k_i for τ_2 and τ_3 at $T_0 = 300\,\text{K}$ are given in Tables 3.3 and 3.4, respectively. Here ψ_x is the fraction of the gas component x, and p is the total gas pressure in torr.

Using [6.20] the temperature dependence of the relaxation time can be obtained. The temperature dependence of τ_2 is, to a good approximation, given by

$$\tau_2(T) = \tau_2^0 \left(\frac{T_0}{T}\right)^{1/2} \exp\left[h\nu_2\left(\frac{1}{kT} - \frac{1}{kT_0}\right)\right] \ , \tag{6.69}$$

where τ_2^0 is the relaxation time at room temperature T_0. Similarly, for τ_3 it can be shown that

$$\tau_3(T) = \tau_3^0 \left(\frac{T_0}{T}\right)^{1/2} \exp\left[h(\nu_3 - \nu_1 - \nu_2)\left(\frac{1}{kT} - \frac{1}{kT_0}\right)\right] \ , \tag{6.70}$$

where τ_3^0 is the relaxation time at room temperature. The relaxation time τ_{12} depends mostly on the collisions with CO_2 molecules and therefore we neglect its dependence on N_2 and He. At $T = T_0$ one finds [6.23] $\tau_{12} = 6.5 \times 10^{-7}$ s at 1 torr. Further, it can be shown [6.20] that the temperature dependence, to a good approximation, is given by

$$\tau_{12} = 6.5 \times 10^{-7} \left(\frac{T_0}{T}\right)^{3/2} (p\psi_{CO_2})^{-1} . \tag{6.71}$$

The relaxation time τ_{43} depends only on N_2. At $T = T_0$ one finds [6.23] $\tau_{43} = 4.9 \times 10^{-5}$ s at 1 torr.

Since the vibrational energy transfer from N_2 to CO_2 is practically at resonance it can be shown that the dependence of τ_{43} on T is

$$\tau_{43} = 4.9 \times 10^{-5} \left(\frac{T_0}{T}\right)^{3/2} (p\psi_{N_2})^{-1} . \tag{6.72}$$

The constants τ_2, τ_{43}, A_3 and A_{12} can now be calculated from the available data. Calculating the cross section for stimulated emission we consider the laser line P(20) at line center so that $S = 2/\pi\Delta\nu_P$ where $\Delta\nu_P$ is given by (3.30).

Example

In the following we consider a small TEA laser with a mixture $CO_2 : N_2 : He = 1 : 1 : 3$. The discharge length is 13 cm, the cavity 24 cm and the electrode gap 0.56 cm. The current pulse and optical output pulse are shown in Figs. 6.14 and 15, respectively. The outcoupling mirror reflectivity is 78 %.

Using the above analysis we now calculate for this experiment the output pulse and all temperatures. To determine the vibrational excitation rates by the electrons we proceed as follows. As we have seen in Sect. 6.4, the excitation cross sections depend on the electron energy. The electron energy distribution, in turn, is a function of the peak value of the electric field divided by the total gas molecule number density. In the case of a self-sustained discharge at atmospheric pressure, a so-called TEA (transversely excited atmospheric) CO_2 laser, the E/N value becomes

$$E/N = (13 \, \text{kV/cm})/(2.5 \times 10^{19} \, \text{molecules/cm}^3) = 5.2 \times 10^{-16} \, \text{V cm}^2 .$$

From this quantity the average electron energy and vibrational excitation can be estimated according to Fig. 6.3. In a $1 : 1 : 8$ mixture of $CO_2 : N_2 : He$ the peak average electron energy is about 2 eV. This is about the same as for a $1 : 1 : 3$ mixture. The effective electron excitation rate is equal to an effective cross section times electron velocity $\langle \sigma v \rangle$. From Fig. 6.1 we estimate the cross sections for the excitation of the three CO_2 vibrations to be equal. The vibrational excitation cross section for N_2 is about 5 times that of the ν_3 vibration (Figs. 3.3 and 4). We find for the χ parameters approximately

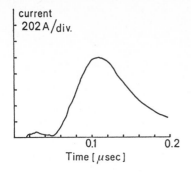

current
202 A/div.

0.1   0.2
Time [μsec]

Fig. 6.14. Experimentally observed discharge current of a small TEA system. The discharge length is 13 cm and the electrode gap 0.56 cm

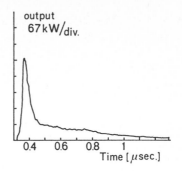

output
67 kW/div.

0.4 0.6 0.8 1
Time [μsec.]

Fig. 6.15. Optical output pulse of a small TEA system with a cavity length of 24 cm. The discharge length is 13 cm and the electrode gap 0.56 cm. The laser mixture is $CO_2 : N_2 : He = 1 : 1 : 3$

$$\chi_1 = 4 \times 10^{-9} \; ; \quad \chi_2 = 4 \times 10^{-9} \; ; \quad \chi_3 = 4 \times 10^{-9} \; ,$$
$$\chi_4 = 2 \times 10^{-8} \, [s^{-1} cm^3] \; .$$

The electron density N_e is related to the current density by

$$v_d e N_e = j \; , \tag{6.73}$$

where j is the total current divided by the cross section of the discharge and v_d the drift velocity. The drift velocity depends on E/N and is shown in Fig. 6.6. The current pulse in this low inductance discharge device as illustrated by Fig. 6.14 more or less resembles the analytic function

$$j(t) = j_0 t \exp(-t/\tau_p) \; . \tag{6.74}$$

The parameters j_0 and τ_p are chosen in such a way that the peak current and the time integral (total charge) correspond to the experimental values of Fig. 6.14. This current pulse together with the calculated temperatures T, T_1, T_2, T_3, and T_4 are shown in Fig. 6.16. The optical pulse and the gain are shown in Fig. 6.17.

It is interesting to see that the temperature T_2 follows the temperature T_1 very closely. During the optical pulse a large change in each of the vibrational temperatures of CO_2 is seen as expected. From Fig. 6.17 it is seen that the fast rise of the optical pulse causes the collapse of gain, which, in turn, terminates the pulse. The first peak is usually called the "gain switched peak". This phenomenon is a consequence of the fact that the time required to establish a significant gain is shorter than the cavity-field build-up time. After the pulse the gain recovers by energy transfer from the vibrationally excited nitrogen to the ν_3 vibration of carbon dioxide. The restored gain

167

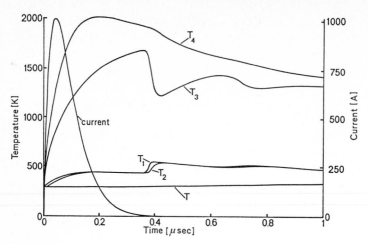

Fig. 6.16. The calculated temperature profiles for the indicated discharge current pulse. Note that T_2 closely follows T_1 except for the regions with large energy transfer

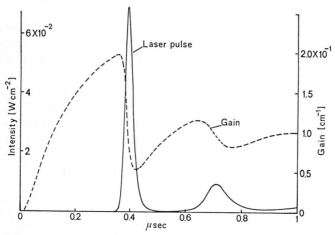

Fig. 6.17. The calculated laser pulse and gain of medium for the discharge current pulse shown in Fig. 6.16

again causes pulse forming but much less pronounced because this pulse starts from the residual field of the first pulse and not from noise.

6.6 Double-Discharge UV Preionized Systems

In the early developments of uv preionization for stable transverse discharges several techniques were successful [6.24–27]. They all had in common the use of trigger discharges to generate a uniform ionization layer near the cathode before the onset of the main discharge; a so-called "double-discharge" system. The electrodes were usually made of solid material (aluminum or brass) shaped according to a so-called Rogowski profile [6.28] to avoid large field concentrations near their edges. In principle, for those systems the following sequence of events is thought to occur. Just before the application of the high-voltage main discharge, a corona discharge is initiated between the cathode and the insulated trigger electrodes creating a thin electron gas layer near the cathode. These electrons produced by uv photoemission at the cathode by the corona discharge are released into the laser volume resulting in a fast change of discharge impedance. At this point the main discharge is initiated. As long as sufficient preionizing charges are present, the main discharge forms a glow, otherwise arc formation follows. Nevertheless the discharge always favors arc formation so that additional requirements are to keep the main discharge shorter than the build-up time of the arc discharge and to set an upper limit to the discharge voltage. As a result, the early techniques were limited to operation in small volumes (small apertures) at input powers less than 200 J/liter with gas mixtures containing low CO_2 concentrations.

The principles of the double-discharge technique have been further improved by *Richardson* et al. [6.29]. In their system the preionization of the discharge volume is obtained with the aid of an auxiliary multiple spark discharge from point electrodes which are situated close behind a perforated mesh anode. The main discharge occurs between the cathode consisting of a solid aluminum electrode approximately shaped as a Rogowski profile and an anode consisting of a fine stainless steel mesh tightly stretched near the edges over an insulating form. The main discharge is energized from a two-stage Marx generator charged at 55 and 75 kV for the 5 and 7.5 cm discharge modules, respectively, as shown in the circuit diagram of Fig. 6.18.

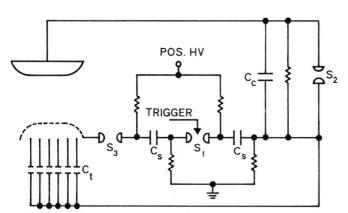

Fig. 6.18. Schematic of Marx-bank excitation circuit for double-discharge TEA laser [6.29]. (C_s: storage capacitor, 0.1 μF; C_t: trigger capacitors, 160 pF; C_c: cathode capacitor, 100 pF; S_1, S_2 and S_3 pressurized nitrogen spark gaps)

The circuit allows time for preionization of the main gas volume prior to the initiation of the main discharge. It consists of six separate strings of 100-point electrodes. Each string is capacitively coupled to the Marx generator. The sequence of events is as follows. After the spark gap S_1 is externally triggered, the Marx bank fires and the voltage builds up between the point electrodes and the anode mesh. The initial coronal emission from the point electrodes provides a thin layer of electron gas in the vicinity of the mesh. As this voltage increases, multiple spark discharge arcs between the anode and point electrodes are formed. These arcs provide the uv source for preionizing the main discharge volume. After sufficient preionization the spark gap S_2 fires and the main current pulse occurs between the main cathode and mesh anode.

Time-resolved studies of the discharge, both electronically and optically by means of a streak camera have shown that the appearance of arc discharges coincides with partial ionization of the medium between the main cathode and mesh anode. The onset of this ionization extends almost completely in the anode-cathode region within 10 ns. Volumetric photoionization by uv radiation is consistent with these observations, although other phenomena such as photoemission at the cathode may play a lesser role.

The development of this discharge technique is capable of exciting a gas mixture with up to 30 % CO_2 with input energies above 300 J/liter, a conversion efficiency of 10 %, and an average small signal gain of 4.3 % per cm. For an active volume of a $3:2:5 = CO_2:N_2:He$ gas mixture with an electrode separation of 7.5 cm the maximum input is 240 J/liter. The same system with 5 cm electrode separation is capable of operating at a higher input energy of 320 J/liter. When energies in excess of these limits were discharged into the gas, small streamers emanating from the cathode were observed [6.29].

6.7 Uniform-Field Electrode Profile (Chang Profile)

In order to obtain uniform pulsed glow discharges in transversely excited systems it is very important that the electrodes produce a very uniform distribution of the electric field. A high degree of field uniformity is not only required to increase the maximum input energy and applied voltage but also to obtain a discharge width comparable to the electrode distance. In this respect the Rogowski profile mentioned in the previous section is not the best choice because it is made up of three unsmoothly joined segments taken from an infinitely wide analytic profile, i.e.,

$$y = \frac{\pi}{2} + ce^{-x} , \quad y = \frac{\pi}{2} \quad \text{and} \quad y = \frac{\pi}{2} + ce^{x} ,$$

where x and y are the coordinates describing the cross section of a profile. These profiles are not compact and their uniformity is uncertain. It is difficult to improve these profiles because of their empirical approximation.

A very interesting analytic approach to this problem has been proposed by *Chang* [6.30]. He presented analytic expressions for compact uniform-field electrodes involving only hyperbolic functions. The formulas enable one to design electrodes for any electrode separation distance with specified field uniformity which can be fabricated on a numerically controlled milling machine. The following conformal transformation is proposed:

$$\varsigma = \omega + k \sinh \omega \; , \tag{6.75}$$

where $\varsigma = x + iy$ and $\omega = u + iv$, with x and y being the space coordinates of an electrode profile, u and v being the flux and potential functions, respectively, and $k > 0$. In the ω plane the equipotential plane is given by a constant value for v, and a variable value for u. In the ς plane the corresponding equipotential surface is then according to (6.75)

$$x = u + k \cos v \sinh u \; , \tag{6.76a}$$

$$y = v + k \sin v \cosh u \; , \tag{6.76b}$$

where $|v| < \pi$. It is seen that this equipotential plane is symmetric with respect to the y axis, while the equipotential planes for v and $-v$, respectively, are mirror images.

It is instructive to examine first the special case with $v = \frac{\pi}{2}$. The equipotential surface simplifies to

$$y = \frac{\pi}{2} + k \cosh x \; , \tag{6.77}$$

which approaches the Rogowski profile asymptotically as $x \to \pm \infty$. The curvature of this particular profile at $x = 0$ is equal to k, which gives the parameter k a simple geometric meaning.

The question arises, which value of v gives the best uniform field distribution at the surface. To answer this the electric field at the surface is examined. Its value is given by [6.31]

$$E^{-2} = \left| \frac{d\varsigma}{d\omega} \right|^2 = (1 + k \cos v \cosh u)^2 + (k \sin v \sinh u)^2 \; . \tag{6.78}$$

The electric field can also be expressed as a power series expansion in u, i.e.

$$E = a_0(v) + a_2(v)u^2 + a_4(v)u^4 + \dots \; . \tag{6.79}$$

The odd-powered terms are missing due to symmetry. To obtain a "maximally flat" distribution of field near the center of the electrode (where $u = 0$), one requires the coefficient $a_2(v)$ to vanish. This condition, which is equivalent to $\partial^2 E^{-2}/\partial u^2 = 0$ at $u = 0$, leads to the following relation between the parameter k and the value of the potential function for the electrode surface, which we denote by v_m

$$v_m = \arccos(-k) = \tfrac{\pi}{2} + \arcsin k \ . \tag{6.80}$$

Curves A, B and C in Fig. 6.19 show three normalized profiles in the first quadrant with $k = 0.2$, 0.06, and 0.01. The distribution of electric field on each electrode is also plotted in the lower part of the figure. Equation (6.80) and Fig. 6.19 indicate that as $k \to 0$, v_m tends toward $\tfrac{\pi}{2}$ and the electrode becomes very wide. As the value of k increases, the optimized profile becomes more compact.

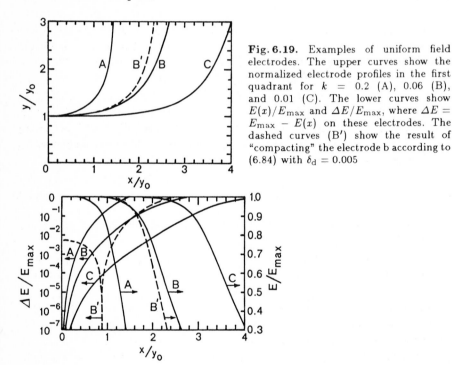

Fig. 6.19. Examples of uniform field electrodes. The upper curves show the normalized electrode profiles in the first quadrant for $k = 0.2$ (A), 0.06 (B), and 0.01 (C). The lower curves show $E(x)/E_{max}$ and $\Delta E/E_{max}$, where $\Delta E = E_{max} - E(x)$ on these electrodes. The dashed curves (B') show the result of "compacting" the electrode b according to (6.84) with $\delta_d = 0.005$

For practical purposes it is desirable to find a relation between the parameter k, which determines the geometric proportions of the uniform field electrode, and a certain practical parameter with a simple and direct meaning, such as the aspect ratio of the electrode pair and the desired uniformity of the field. To do this, one first defines a parameter δ_m as the maximum fractional variation of electric field that can be tolerated within a critical surface area on the electrode. The value of x (or u) at the edge of the critical area will be denoted by x_m (or u_m), and the height of the electrode surface at $x = 0$ by y_0. Since for a maximally flat field design (6.80) the value of $E(u)$ decreases monotonically with $|u|$, we have

$$\delta_m = [E(0) - E(u_m)]/E(0) \ . \tag{6.81}$$

Equation (6.81) combined with (6.76, 78, 80) can be used to calculate the aspect ratio x_m/y_0 of the uniform field electrode in terms of parameters k and δ_m. In practice, it is more natural to calculate k from the specified values of δ_m and x_m/y_0. This can be done either by a graphical method or by using the following approximate formula [6.30]

$$\left(\frac{l(1 - 0.64\,l)}{1 - 0.64\,l - 0.36\,l^2} \right)^6 \left[\left(\cosh \frac{\pi x_m}{2y_0} \right) - 1 \right]^2 = \frac{1}{(1 - \delta_m)^2} - 1 \ , \tag{6.82}$$

where $l = k^{1/3}$. For $\delta_m < 0.01$ and $x_m/y_0 > 0.05$ the error in the approximate value of k calculated from (6.82) is always less than 10 %, while the actual aspect ratio of the resulting profile deviates by no more than 5 % from the specified value. The errors diminish as x_m/y_0 becomes very large. A value of this which is slightly different from that given by (6.80), such that $\cos v < -k$, would lead to a somewhat more compact electrode profile. This may be desirable for some applications even though the distribution of $E(u)$ on the electrode surface is no longer maximally flat. It can be shown from (6.78) that for $\cos v < -k$, $E(u)$ has double maxima located at

$$u = \pm u_d = \pm \text{arccosh} \left[(-\cos v)/k \right] \ , \tag{6.83}$$

while $u = 0$ becomes a local minimum for $E(u)$. The fractional depression of $E(u)$ defined as $\delta_d = [E(u_d)/E(0)] - 1$ is given by

$$(1 - \delta_d)^2 - 1 = (k + \cos v)^2 [(1 - k^2)(1 - \cos^2 v)]^{-1} \ . \tag{6.84}$$

If $\delta_d < \delta_m$, then, for $|u| \leq |u_d|$, $E(u)$ is still uniform within the specified limit. To design such a "compacted" profile, one can specify δ_m and x_m/y_0 and determine the value of k from (6.82) as before, then specify δ_d and obtain a modified value of v from (6.75). Finally, these values of k and v can be used in (6.76) to calculate the electrode profile. An example is shown in Fig. 6.19 by dashed curves. Here the electrode B$'$ is compacted to the extent $\delta_d = 0.005$. It is apparent that electrode B$'$ is not only more compact than electrode B but also has a larger surface area over which E is constant within $\frac{1}{2}$ %.

So far, only two-dimensional profiles have been discussed. Practical electrodes must of course have a three-dimensional profile of finite length. Since conformal transformations cannot directly generate a three-dimensional profile, one has to be content with a composite profile. Consider an electrode that is long in the z dimension and narrow in the x dimension. The yz profile in the $x = 0$ plane can be designed to have a maximally flat field profile, according to (6.75–82) (with x replaced by z). The value of y along the z-axis can then be regarded as y_0 in (6.82) to generate a transverse profile for each value of z. The parameter δ_m can be specified as independent of z, but the aspect ratio $x_m(z)/y_0(z)$ must be adjusted along the z-axis. A very satisfactory result can be obtained by varying the aspect ratio according to the formula [6.30]

$$\frac{x_m(z)}{y_0(z)} = \frac{x_m(0)}{y_0(0)} \left(\frac{y_0(0)}{y_0(z)} \right)^{3/2} . \tag{6.85}$$

6.8 Minimum-Width Electrode (Ernst Profile)

Although Chang's profiles provide a field uniformity within predescribed limits in a central region between two electrodes, there is also a relatively large transition region where the field continuously and slowly decreases to a value far below the breakdown strength of the laser gas. This transition region makes the electrode width much larger than the discharge width so that only a fraction of the gas in the laser is excited by the discharge. Another disadvantage, even more severe, is the reduced penetration of the preionizing uv radiation by absorption, because its source must be kept away from the discharge at a distance where the field strength is sufficiently low. In practice, these profiles limit, for that reason, the maximum separation distance or aperture which, in turn, limits the maximum laser energy. A reduction of the electrode width for a given discharge width requires therefore a field distribution in the transition region that decreases faster with distance than that given by Chang's analytic expression. One may then ask what is the minimum width of the electrodes for a given discharge width. This problem has been treated by *Ernst* [6.32], who considered Chang's profile as a first-order approximation of a serial expansion in terms of sine-hyperbolic-functions. He used the conformal transformation

$$\varsigma = \omega + k_0 \sinh \omega + k_1 \sinh 2\omega + k_2 \sinh 3\omega + \dots . \tag{6.86}$$

It is instructive to study the effect of each addional term in this expansion. If only the third term, $k_1 \sinh 2w$, is added, the equipotential surface is analogous to (6.76) given by

$$x = u + k_0 \sinh u \cos v + k_1 \sinh 2u \cos 2v , \tag{6.87a}$$

$$y = v + k_0 \cosh u \sin v + k_1 \cosh 2u \sin 2v , \tag{6.87b}$$

where v is again constant and u the running variable. To calculate a profile, the three parameters v, k_0, and k_1 must be chosen. This is again done under the conditions of the desired aspect ratio x_m/y_0 and the best uniformity of the field. As compared with Chang's approach one more parameter can be specified. The field strength is now given by [6.31]

$$E^{-2} = \left| \frac{d\varsigma}{d\omega} \right|^2 = |1 + k_0 \cosh \omega + 2k_1 \cosh 2\omega|^2 = f^2(u) + g^2(u) \quad (6.88)$$

with

$$f(u) = 1 + k_0 \cosh u \cos v + 2k_1 \cosh 2u \cos 2v , \quad (6.89a)$$

$$g(u) = k_0 \sinh u \sin v + 2k_1 \sinh 2u \sin 2v , \quad (6.89b)$$

Using the power series expansion according to (6.79) the coefficients are

$$a_2 = -\{f(0)f^{(2)}(0) + [g^{(1)}(0)]^2\}/f^3(0) , \quad (6.90)$$

$$a_4 = -\{f(0)f^{(4)}(0) + 3[f^{(2)}(0)]^2 + 4g^{(1)}(0)g^{(3)}(0)\}/f^3(0) , \quad (6.91)$$

where the exponent in brackets denotes the number of differentiations with respect to u. For the best uniformity in the central region the coefficients a_2 and a_4 must vanish. For a given value of a_0, which is a measure of the aspect ratio, the parameters k_1 and v are calculated and indicated in Table 6.2. It is seen that the optimized values of v differ only slightly from $\frac{\pi}{2}$. In practice, v can be taken equal to $\frac{\pi}{2}$ and substituting this value into (6.90) the optimized value of k_1 is simply given by

$$k_0^2 - 8k_1(1 - 2k_1) = 0 , \quad (6.92)$$

or approximately

$$k_1 = \tfrac{1}{8}k_0^2 . \quad (6.93)$$

Using (6.92) the resulting profile is then

$$x = u - \tfrac{1}{4}[1 - (1 - k_0^2)^{1/2}]\sinh 2u , \quad (6.94a)$$

$$y = \tfrac{\pi}{2} + k_0 \cosh u . \quad (6.94b)$$

In Fig. 6.20 this profile calculated by *Ernst* for $k = 0.02$ is compared with Chang's profile. [The exact profile with the optimized values of v and k_1 differs only slightly from the approximate curve according to (6.94).] All x and y values in Fig. 6.20 have been normalized to $y_0 = \frac{\pi}{2} + k_0$.

The results, as shown in Fig. 6.20, demonstrate that the addition of one extra term in the profile expansion (6.86) reduces the electrode width

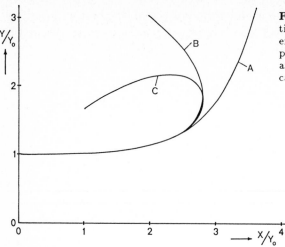

Fig. 6.20. The shape of the optimized profiles for three different cases. Curve A is Chang's profile, curve B relates to (6.94) and curve C to (6.95). For all cases $k_0 = 0.02$

by roughly 15 %. Moreover, the uniformity of the field distribution over the electrode surface is greatly improved. This is shown in Fig. 6.21 where $\Delta E = E(0) - E(x)$ divided by its maximum value $E(0)$ is plotted versus x/y_0. The curve A is for the Chang profile, and the curves B and B' are calculated according to the approximate solutions (6.94a and b), and the exactly optimized parameters of the profile according to (6.87a and b). It is seen that the approximate solution differs only slightly from the exact solution. Next, Ernst also included the fourth term, $k_2\sinh 3\omega$, in (6.86) to obtain an analytic expression for the profiles. In the subsequent expression there is an extra term k_2 to be used for optimizing the field uniformity. This

Table 6.2. Optimized values of k_1 and v for the profile of (6.87) with given values for k_0 [6.32]

k_0	k_1	v
0.001	0.1250002×10^{-6}	1.570796
0.0015	0.2812508×10^{-6}	1.570796
0.002	0.5000025×10^{-6}	1.570796
0.003	0.1125013×10^{-5}	1.570796
0.005	0.3125098×10^{-5}	1.570796
0.007	0.6125375×10^{-5}	1.570796
0.01	0.1250156×10^{-4}	1.570795
0.015	0.2813292×10^{-4}	1.570793
0.02	0.5002505×10^{-4}	1.570788
0.03	0.1126271×10^{-3}	1.570769
0.05	0.3134887×10^{-3}	1.570670
0.07	0.6163444×10^{-3}	1.570445
0.1	0.1266441×10^{-2}	1.569746
0.15	0.2901734×10^{-2}	1.567006
0.2	0.5315494×10^{-2}	1.560798
0.3	0.1400316×10^{-1}	1.514851

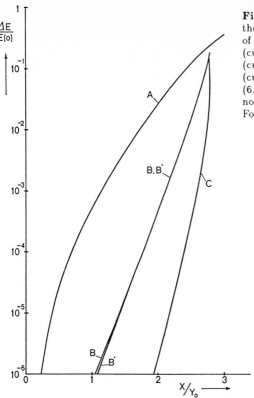

Fig. 6.21. The normalized deviation of the electric-field strength at the surface of the electrodes for Chang's profile (curve A), for the profile from (6.94) (curve B), for the profile from (6.87) (curve B′) and for the profile from (6.95) (curve C), plotted versus the normalized position at the electrode. For all cases $k_0 = 0.02$

means in analogy to the above analysis that now also the next term $a_6(v)$ in the series expansion of E given by (6.79) can vanish with the proper choice of the parameter v, k_1, and k_2. It is found that v approaches $\frac{\pi}{2}$ even more closely than in the former case. For instance, for $k_0 = 0.02$ the optimized values of k_1, k_2 and v are: $k_1 = 0.5000125 \times 10^{-4}$, $k_2 = 0.8334357 \times 10^{-7}$, and $v = 1.570796$. So, the profile is given by

$$x = u - k_1 \sinh 2u \ , \tag{6.95a}$$

$$y = \tfrac{\pi}{2} + k_0 \cosh u - k_2 \cosh 3u \ . \tag{6.95b}$$

Using $v = \pi/2$ the optimized values for k_1 and k_2 as a function of k_0 can be simply obtained from the conditions $a_2 = 0$ and $a_4 = 0$. This gives

$$k_1 = \tfrac{1}{4}\{1 - [1 - (k_0 - 9k_2)^2]^{1/2}\} \ , \tag{6.96}$$

$$k_2 = \tfrac{5}{81} k_0 \{1 - [1 - \tfrac{9}{25}(1 - 8k_1/k_0^2 + 64k_1^2/k_0^2)]^{1/2}\} \ . \tag{6.97}$$

The values of k_1 and k_2 can be easily found after some iteration steps. A good approximation is

$$k_1 = \tfrac{1}{8}k_0^2 \ , \tag{6.98}$$

$$k_2 = \tfrac{1}{90}k_0^3 \ . \tag{6.99}$$

A plot of this profile, with $k_0 = 0.02$, is curve C in Fig. 6.20, and its field-strength distribution is curve C in Fig. 6.21. A further improvement of the uniformity and compactness of the electrode profile can be obtained by using more terms in the expansion of (6.86). However, this will not lead to any substantial improvement over what has already been obtained.

Finally, we remark that for the realization of an electrode the analytic form given by (6.95) must be truncated on the "back surface", where the field is much smaller than on the "front surface". The truncation will then have little influence on the calculated field uniformity.

6.9 Dielectric Corona Preionization

As mentioned in Sect. 6.6, the historical development of self-sustained discharges using uv preionization has shown several sophisticated techniques [6.24–27, 29]. The performance is mainly due to the technical integration and operation of the subsystems forming the laser device. Without underestimating the effect of the other subsystems, the preionizer is of utmost importance from the standpoint of obtaining optimized laser performance and long sealed-off lifetime.

The preionization provides the initial electrons in the main electrode gap that condition the start of a glow discharge between the electrodes. Experiments have indicated that uniform preionization is highly desirable. For this it is very important that the uv source is uniform and extends over an area equal to the length of the electrodes times their separation distance. The most successful technique for achieving this is the use of a dielectric corona discharge. The term corona discharge often refers to a phenomenon which occurs in a gaseous medium in the vicinity of conductors of small radius of curvature subjected to high voltages. The high electric field along the surface which is usually non-uniform, produces a partial breakdown of the surrounding gas. A similar situation may arise when a dielectric plate is subjected to a high electric field. The field polarizes the dielectric plate and a high field is present at its surface. A corona discharge may then be produced in the surrounding gas near the dielectric.

This basic mechanism has been successfully exploited in two different laser constructions. The first involves the use of a suitable dielectric sheet subjected to a high-voltage gradient across its thickness at one end. The method was pioneered by *Ernst* and *Boer* [6.33–36]. The second method developed by *Hasson* and co-workers [6.37–39] involves the use of two knife edges separated by a gap with a dielectric sheet backing. The knife-edge

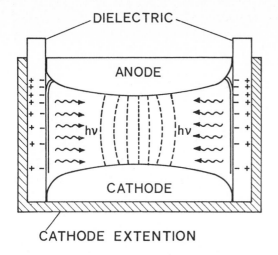

DIELECTRIC

ANODE

hv   hv

CATHODE

CATHODE EXTENTION

Fig. 6.22. Schematic drawing of the corona discharge along the dielectric material and the uv-radiation that preionizes the main discharge

method with very high fields has a stronger interaction with the surrounding gas, and so stronger dissociation, than the former method. Therefore the first method, which we shall discuss in more detail in Sect. 6.10, has greater potential for sealed-off operation.

The physical principles involved in producing a uniform preionization in a laser are explained with the aid of Fig. 6.22. This figure shows a cross section of the laser cavity perpendicular to the optical axis. The contours of the anode and cathode are according to a uniform electric field distribution at their surfaces. Through an electric circuit a high-voltage pulse is suddenly applied between the electrodes. Thus a high electric field, with a rapid rise appears between them. Because the cathode has a metalic extension bent around the two dielectric plates and closely approaching the anode, the dielectric plates are subjected to a very strong electric field. They become polarized with surface charges, as shown in Fig. 6.22. The high voltage in the surrounding gas produces a surface corona discharge. The ions formed in the surrounding gas are attracted by the dielectric plates and the electrons are easily pulled from the gas layer by the strong electric field in the corner formed by the dielectric and the anode. Once sufficient electrons are present in the gas layer, streamers are formed which rapidly develop into avalanches. Because of the non-conductivity of the dielectric, the surface discharge is uniform in the direction of the optical axis, i.e. perpendicular to the figure. The plasma produced near the dielectric plates forms a channel with a width equal to the length of the electrodes for the passage of current from the anode to the cathode.

The corona surface current, in turn, emits the uv photons. The light emission from current surface-spark discharges have been extensively studied by *Beverly* [6.40]. He argued that the initial rate of energy input into the surface discharge channel and hence its light emission is largely deter-

179

mined by the initial rate change of the current with time. A rapid change of current requires a low inductance driving circuit, because the initial current rise is

$$\frac{di}{dt} = \frac{V}{L} \; , \tag{6.100}$$

where V is the imposed voltage and L the inductance of the circuit. However, the current must not be too high. In the strong discharge regime, i.e. $V/L > 10^{10}$ V/H, radiation emitted by the plasma heats and vaporizes surface constituents [6.41]. Erosion due to radiation vaporization of the dielectric limits the life expectancy of the uv source. Moreover it may happen that in this regime sputtered material from the cathode condenses on the dielectric so that a conducting layer is formed. For these reasons the weak discharge regime, i.e. $V/L < 10^{10}$ V/H, is preferred for laser operation [6.41].

The uv photons emitted by this uniform dielectric corona discharge illuminate the electrode gap uniformly in the direction of the optical axis. This is not necessarily the case in the direction across the width of the plates. A gradient in the direction of the current does not influence the quality of the resulting main discharge. In fact, it has been reported that an excellent glow discharge between the electrodes without streamers is observed [6.34]. Such a discharge is a necessary condition for high optical quality of the laser beam.

The mechanism of the corona preionization is also reflected by the observed waveform shown in Fig. 6.23. The first current pulse is identified with the preionization process. The preionization current pulse which is the sum of the surface charge plasma current and the displacement current in the di-

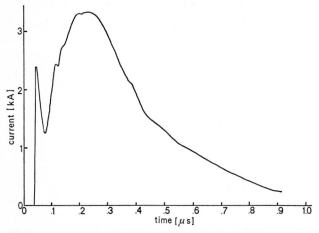

Fig. 6.23. The current as a function of time for a 5 cm electrode gap CO_2 system. The Marx generator voltage is 95.5 kV and the gas mixture $CO_2 : N_2 : He = 1 : 1 : 3$ [6.36]

electric is approximately 25 ns wide. The second pulse is the main discharge. To optimize the system the current distribution over these two pulses must be balanced. If, for instance, the preionization current pulse emits more light than is required, the preionization energy is too large at the expense of the main discharge energy that provides the laser excitation. Thus the two energies must be changed by the current pulses. The balancing process between the two current pulses depends on several physical variables: dielectric material, dielectric length from anode to cathode, dielectric thickness, V/L value of the circuit, gas pressure, and gas composition. Experimental investigations for each aperture and operation range must be done. Extensive studies concerning the general effect of these variables on the radiation emission are reported in [6.41]. The main conclusions can be summarized as follows:

Voltage

Up to a certain voltage, which depends on pressure and gas composition, a normal uniform corona is observed. As the voltage is further increased the regime of surface sparking is reached. The sparking regime should be avoided. The spark is unstable in location and meanders over the dielectric surface. The sparking threshold voltage is rather low for pure helium. Nitrogen and carbon dioxide (in this sequence) increase the threshold level of surface sparking. As expected, this threshold increases with gas pressure.

Gas Composition

A spectral investigation of the corona discharge in pure CO_2 shows little spectral light intensities. However, the spectrum obtained for pure nitrogen contains considerably more intensity in the range from 100 to 200 nm. Absorption measurements in CO_2 have shown that the absorption is strong below 115 nm. The absorption then decreases to a few $cm^{-1} atm^{-1}$ in a 10 nm band centered around 120 nm. Above 120 nm the absorption increases strongly, but it falls again to about $2 cm^{-1} atm^{-1}$ at 160 nm and continues to decrease thereafter, reaching a negligible level at about 195 nm [6.42]. Thus only uv photons in the windows $117 nm < \lambda < 124 nm$ and $\lambda > 160 nm$ can penetrate CO_2 sufficiently. This is in agreement with *Kaminski*'s investigations [6.41] that rich CO_2 mixtures have very little uv activity below 155 nm in contrast with poor, say 10 %, CO_2 mixtures. It is the main reason why corona preionized CO_2 lasers have the best performance for gas mixtures with low CO_2 percentages.

Pressure

The uv light produced by the corona for constant voltage decreases as pressure increases. The reason is that with increasing pressure and constant voltage the current change, di/dt, decreases. As stated before, the change of current with time determines largely the light emission. The greater di/dt

the more intense is the uv radiation. However, the discharge voltage of a laser system is more or less proportional to total gas pressure. According to (6.100) the same can then be said for the current change. Therefore, in practice, it is not expected that higher pressure will have an unfavorable effect on the light emission.

Dielectric Thickness

The capacity of the dielectric plate is inversely proportional to its thickness. Thus the displacement current and consequently also di/dt increase with decreasing thickness. This, in turn, effects the energy deposition into the corona plasma and the uv-light intensity. However, the thickness reduction is limited by the formation of a surface arc at increasing corona current.

Gap Distance

The width of the dielectric in the direction of the electrode distance influences for constant voltage the light emission; namely, as the gap is decreased the light intensity increases. However, this parameter is not important in practice because the voltage increases for an operating system more or less linearly with the gap distance so that the field of the corona discharge is constant.

Material Choice

Important material properties are the dielectric constant and the dielectric strength. In comparison with other ceramics it is found that Zerodur is a very attractive candidate [6.41]. All other parameters being constant it was observed that the light production is superior. As before, the reason is that di/dt is greater for Zerodur than for pyrex glass or Macor.

6.10 Single-Discharge Corona Preionized Systems

In the previous section we described the performance of preionization by means of a dielectric surface discharge in the surrounding laser gas. Realizations of the basic idea of a rapid corona discharge in a self-sustained TEA laser have been described in the literature [6.33–39]. It has been shown that this type of self-sustained TEA laser is superior to other types. This is because of its uniform preionization, its potential of relatively large electrode gap, and the simplicity of the single discharge. It has been shown that the technique can be used even for an electrode gap of 10 mm [6.36]. According to work of *Ernst* and *Boer* [6.36] the output ranges from 60 J/liter for 2 cm, up to 40 J/liter for a 10 cm electrode gap. The lower output per volume for larger electrode gap is due to the uv absorption loss from the corona to the discharge area. In fact, the high output energy at 10 cm electrode gap is obtained in a relatively poor CO_2 mixture of $CO_2 : N_2 : He = 1 : 1 : 10$.

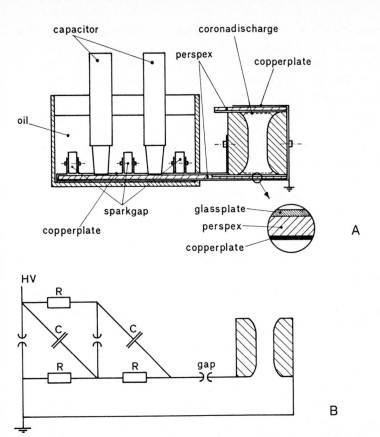

capacitor   coronadischarge

Fig. 6.24. (a) Scheme of the construction of the system and details of the side-wall construction of the laser. (b) Electrical scheme of the laser system [6.36]

The uniform preionization of this type of self-sustained discharge allows for high excitation rate, i.e. high input energy, comparable to those of uv or electron-beam sustained systems. Also the maximum small-signal gain is comparable.

It is essential that the construction consists of uniform field electrodes and a very rapid power source. A typical construction, as shown in Fig. 6.24a, was originally described by *Ernst* [6.36]. The electrodes are fabricated according to the analytic expressions (6.76a,b) for an electrode gap of 10 cm. A k value of 0.015 and $v = \arccos{(-k)}$ are used. The length of the profiles is 60 cm and their width is 364 or 346 mm. For outcoupling a plane-parallel uncoated germanium flat is used. The glass plates, as shown in the figure, are glued to the electrodes. It was experimentally found that the thickness of the glass plates depends on the width of the electrodes. The smaller the width the thicker the plate. If the glass plate is too thin, di/dt of the corona discharge apparently degrades into a surface sparking, as discussed in the previous section. Instead of changing the thickness, the capacity for the displacement current can also be varied by placing perspex plate between the glass plate and the copper plate. For the width of the electrodes of 364 mm the thickness of the glass plate and perspex plate is 13 mm and for the width of 346 mm it has an optimum of 23 mm. The rapid discharge is obtained by using a Marx generator with low inductance. The electric scheme is shown in Fig. 6.24b. It has two stages and an additional gap, which provides for the insulation of the laser head from the Marx generator when the first spark gap fires. For a sparking discharge the oscillation frequency of the current through the system gives a rough indication of the self-inductance of the system. It was estimated as about 200 nH.

Figure 6.25 shows the output energy of the system with the larger electrode width of 364 mm for a $CO_2 : N_2 : He = 1 : 1 : 10$ mixture as a function of the capacity of the Marx generator. The maximum output energy we measured was 148 J (accuracy 10%), for a Marx-generator voltage per stage of 50 kV. For a voltage of 45 kV the maximum output energy was 135 J. The capacitance per stage of the Marx generator for that case was 360 nF. Figure 6.26 shows the measured output beam diameter. The data were obtained by moving a 2 mm slit over the beam. The width of the output beam appears to be 55 mm, which means an active volume of $5.5 \times 60 \times 10 \, cm^3 = 3.3$ liter.

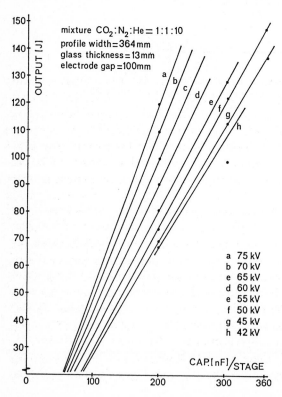

mixture $CO_2 : N_2 : He = 1 : 1 : 10$
profile width = 364 mm
glass thickness = 13 mm
electrode gap = 100 mm

a 75 kV
b 70 kV
e 65 kV
d 60 kV
e 55 kV
f 50 kV
g 45 kV
h 42 kV

Fig. 6.25. Output energy as a function of the capacity per stage of the Marx generator for a $CO_2 : N_2 : He = 1 : 1 : 10$ mixture. The voltage per stage of the Marx generator has been used as a parameter [6.36]

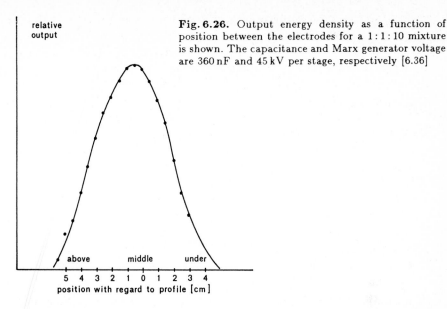

Fig. 6.26. Output energy density as a function of position between the electrodes for a $1:1:10$ mixture is shown. The capacitance and Marx generator voltage are 360 nF and 45 kV per stage, respectively [6.36]

This results in an output energy per unit volume of 40 J/liter. The efficiency calculated with an input energy of 729 J is 18.5 %.

6.11 Electron-Beam Controlled Systems

For electron-beam controlled systems the ionization necessary to maintain the discharge is produced by an external electron beam. The electron beam controls the level of ionization in the discharge and therefore the conductivity of the discharge while an independently controlled main discharge provides the excitation of the vibrational levels in the laser process. Since the electric field of the main discharge in the conductive plasma can be chosen independently, this field will be adjusted for optimum excitation rates into the vibrational levels. This technique for CO_2 systems was pioneered by *Fenstermacher* et al. [6.43–44], and later also by *Daugherty* et al. [6.45] and *Basov* et al. [6.46].

A cross-sectional view of an electron-beam controlled CO_2 laser system is shown in Fig. 6.27. The ionizing energetic electron beam is typically in the range of 100–300 KeV. The field across the laser gas accelerates the resulting charges and provides the vibrational excitation of the molecules. This discharge is non-self-sustained because the electric field will be chosen in such a way that the resulting electron velocity distribution is optimum for pumping the upper laser level. The optimum E/N is about $(1-2) \times 10^{-16}\,\mathrm{V\,cm^2}$, as can be seen in Fig. 6.11. This choice of field does not provide sufficient ionization to compete with the recombination process. An additional advan-

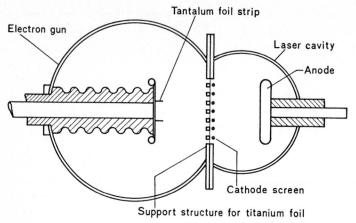

Fig. 6.27. Cross-sectional view of an e-beam controlled CO_2 system. The optical beam is perpendicular to the figure [6.47]

tage of the separation of the source of ionization from the laser excitation is the avoidance of self-sustained breakdown and arc formation.

The ionizing electron beam is produced by a pulsed cathode diode mounted in a vacuum chamber shown on the left of Fig. 6.27. This cathode may be a thermionic cathode, a plasma cathode, or a cold cathode. However, because of its simplicity and ease of use, the cold cathode is preferred and will be discussed here. As an example, we describe the system built at the University of Twente, Enschede, The Netherlands [6.47].

The emitting part of the cathode consists of two tantalum foil strips placed perpendicular to the backing plate. This backing plate is surrounded by a cylinder which serves to prevent emission from the edges of the plate as well as to focus the spreading electron beam for impinging of the titanium foil window of 50 μm thickness. Titanium is used for the foil because of its favorable combination of transmission and strength. *Mesyats* et al. [6.48] found that the mechanism responsible for electron emission from these foil strips are the explosions of whisker microprotrusions at the tips of the strips. As the applied voltage rises, field emission occurs from the microtips of the whiskers. An estimate of the electric field at the tip of the blade is given by the ratio of the applied voltage and the anode-cathode spacing, multiplied by two enhancement factors. The first factor comes from the geometry of the foil strips on the backing plate, whereas the second one arises due to the geometry of the microprotrusions relative to the edge of the strip. Both factors may have values of 100 to 1000. If the applied voltage is about 200 kV and the anode-cathode spacing is 6.5 cm the average field is of the order of 10^4 V/cm. The field at the tip of the microprotrusion can then be estimated in the range of 10^7 to 10^8 V/cm. For such fields, field emission takes place quite readily.

The initial field emission at the microtip may produce an emission current in the order of 10^6 A/cm². Such currents give rise to heating and vaporization processes which increase the current densities further to values of the order of 10^9 A/cm². This, in turn, produces even more heating. The increased metal vapor in the vicinity of the microprotrusion produces a gas cloud that will then be ionized by the emitted electrons in the strong field. The ions formed, in turn, give secondary-electron emission from the cathode, so that the plasma produced from the metal vapor acts as the source of the electron beam.

The high-energy electrons from the beam are accelerated through the 50 μm titanium foil window separating the vacuum chamber from the laser system. After passing through the multiple-grid cathode structure these electrons provide the ionization. A constant electric field of $E/N = (1.2 - 2.4) \times 10^{-16}$ V cm² is applied to the laser gas between cathode and anode shown on the right of Fig. 6.27.

6.11.1 Recombination-Limited Plasma

The energetic electrons entering the laser cavity ionize the laser gas and produce free secondary electrons with a mean energy in the range of a few eV. The electron beam sustains the discharge in the active laser volume. The number of produced secondary electrons per cm³ and per second can be obtained from the relation

$$S = \frac{j_b \varrho p (dE/dx)}{e W_{ion}} , \qquad (6.101)$$

where j_b is the current density of the beam, $\varrho \sim 10^{-3}$ g cm^{-3} atm^{-1} is the gas density at one atmosphere, p the gas pressure in atmospheres, dE/dx the stopping power in the laser gas equal to 2.4×10^6 eV cm² g^{-1} [6.49], W_{ion} is the mean energy required to produce an ion pair in the laser mixture of about 30 eV.

The density n of the free electrons of this type of discharge can be described by the simple relation

$$\frac{dn}{dt} = S - \gamma n^2 - \beta n . \qquad (6.102)$$

The second term on the right represents the recombination loss due to the existing densities of ions and electrons. γ is the electron-ion recombination coefficient. Since the plasma may be considered neutral, the densities of ions and electrons are equal so this term contains n^2. The last term on the right describes the loss of electrons due to attachment to neutral particles. In this simple model we have assumed for the different particles the same values of γ and β. The value of $\beta = a v_d$ depends on the E/N value, as described in

187

Sect. 6.4.1. For optimized conditions with respect to vibrational excitation of the active medium the E/N value is about $(1.2 - 2.4) \times 10^{-16}\,\mathrm{V\,cm^2}$, as can be seen in Fig. 6.11. For this value of E/N the attachment is much smaller than the recombination rate so that the last term in (6.102) can be neglected. The solution of (6.102) can be obtained in the form

$$ n = \left(\frac{S}{\gamma}\right)^{1/2} \tanh\left[(S\gamma)^{1/2}t\right] \; . \tag{6.103}$$

Upon the removal of S we find the solution of (6.102), now describing the electron density decays from an initial density value n_0, as

$$ n = \frac{n_0}{1 + n_0\gamma t} \; . \tag{6.104}$$

It is seen from (6.103) that the approximate time constant τ for approaching the steady state is given by

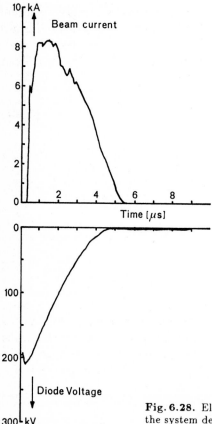

Fig. 6.28. Electron-beam current and diode voltage for the system described in the text

$$\tau = (S\gamma)^{-1/2} \ . \tag{6.105}$$

The system to be discussed here [6.47] has a maximum total beam current of $8\,\text{kA}$ (Fig. 6.28). The beam cross section in the active medium is $100 \times 10\,\text{cm}^2$, whereas the transmission through the foil and the support structure is about 0.5. From this we estimate $j_b = 4\,\text{A/cm}^2$. Using (6.101) S is about $2 \times 10^{21}\,\text{cm}^{-3}\text{s}^{-1}$. For the typical values of E/N the recombination coefficient can be estimated as $\gamma = 10^{-7}\,\text{cm}^3\,\text{s}^{-1}$. The time constant τ according to (6.105) is then about $70\,\text{ns}$. The steady state free electron density is approximately $n_0 = (S/\gamma)^{1/2} \sim 10^{14}\,\text{cm}^{-3}$. In the laser discharge the current density j_L is

$$j_L = n_0 e v_d \ , \tag{6.106}$$

where v_d is the electron drift velocity depending on E/N and gas mixture. The drift velocity is according to (4.17) given by

$$v_d = \bar{\mu}_e E / \varrho \tag{6.107}$$

where $\bar{\mu}_e$ is the electron mobility for unit density which for the usual gas mixtures is about $1.2\,(\text{s}^{-1}\,\text{cm}^{-1}\,\text{g}\,\text{V}^{-1})$.

The impedance of the discharge can now be calculated from (6.106) by using $n_0 = (S/\gamma)^{1/2}$, (6.101 and 107)

$$Z_L = \frac{d_L}{\bar{\mu}_e} \left(\frac{\varrho \gamma W_{\text{ion}}}{e p \frac{dE}{dx} \varepsilon A} \right)^{1/2} \frac{1}{\sqrt{J_b}} \ , \tag{6.108}$$

where $A = 10 \times 100\,\text{cm}^2$ is the cross section of the discharge, $\varepsilon = 0.5$ the transmission factor of the e-beam through the foil and the support structure, J_b the total e-beam current, and $d_L = 10\,\text{cm}$ is the anode-cathode distance. Using the above defined constants we obtain for the $CO_2 : N_2 : He = 1 : 2 : 5$ gas mixture at one atmosphere

$$Z_L = \frac{33}{\sqrt{J_b}} \Omega \ . \tag{6.109}$$

Although the specific impedance of the discharge depends on the e-beam current density in the discharge this density may be as low as $0.1\,\text{A/cm}^2$ to provide sufficient preionization for the conductive plasma and to sustain the discharge for pumping the relevant vibrational levels. However, the larger e-beam density the faster the pump energy is deposited into the active medium.

6.11.2 Cold Cathode

The electron emission occurs via explosions of whisker microprotrusions at a few tips of the tantalum strips. After the whisker explosion the space between the cathode plasma and the anode is space charge limited. Along the strips the emission starts at a few tips and expands with time so that the effective diode distance between anode and cathode decreases with $v_p t$, where v_p is the expansion velocity of a microprotrusion. For the large cathode surface the field depends mainly on the coordinate perpendicular to the anode. The diode can then be considered as a simple parallel planar gap having a current density given by

$$j_b = \frac{2.33 \times 10^{-6} V_g^{3/2}}{(d_g - v_p t)^2} \ , \tag{6.110}$$

where d_g is the anode-cathode spacing and V_g the diode voltage. The diode behavior as a function of time has been studied by *Mesyats* et al. [6.48]. The surface area A and the number m of microprotrusions can be estimated by considering the capacitance. If the produced plasma expands as a sphere with velocity v_p we find the capacitance as

$$C = 4\pi\varepsilon m(r_0 + v_p t) \ , \tag{6.111}$$

where r_0 is the finite initial radius of plasma cloud. Alternately, the capacitance approximates to that of a plane capacitor given by

$$C = \frac{A\varepsilon}{d_g - v_p t} \ . \tag{6.112}$$

Equating the two expressions for C we obtain for the effective area

$$A = 4\pi m(d_g - v_p t)(r_0 + v_p t) \ . \tag{6.113}$$

The total diode current J_b is given by $A j_b$ or by using (6.110, 113)

$$J_b = 2.9 \times 10^{-5} m V_g^{3/2} \frac{r_0 + v_p t}{d_g - v_p t} \ . \tag{6.114}$$

It is seen that for small times, $t \ll d_g / v_p$, the current rises linearly. The values for v_p and m can be obtained from experiments.

From the measured current and voltage characteristics, as shown in Fig. 6.28, we deduce for times, $t \gg r_0 / v_p$, the quantity

$$\frac{V_g^{3/2}}{J_b} = \frac{d_g}{K v_p} \frac{1}{t} - \frac{1}{K} \ , \tag{6.115}$$

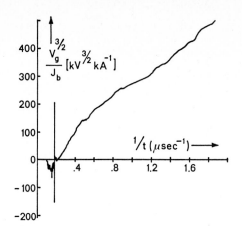

Fig. 6.29. The computed $V_g^{3/2} J_b^{-1}$ curve from the experimentally observed voltage and current profiles. For small $1/t$ values this curve is approximately linear

where $K = 2.9 \times 10^{-5}\,m$. This expression is plotted versus $1/t$ in Fig. 6.29. It is seen that for small values of $1/t$ the function is approximately linear. The short-circuiting of the diode occurs for $1/t = 0.2$ or $\tau_s = 5\,\mu s$. From the anode-diode spacing $d_g = 6.5\,cm$ we find $v_p = d_g/\tau_s = 1.3 \times 10^6\,cm\,s^{-1}$. Further we obtain from Fig. 6.29 the value $K = 3.5 \times 10^{-4}\,AV^{-3/2}$ and from this $m = 12$. Thus the experimental analysis shows that the micropro- trusions along the two tantalum strips have an average separation of about 16 cm.

6.11.3 Simulation of e-Beam Sustained Discharge

The equivalent circuits for both the e-beam and the sustainer discharge are shown in Fig. 6.30.

The impedance Z_g of the diode is obtained from (6.115) and that of the laser discharge, Z_L, from (6.109). The time dependence of V_L, J_L, V_g, J_b, and laser input energy for the one atmosphere mixture are shown in

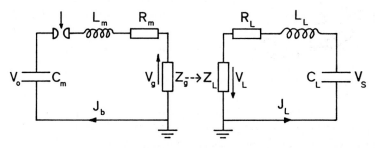

Fig. 6.30. The equivalent RLC circuits for an e-beam sustained discharge. The system built at the University of Twente, The Netherlands, has the following parameters: $C_m = 0.128\,\mu F$ is a 5-stage Marx generator of the e-beam; $L_m = 3\,\mu H$; $R_m = 5\,\Omega$, necessary to protect the e-beam and diode in the case of short circuiting; $V_0 = 225\,kV$ or $45\,kV$ per stage, $C_L = 4.5\,\mu F$; $L_L = 0.5\,\mu H$; $R_L = 0.2\,\Omega$, necessary to protect the capacitor in the case of short circuiting; $V_s = 50\,kV$

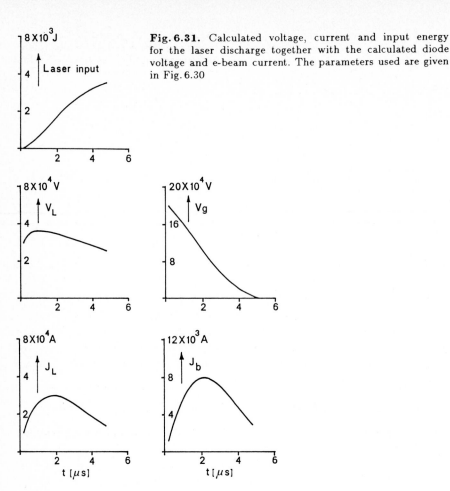

Fig. 6.31. Calculated voltage, current and input energy for the laser discharge together with the calculated diode voltage and e-beam current. The parameters used are given in Fig. 6.30

Fig. 6.31 as a function of time. The main discharge starts and terminates with J_b. The laser efficiency depends strongly on E/N. Low values of E/N are inefficient, as seen in Fig. 6.11. This means that an optimum RLC circuit must be designed for the discharge. The magnitude of L_L is chosen to have the most effective E/N in the laser cavity during the discharge. It is seen in Fig. 6.31 that V_L is between 30 and 40 kV or the field is between 3 and 4 kV cm^{-1} atm^{-1}. The discharge may terminate by a crowbar to short circuit the system or as in the present example by the end of the e-beam current. In the latter case, a part of the electric energy remains in the sustainer capacitor and can be used for the next shot.

6.11.4 Optimized Output from the e-Beam Sustained System

The efficiency of the vibrational excitation depends on the electron energy distribution which, in turn, is governed by the E/N value of the discharge.

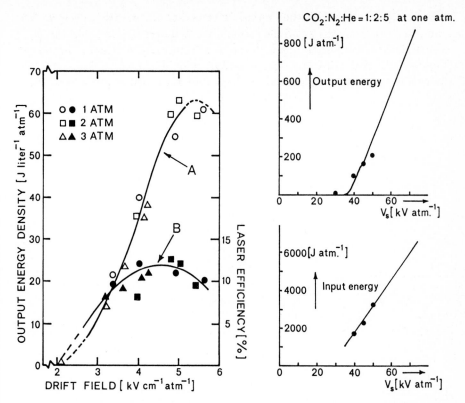

Fig. 6.32. Single-pulse energy density (curve A) and the efficiency (cure B) for the $CO_2 : N_2 : He = 1:2:3$ gas mixture vs normalized drift field for 1, 2, and 3 atm. [6.51]

Fig. 6.33. Input and output energies per unit atmosphere of an e-beam sustained system. The lines are calculated and the dots are experimental values [6.47]

For usual gas mixtures theory predicts E/N values in the range $(1.2 - 2.4) \times 10^{-16}$ V cm^2 or $(3-6)$ kV cm^{-1} atm^{-1}, as can be seen in Fig. 6.11.

This has indeed been confirmed by experimental studies with a variable drift field over the discharge [6.51]. Using a crowbar in the laser discharge circuit the discharge time can be fixed and for the chosen period a more or less constant E field can be chosen. The experiments show that the pressure-normalized output energy density and the efficiency on E/N are independent of pressure. This means that the pressure-normalized input energy density is likewise independent of pressure. The results for a 1.6 μs discharge are shown in Fig. 6.32 [6.50]. The data indicate that the maximum in output energy density occurs between 5 and 6 kV cm^{-1} atm^{-1}. The obtained efficiency curve does not depend on the pulse length provided the total input energy does not substantially raise the translational temperature so that the depopulation of the lower laser level is hampered.

Thus, the experimentally obtained efficiency versus E/N according to Fig. 6.32 can be used to predict the performance of an e-beam sustained

193

system. This is done for the system described in the previous section. The input energy from the sustainer bank is plotted versus V_S. Using the E/N dependence according to Fig. 6.32 the total integrated efficiency can be calculated and from this the output power. This is shown in Fig. 6.33. The solid lines have been calculated, and the dots represent experimental data.

7. AM Mode Locking of TEA Lasers

For laser systems generating only axial modes, i.e. modes with Gaussian energy distribution, a phase correlation of the modes and an equal frequency spacing between the modes result in an oscillating pulse in the cavity; the outcoupled beam appears as a train of pulses. The phenomenon can be understood by considering either the interference of the modes in the frequency domain or the self-consistency of an oscillating pulse in the time domain. The physics to describe these two approaches can be discussed by means of Fig. 7.1. In the upper part the frequency distribution of a multi-mode axial beam over the line profile is indicated. In this frequency domain the field consists of a number of discrete axial modes with frequency spacing equal to $c/2L$. The frequency spacing is nearly constant and the variations are due to the dispersion of the medium. In the case of a CO_2 laser the line profile is homogeneously broadened so that at most a few axial modes near the center of the gain curve oscillate. (At the onset of laser action many axial modes develop from the noise, but after some time, as a result of strong mode competition in a homogeneous medium, only one, the strongest near the line center, survives.) In the presence of a modulating element driven at a frequency near the mode spacing, $f_m \simeq c/2L$, side bands are created on each oscillating mode, which overlap with adjoining axial modes. These side

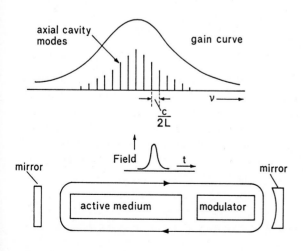

Fig. 7.1. (a) The frequency distribution of axial modes with respect to the line of a transition; (b) schematic round trip of a laser pulse through the cavity

bands are amplified by the medium because of their resonance behavior. In this way many modes are forced to oscillate at equal frequency spacing. The modes are then strongly coupled in both amplitude and phase. The resulting field of these modes shows a short circulating optical pulse inside the cavity because within each period there is, at any place, a moment when all mode fields interfere constructively to reach a maximum.

Looking at the time domain, the field starts from the noise with an arbitrary distribution which circulates inside the cavity with a repetition period equal to the round-trip time $2L/c$. The field loses energy due to outcoupling and gain due to the amplification of the active medium. In the presence of an intracavity modulating element with a modulation period ν_{m}^{-1} equal to the round-trip time $2L/c$, the circulating field distribution will be reshaped repeatedly on each transit through the modulator. The resulting modulation effects will accumulate over many passes, and finally the remaining field passes the modulator when its transmission is maximum. In other words, a short circulating optical pulse is formed within the cavity.

The active intracavity modulation can be either on the amplitude (AM) or on the frequency (FM). In straightforward AM mode locking the time-varying transmission locks all the oscillating modes in phase with each other, producing a circulating pulse which passes the modulator at maximum transmission. In the case of FM modulation the circulating energy that passes the FM modulator receives a Doppler shift proportional to $d\phi/dt$. After repeated shifts on successive passes this energy is pushed outside the gain band of the active medium. Only the part of the circulating energy that passes through the modulator at the time when the intracavity modulation is stationary at either side of its extrema will remain at the same frequency of the gain band. In this way pulse forming is obtained. In this chapter we only treat AM mode locking. The analysis can be done either in the time domain or in the frequency domain. We shall start with the time domain and continue with a study in the frequency domain. The results are the same.

7.1 Acousto-Optic Modulation

An acousto-optic device is generally used to obtain the necessary periodic loss modulation for amplitude modulation in CO_2 lasers. Such a modulator, which is located in the laser cavity near one of the mirrors, consists of a germanium crystal. A suitable acoustic transducer is mounted at the end of the crystal. In general, to avoid the 36 % Fresnel reflections and failure of the antireflection coatings an acousto-optic modulator with Brewster angles is used. A schematic drawing of the modulator is shown in Fig. 7.2. The acoustic transducer, which is supplied by a power oscillator, transmits an acoustic wave in the germanium crystal. The acoustic wave reflects at the end of the crystal, so that at acoustic resonance a standing wave is obtained.

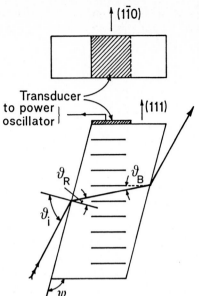

This standing acoustic wave produces compressions and rarefactions in the crystal, resulting in a phase grating. For the proper operation of the modulation we have to ensure that the incident optical beam impinges on the surface at the Brewster angle, i.e. $\tan \vartheta_i = n$, and that the radiation inside the crystal is incident on the phase grating at the Bragg angle. The latter condition can be analyzed by considering in detail the interaction of an optical beam and a sound wave in a crystal [7.1,4].

Acoustic scattering of light is associated with the change in the index of refraction, which is caused by the acoustic wave in the crystal. The frequency of the acoustic waves under discussion will be assumed to be sufficiently high, so that the wavelength is very small compared to the cross-sectional dimensions of the wave. The scattering process is especially simple when we consider the interaction of plane monochromatic optical and acoustic waves. The incident light is denoted by a wave vector or propagation constant $k_1 = \omega_1/c_1$, in which ω_1 is the optical angular frequency and c_1 is the light velocity in the germanium crystal. The direction of the incident light is perfectly defined since the wave front is planar and infinitely wide. The sound wave is also denoted by a wave vector or propagation constant $k_s = \omega_s/c_s$, in which ω_s is the acoustic angular frequency and c_s the sound velocity in the germanium crystal. The interaction of the optical wave and the sound wave occurs if conservation of energy and momentum is achieved. Figure 7.3 illustrates these conditions. Conservation of energy yields

$$\frac{h}{2\pi}\omega_1 = \frac{h}{2\pi}(\omega_1 \pm \omega_s) \ , \tag{7.1}$$

197

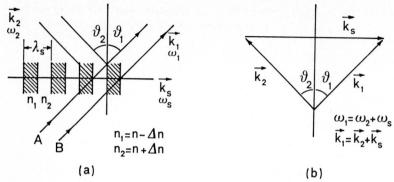

Fig. 7.3. **(a).** Schematic drawing of the variation of refractive index due to a standing acoustic wave. **(b)** The condition for maximum interaction is shown

and conservation of momentum

$$\frac{h}{2\pi}\boldsymbol{k}_1 = \frac{h}{2\pi}(\boldsymbol{k}_2 \pm \boldsymbol{k}_s) \ . \tag{7.2}$$

So we must satisfy the (Bragg) conditions

$$\omega_1 = \omega_2 \pm \omega_s \ , \tag{7.3}$$

$$\boldsymbol{k}_1 = \boldsymbol{k}_2 \pm \boldsymbol{k}_s \ . \tag{7.4}$$

It should be noted that $\omega_s \ll \omega_1, \omega_2$ so that $\omega_1 \simeq \omega_2$. The last equation can be written as

$$|\boldsymbol{k}_1| \sin \vartheta_1 + |\boldsymbol{k}_2| \sin \vartheta_2 = |\boldsymbol{k}_s| \ , \tag{7.5}$$

$$|\boldsymbol{k}_1| \cos \vartheta_1 = |\boldsymbol{k}_2| \cos \vartheta_2 \ . \tag{7.6}$$

Since it follows from (7.3,6) that

$$\frac{\cos \vartheta_1}{\cos \vartheta_2} = \frac{|\boldsymbol{k}_2|}{|\boldsymbol{k}_1|} \simeq 1 \ ,$$

$\vartheta_1 = \vartheta_2$, which is called the *Bragg angle* ϑ_B. Using (7.5) we find

$$2\lambda_s \sin \vartheta_B = \lambda_1 \ , \tag{7.7}$$

where λ_s and λ_1 are the length of the sound wave and the incident light wave, respectively. This equation is known as *Bragg's law*.

From this scattering process under the Bragg condition the transmitted and scattered light intensities can be calculated. This can be done as follows.

In the scattering process, mentioned above, a wave of angular frequency ω_1 changes to $\omega_1 + \omega_s$ and the propagation vector $\boldsymbol{k}_1$ changes to $\boldsymbol{k}_1 + \boldsymbol{k}_s$. Thus the energy is increased by $h\omega_s/2\pi$ and the momentum by $h\boldsymbol{k}_s/2\pi$. The interpretation of the scattering process is that a photon energy $h\omega_1/2\pi$ and momentum $h\boldsymbol{k}_1/2\pi$ is changed into a new photon of energy $h(\omega_1 + \omega_s)/2\pi$ and momentum $h(\boldsymbol{k}_1 + \boldsymbol{k}_s)/2\pi$. The conservation of energy and momentum is achieved by the absorption of a phonon of energy $h\omega_s/2\pi$. In an analogous way the scattering process with a final frequency $\omega_1 - \omega_s$ and a final propagation vector $\boldsymbol{k}_1 - \boldsymbol{k}_s$ creates a phonon of energy $h\omega_s/2\pi$ and momentum $h\boldsymbol{k}_s/2\pi$.

In our case we are dealing with a standing acoustic wave in the germanium crystal. It means that the scattering process varies periodically with twice the sound-wave frequency. At the moment the standing soundwave reaches its maximum amplitude, the diffraction ratio is maximum; and at the moment the amplitude is zero, the diffraction ratio is zero. Now we consider the moment of maximum diffraction. The phase grating which we have created with the standing acoustic wave in the germanium crystal can be composed of many adjacent thin strips, each having a width dL and each producing a different optical phase shift of amplitude $d\phi$, which varies in the y direction as $\sin(2\pi y/\lambda_s)$. In Fig. 7.4 such a situation is plotted.

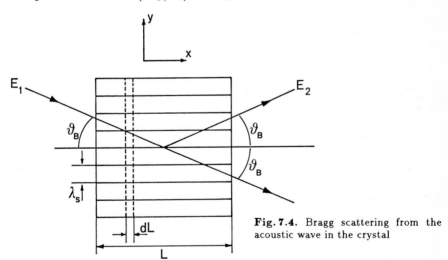

Fig. 7.4. Bragg scattering from the acoustic wave in the crystal

Now we consider only one thin strip of width dL. The incident light beam, which has a complex amplitude E_1, passes through such a thin grating at the Bragg angle. Since the index of refraction varies in the y direction, the wave number k_1 also varies in the y direction. Hence we can write the change of the wave number k_x in the x direction

$$\Delta k_x = k_1 \Delta n \cos \vartheta_B \sin \frac{2\pi y}{\lambda_s} \, , \qquad (7.8)$$

and for the change of the wave number k_y in the y direction

$$\Delta k_y = k_1 \Delta n \sin \vartheta_B \sin \frac{2\pi y}{\lambda_s} \ , \tag{7.9}$$

where Δn is the maximum amplitude of the refractive index. After passing through the thin strip with width dL the coordinates of the wave front of the incident beam have changed by dx and dy so that

$$E_1 \exp\left[i(\Delta k_x dx + \Delta k_y dy)\right] \ , \tag{7.10}$$

which equals

$$E_1 \exp\left[i\left(k_1 \Delta n \cos \vartheta_B \sin \frac{2\pi y}{\lambda_s} + k_1 \Delta n \sin \vartheta_B \sin \frac{2\pi y}{\lambda_s} \tan \vartheta_B\right)dL\right] \tag{7.11}$$

with $dx = dL$ and $dy = \tan \vartheta_B dL$. Equation (7.11) can be written as

$$E_1 \exp\left(\frac{ik_1 \Delta n \sin \frac{2\pi y}{\lambda_s}}{\cos \vartheta_B} dL\right) \ . \tag{7.12}$$

Because $[k_1 \Delta n \sin (2\pi y/\lambda_s)/\cos \vartheta_B]dL \ll 1$, we write for the light amplitude

$$E_1\left(1 + i\frac{k_1 \Delta n}{\cos \vartheta_B} \sin \frac{2\pi y}{\lambda_s} dL\right)$$

or alternatively

$$E_1 + E_1 \frac{k_1 \Delta n dL}{2\cos \vartheta_B}\left[\exp\left(\frac{i2\pi y}{\lambda_s}\right) - \exp\left(\frac{-i2\pi y}{\lambda_s}\right)\right] \ .$$

The last two terms on the right-hand side of this equation are upper and lower side bands, which diverge from the carrier E_1 due to the spatial frequency components $1/\lambda_s$ in the y direction. Only the upper side band persists since it constructively interferes with the upper side band emanating from adjacent strips because of the Bragg condition. The lower side band is eliminated by destructive interference. This means that if we want to know the total diffraction of the optical beam, we only have to integrate the upper new wave component. In our case

$$\int_0^L \frac{E_1 k_1 \Delta n}{2\cos \vartheta_B} e^{i2\pi y/\lambda_s} dL = E_2 \ . \tag{7.14}$$

The increase of the amplitude per unit of length of·the 'new' wave will be

$$\frac{d|E_2|}{dL} = \frac{|E_1|k_1 \Delta n}{2\cos \vartheta_B} \ . \tag{7.15}$$

The condition for the conservation of energy requires

$$|E_0|^2 = |E_1|^2 + |E_2|^2 \ . \tag{7.16}$$

Differentiation of this equation gives

$$|E_1|\frac{d|E_1|}{dL} = -|E_2|\frac{d|E_2|}{dL} \ . \tag{7.17}$$

From (7.15, 17) we obtain

$$\frac{d|E_1|}{dL} = \frac{-|E_2|k_1 \Delta n}{2\cos \vartheta_B} \ . \tag{7.18}$$

Using (7.15, 18) the equations

$$\frac{d^2|E_2|}{dL^2} + |E_2|\left(\frac{k_1 \Delta n}{2\cos \vartheta_B}\right)^2 = 0 \quad \text{and} \tag{7.19}$$

$$\frac{d^2|E_1|}{dL^2} + |E_1|\left(\frac{k_1 \Delta n}{2\cos \vartheta_B}\right)^2 = 0 \ . \tag{7.20}$$

are derived. Equations (7.19, 20) have simple solutions incorporating the conditions that $|E_2| = 0$ and $|E_1| = E_0$ for $L = 0$, yielding

$$|E_2| = E_0 \sin\left(\frac{k_1 \Delta n L}{2\cos \vartheta_B}\right) \quad \text{and} \tag{7.21}$$

$$|E_1| = E_0 \cos\left(\frac{k_1 \Delta n L}{2\cos \vartheta_B}\right) \ . \tag{7.22}$$

In terms of intensities we obtain

$$\frac{I_2}{I_0} = \sin^2\left(\frac{k_1 \Delta n L}{2\cos \vartheta_B}\right) \ , \tag{7.23}$$

$$\frac{I_1}{I_0} = \cos^2\left(\frac{k_1 \Delta n L}{2\cos \vartheta_B}\right) \ . \tag{7.24}$$

Since we consider a standing acoustic wave Δn has a time dependence given by

$$\Delta n = \Delta n_{\max} \cos\left(\omega_m t/2\right) \ . \tag{7.25}$$

The Bragg angle can be expressed in material parameters, according to (7.7),

$$\sin \vartheta_B = \frac{\lambda_1}{2\lambda_s} = \frac{\lambda_0 \nu_s}{2nV_s} \ ,$$

where $\lambda_0 = 10.6 \, \mu$m is the vacuum wavelength, $n = 4$ is the refractive index of germanium, ν_s is the modulation frequency, and V_s is the sound velocity. In the case that the acoustic wave is in the [111] direction of the germanium crystal with a velocity of 5.5×10^3 m/s, the modulation frequency is 40 MHz, the Bragg angle becomes $\vartheta_B = 0.552°$. Assuming that the crystal is a parallelepiped, the relation between the angles indicated in Fig. 7.2 is $\psi = 90° + \vartheta_R - \vartheta_B$, where ϑ_R is the Brewster angle in germanium. In the case of the present example $\psi = 103.5°$. The size of the crystal is $80 \times 40 \times 20$ mm^3. The transducer is a lithium niobate crystal ($36°$ rotated y cut), 1.5×1.5 cm^2, bounded to one surface of the Ge crystal, so that the resultant acoustic wave propagates in the [111] crystallographic direction. The transducer was bound to the Ge by heating the crystal to $50°$ C, dropping some phenylsalicylate onto the heated surface, applying the transducer and cooling the total assembly for one hour. The electrodes were attached to the transducer and the germanium crystal by means of a conducting silver paint.

Using a rf power oscillator that requires an input impedance of $50 \, \Omega$, an autotransformer system is installed to obtain the maximum power transfer. Depending on the required modulation depth the power should be of the order of several watts.

The attenuation of the transmitted beam, as given by (7.24), can be measured if we irradiate the modulated crystal under the Bragg condition with a cw CO_2 laser. The attenuation can be described in terms of modulation depth parameter to be discussed later on and defined as $\alpha_a = -\frac{1}{2}\ln(I_1/I_0)$. In Table 7.1 some results are given from measurements of the maximum values of the parameter α_a, i.e., $\Delta n = \Delta n_{max}$, as a function of the input power P of the transducer.

In the case of mode locking the attenuation is only a few percent. For the analytical treatment of mode locking it is advantageous to describe the transmitted field in terms of an absorption coefficient given by a complex

Table **7.1.** The maximum value of the modulation depth parameter as a function of the input power of the transducer

P[W]	α_a
2.2	0.04
3.4	0.21
5.0	0.40
6.5	0.85

index of refraction. Thus we represent E_1 by

$$E_1 = E_0 e^{-ik'z} ,$$ (7.26)

where k' can be expressed in terms of the complex susceptibility, $\chi = \chi' - i\chi''$ i.e.,

$$k' = k_1 \left(1 + \frac{\chi}{n^2}\right)^{1/2}$$ (7.27)

Since $|\chi' - i\chi''| \ll 1$ we write

$$k' = k_1 \left(1 + \frac{1}{2}\frac{\chi'}{n^2} - \frac{1}{2}i\frac{\chi''}{n^2}\right)$$ (7.28)

so that after traversing the modulator the amplitude of the field is given by

$$|E_1| = E_0 \exp\left(-\frac{1}{2n^2}k_1\chi''L\right) .$$ (7.29)

In order to calculate χ'' we compare (7.22) and (7.29). Using, in addition, (7.25) we find

$$\cos\left(\frac{Lk_1\Delta n_{\max}\cos(\omega_m t/2)}{2\cos\vartheta_B}\right) = \exp\left(-\frac{1}{2n^2}k_1\chi''L\right) .$$

Since the arguments of the cosine and exponential functions are small we can expand both functions, and obtain approximately

$$1 - \frac{L^2 k_1^2 \Delta n_{\max}^2 \cos^2(\omega_m t/2)}{8\cos^2\vartheta_B} = 1 - \frac{1}{2n^2}Lk_1\chi_M'' .$$

From the last result we obtain for the susceptibility

$$\chi_M'' = \Delta\chi''(1 + \cos\omega_m t) \quad \text{where}$$ (7.30)

$$\Delta\chi'' = \frac{n^2 Lk_1 \Delta n_{\max}^2}{8\cos^2\vartheta_B} .$$

The amplitude transmission through the modulator is then derived by substituting (7.30) into (7.29), i.e.,

$$|E_1(t)| = E_0 \exp\left[-\alpha_a \cos^2\left(\frac{\omega_m t}{2}\right)\right] , \quad \text{where}$$ (7.31)

$$\alpha_a = \frac{Lk_1\Delta\chi''}{n^2} .$$ (7.32)

7.2 Time-Domain Analysis of AM Mode Locking

In this section we shall follow the analysis of *Kuizenga* and *Siegman* [7.5]. Their treatment of the pulse forming by AM mode locking for a homogeneously broadened system starts with the assumption that the pulse is Gaussian and that necessary approximations to the line profile and intracavity modulator can be made to keep the pulse Gaussian in shape. For a homogeneously broadened line with a Lorentzian profile it can be shown that the complex gain factor (including the amplification of field amplitude and its phase shift) for the field after a round trip through the active medium is

$$G(\omega) = \exp\left(\frac{g}{1 + 2i(\omega - \omega_L)/\Delta\omega_N}\right) , \tag{7.33}$$

where e^g is the amplitude gain factor at line center ω_L, and $\Delta\omega_N$ is the line width at half maximum. The term $g[1 + 2i(\omega - \omega_L)/\Delta\omega_N]^{-1}$ describes the complex Lorentzian gain profile. It will be seen later that the frequency distribution of the pulse is much smaller than $\Delta\omega_N$. In this frequency region we may expand the line profile about the pulse center frequency ω_0 as

$$G(\omega) = \exp\left\{\frac{g}{1 + 2i\eta}\left[1 - \left(\frac{2i}{1 + 2i\eta}\right)\left(\frac{\omega - \omega_0}{\Delta\omega_N}\right)\right.\right.$$
$$\left.\left. - \frac{4}{(1 + 2i\eta)^2}\left(\frac{\omega - \omega_0}{\Delta\omega_N}\right)\right]^2\right\} , \tag{7.34}$$

where $\eta = (\omega_0 - \omega_L)/\Delta\omega_N$ is the shift of the pulse center frequency normalized to the line width. It is seen that the line profile has now become Gaussian in shape.

In the following we assume a Gaussian pulse with respect to time, having a uniform intensity profile over its cross section. When this pulse goes through the above described medium it will remain Gaussian. The same can be said for traversing the modulator. This is crucial for the present analysis. Such an optical pulse can be described most generally by

$$E(t) = E_0\exp(-\alpha t^2)\exp[i(\omega_0 t + \beta t^2)] , \tag{7.35}$$

where the term α determines the Gaussian envelope of the pulse and the term βt is the linear frequency shift during the pulse. A complex constant γ can be defined as

$$\gamma = \alpha - i\beta \tag{7.36}$$

so that

$$E(t) = E_0 \exp(-\gamma t^2) \exp(i\omega_0 t) . \tag{7.37}$$

The Fourier transform of this pulse is

$$E(\omega) = \sqrt{\frac{\pi}{\gamma}} E_0 \exp[-(\omega - \omega_0)^2/4\gamma] \ . \tag{7.38}$$

The ideal mode locking is when the pulse passes through the modulator at the instant of maximum transmission. Looking at (7.31) this occurs for each instant $\cos(\omega_m t/2) = 0$. With the assumption that the pulse is short compared to the modulation period and introducing for the modulator a time shift π/ω_m so that $\cos^2(\omega_m t/2)$ will be replaced by $\sin^2(\omega_m t/2)$ the transmission near its maximum can be approximated as

$$|E_1(t)| = E_0 \exp\left[-\alpha_a \left(\frac{\omega_m t}{2}\right)^2\right] \ . \tag{7.39}$$

For the more general case where the pulse passes through the modulator at the phase angle θ to the ideal case, the transmission in terms of a time relative to the moment of reaching this phase angle θ is then approximated by

$$|E_1(t)| = E_0 \exp\left[-\alpha_a \sin^2\theta - \alpha_a \sin 2\theta \left(\frac{\omega_m t}{2}\right)\right.$$
$$\left. - \alpha_a \cos 2\theta \left(\frac{\omega_m t}{2}\right)^2\right] \ . \tag{7.40}$$

It is seen that with the above approximation the transmission of the modulator is Gaussian so that the Gaussian pulse through the modulator will remain Gaussian.

7.2.1 Self-Consistency of the Ideal, Circulating Pulse

We now investigate the sequence of the pulse going through the active medium and the modulator. This is illustrated in Fig. 7.5. First, we consider the ideal situation where the pulse is on line center, i.e. $\omega_L = \omega_0$, or $\eta = 0$, and passes the modulator at minimum loss. If $E_1(t)$ is the pulse entering the active medium, then the Fourier transform of the pulse coming out is given by

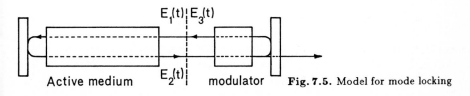

$E_1(t)$ | $E_3(t)$

$E_2(t)$

Active medium   modulator   Fig. 7.5. Model for mode locking

205

$$E_2(\omega) = G(\omega)E_1(\omega)$$

$$= E_0\sqrt{\frac{\pi}{\gamma}}\exp\left[g - 2ig\left(\frac{\omega - \omega_0}{\Delta\omega_N}\right) - 4g\left(\frac{\omega - \omega_0}{\Delta\omega_N}\right)^2 - \frac{(\omega - \omega_0)^2}{4\gamma}\right]$$

$$(7.41)$$

$$E_2(\omega) = E_0\sqrt{\frac{\pi}{\gamma}}\exp[g - A(\omega - \omega_0)^2]\exp[-iB(\omega - \omega_0)] , \qquad (7.42)$$

$$A = \frac{1}{4\gamma} + \frac{4g}{\Delta\omega_N^2} \qquad (7.43a)$$

and

$$B = \frac{2g}{\Delta\omega_N} . \qquad (7.43b)$$

Transforming into the time domain, the pulse becomes

$$E_2(t) = \frac{E_0}{2\sqrt{\gamma A}}\exp\left[g - \frac{(t - B)^2}{4A}\right]\exp(i\omega_0 t) . \qquad (7.44)$$

Next, we treat the effect of the modulator. The peak of the pulse goes through the modulator at the time $t = B$, and hence by using (7.39) the amplitude of the pulse apart from a phase shift is after a round trip

$$E_3(t) = E_2(t)\exp\left[-2\alpha_a\frac{\omega_m^2}{4}(t - B)^2\right] . \qquad (7.45)$$

Finally the total round trip through the resonator is completed by including an additional time delay $2L/c$ for the travel time and an effective amplitude reflectivity r of a mirror to include all losses in the cavity. The pulse is then given by

$$E_4(t) = rE_3(t - 2L/c) . \qquad (7.46)$$

To obtain a self-consistent solution, the envelope of the pulse must go through the modulator at the same modulation phase every time. Hence the total round-trip time T_m for the pulse is equal to $2\pi/\omega_m$. The self-consistency requirement now becomes

$$E_1(t - T_m)e^{-i\phi} = E_4(t) , \qquad (7.47)$$

where the phase angle ϕ is included to allow for a possible phase shift of the optical signal with respect to the pulse envelope. Using (7.37, 44, 45, and 46) the self-consistency yields

206

$$E_0 \exp[-\gamma(t - T_m)^2] \exp[i\omega_0(t - T_m)] \exp(-i\phi)$$

$$= \frac{rE_0}{2\sqrt{\gamma A}} \exp\left(g - \frac{(t - B - 2L/c)^2}{4A}\right)$$

$$\times \exp\left[-2\alpha_a \frac{\omega_m^2}{4}(t - B - 2L/c)^2\right] \exp[i\omega_0(t - 2L/c)] \; . \tag{7.48}$$

From this equation it follows that

$$T_m = 2L/c + B \; , \tag{7.49a}$$

$$\gamma = \frac{1}{4A} + \frac{1}{2}\alpha_a \omega_m^2 \; , \tag{7.49b}$$

$$e^{-i\phi} = \frac{r}{2\sqrt{\gamma A}} \exp(g + i\omega_0 B) \; . \tag{7.49c}$$

The last three equations together with (7.43) now essentially solve the problem. From these equations the desired analytical expressions for the modulation frequency, the pulse width, and the band width of the pulse can be obtained. First, we consider (7.49b). Using (7.43a) γ can be calculated as

$$\gamma = \frac{1}{4}\alpha_a \omega_m^2 \pm \frac{1}{2}\left[\frac{1}{4}(\alpha_a \omega_m^2)^2 + \frac{\alpha_a \omega_m^2 \Delta\omega_N^2}{8g}\right]^{1/2} \; . \tag{7.50}$$

We notice from (7.50) that γ is real. Thus $\gamma = \alpha$ and $\beta = 0$. Since α must always be positive we retain the positive sign in (7.50). The modulation frequency ω_m will be much less than the line width $\Delta\omega_N$ and for practical values of g and α_a it is then found that the second term under the square root sign of (7.50) is much larger than the first term. This means that α can be approximated to

$$\alpha = \frac{\omega_m \Delta\omega_N}{4}\sqrt{\frac{\alpha_a}{2g}} \; . \tag{7.51}$$

The pulse width, defined as the time between half-intensity points, can be obtained from (7.37) by using (7.51)

$$\tau_p = \frac{1}{\pi}(2\ln 2)^{1/2}\left(\frac{2g}{\alpha_a}\right)^{1/4}\left(\frac{1}{\nu_m \Delta\nu_N}\right)^{1/2} \; , \quad \text{where} \tag{7.52}$$

$$\nu_m = \frac{\omega_m}{2\pi} \quad \text{and} \quad \Delta\nu_N = \frac{\Delta\omega_N}{2\pi} \; .$$

The band width or spectral width $\Delta\nu_{pulse}$ of the pulse, defined as the frequency between half-power points of the pulse spectrum, can be obtained from (7.38). With (7.51) we find

$$\Delta\nu_{\text{pulse}} = (2\ln 2)^{1/2}\left(\frac{\alpha_a}{2g}\right)^{1/4}(\nu_m\Delta\nu_N)^{1/2} . \tag{7.53}$$

The frequency spacing between the modes ν_m can be derived from (7.49a). Substituting the value of B from (7.43b) we obtain

$$\nu_m = \left(\frac{2L}{c} + \frac{2g}{\Delta\omega_N}\right)^{-1} . \tag{7.54}$$

The term $2g/\Delta\omega_N$ is the dispersion or linear delay in the Lorentzian line. Finally, from (7.49c) the gain g can be deduced. The values of γ and A can be obtained from (7.43a, 51). However, it turns out that $1/4\gamma \gg 4g/\Delta\omega_N^2$ so that we may use $A = 1/4\gamma$. Then we find to a good approximation

$$g = -\tfrac{1}{2}\ln R , \tag{7.55}$$

where $R = r^2$ is the effective power reflectivity of a mirror and includes all losses. Thus, the gain is practically the same as for a cavity without modulator. This is understood from the fact that the pulse duration is much smaller than the modulator period and that the pulse center passes at the instant of full transmission. Equating the phase part of (7.49c) we find $\phi = 2g\omega_0/\Delta\omega_N$ which is some large angle. Its value does not affect any of the pulse parameters. We can neglect this angle.

7.2.2 Self-Consistency of the Circulating Pulse with Detuning

In the above analysis we considered the ideal case of a pulse being on line center $(\omega_0 = \omega_L)$ and passing through the modulator at the instant of maximum transmission. For these conditions the ideal modulation frequency is given by (7.54). However, if the modulation frequency is detuned from the ideal case the pulse might shift off line center and might not pass the modulator at the instant of minimum loss. Investigating this problem we again follow the above procedure, except that we now substitute for $G(\omega)$ from (7.34) and we use (7.40) for the transmission at a phase angle θ. Then, from the self-consistency condition the detuned frequency, the gain, the pulse width, etc., can be obtained.

When this pulse passes through the active medium and modulator we have

$$E_3(t) = \frac{E_0}{2\sqrt{\gamma A'}}\exp\left(g' - \frac{(t - B')^2}{4A'}\right)\exp(i\omega_0 t)$$
$$\times \exp\{-2\alpha_a[\sin^2\theta + \tfrac{1}{2}w_m(t - B')\sin 2\theta$$
$$+ \tfrac{1}{4}w_m^2(t - B')^2\cos 2\theta]\} \tag{7.56}$$

where

$$A' = \frac{1}{4\gamma} + \frac{4g}{\Delta\omega_N^2} \frac{1}{(1 + 2i\eta)^3} \; , \tag{7.57a}$$

$$B' = \frac{2g}{\Delta\omega_N} \frac{1}{(1 + 2i\eta)^2} \; , \tag{7.57b}$$

$$g' = \frac{g}{1 + 2i\eta} \; . \tag{7.57c}$$

If we substitute $K = \frac{1}{4}A' + \frac{1}{2}\alpha_a\,\omega_m^2\cos 2\theta$ and include the round-trip time $2L/c$ and effective reflectivity, the pulse after one round trip can be written as

$$E_4(t) = \frac{rE_0}{2\sqrt{\gamma A'}}\exp\left[g' - K\left(t - B' + \frac{\alpha_a\omega_m\sin 2\theta}{2K} - \frac{2L}{c}\right)^2\right]$$

$$\times \exp\left[i\omega_0\left(t - \frac{2L}{c}\right)\right]$$

$$\times \exp\left(-2\alpha_a\sin^2\theta + \frac{(\alpha_a\omega_m\sin 2\theta)^2}{4K}\right) \; . \tag{7.58}$$

The self-consistency conditions now become

$$T_m = \frac{2L}{c} + B' - \frac{\alpha_a\omega_m\sin 2\theta}{2K} \; , \tag{7.59a}$$

$$\gamma = K = \frac{1}{4A'} + \frac{1}{2}\alpha_a\omega_m^2\cos 2\theta \; , \tag{7.59b}$$

$$e^{-i\phi} = \frac{r}{2\sqrt{\gamma A'}}\exp\left[g' + i\omega_0\left(B' - \frac{\alpha_a\omega_m\sin 2\theta}{2K}\right)\right]$$

$$\times \exp\left(-2\alpha_a\sin^2\theta + \frac{\alpha_a^2\omega_m^2\sin^2 2\theta}{4K}\right) \; . \tag{7.59c}$$

From (7.59a) we must obtain a real value of T_m. This is found for $\eta = 0$. From (7.59b) we then find that γ is real, or $\gamma = \alpha$. The pulse width and band width will be the same as in the previous case with no detuning if α_a is replaced by $\alpha_a\cos 2\theta$ in (7.52, 53). This means that the pulses get longer with detuning for both positive and negative detuning. The modulation frequency is now obtained from (7.59a) by substituting the values of $K = \alpha$ and $B' = B$, i.e.,

$$\nu_m = \left(\frac{2L}{c} + \frac{2g}{\Delta\omega_N} - \frac{2\alpha_a\sin 2\theta}{\Delta\omega_N}\sqrt{\frac{2g}{\alpha_a\cos 2\theta}}\right)^{-1} \; . \tag{7.60}$$

For a given modulation frequency ν_m the phase angle θ for which the pulse passes the modulator can be calculated by means of the last equation. The second and third terms in (7.60) are very small compared to the first one. We may therefore approximate (7.60) by

$$\nu_m = \frac{c}{2L} - \frac{c^2}{2L^2}\frac{g}{\Delta\omega_N} + \frac{c^2}{2L^2}\frac{\alpha_a \sin 2\theta}{\Delta\omega_N}\sqrt{\frac{2g}{\alpha_a \cos 2\theta}} \ . \qquad (7.61)$$

If the detuning is defined as

$$\Delta\nu_t = \frac{c}{2L} - \nu_m - \frac{c^2}{2L^2}\frac{g}{\Delta\omega_N} \ , \qquad (7.62)$$

so that by using (7.61)

$$\Delta\nu_t = -\frac{c^2}{2L^2}\frac{\alpha_a \sin 2\theta}{\Delta\omega_N}\sqrt{\frac{2g}{\alpha_a \cos 2\theta}} \ , \qquad (7.63)$$

it will be zero for $\sin 2\theta = 0$, i.e., the pulse passes the modulator at minimum loss.

Calculating the gain g from equation (7.59c) we note that the term $(\alpha_a \omega_m \sin 2\theta)^2/4K$ is much smaller than g, and hence we find

$$g = \tfrac{1}{2}\ln(1/R) + 2\alpha_a \sin^2\theta \ . \qquad (7.64)$$

Comparing the last equation with (7.55) we note that the increase in gain is due to the absorption loss in the modulator.

7.3 Frequency-Domain Analysis of AM Mode Locking

The analysis in the frequency domain [7.6] starts with a field distribution of axial modes and the assumption that this distribution is Gaussian over the mode frequencies. This treatment is more general and deals also with real nonuniform (Gaussian) beam profiles. The standing field can then be represented by

$$E(z,t) = \sum_n \psi_0 E_n \sin k_n z \exp[i(\omega_n t + \phi_n)] \ , \qquad (7.65)$$

where $\psi_0 \sin k_n z$ is the spatial part of an axial laser mode, E_n is the field of the mode with frequency ω_n and phase ϕ_n, and k_n is the wave number of the mode.

The electric field polarizes the active medium. This can be described in terms of the susceptibility $\chi = \chi' - i\chi''$. The polarization of the active medium will be

$$P_A = \sum_n \psi_0 E_n \varepsilon_0 (\chi_n' - i\chi_n'') \sin k_n z \exp[i(\omega_n t + \phi_n)] \ . \tag{7.66}$$

The absorption caused by the modulator can, as we have discussed in Sect. 7.1, also be described in terms of the susceptibiliy. The polarization of the modulator is then given by

$$P_M = -i\varepsilon_0 \chi_M'' E(z,t) \ , \tag{7.67}$$

where χ_M'' is given by (7.30). Substituting (7.65) into (7.67) we get

$$P_M = \sum_n \psi_0 E_n \varepsilon_0 (-i\chi_M'') \sin k_n z \exp[i(\omega_n t + \phi_n)] \ . \tag{7.68}$$

For the stationary process of mode locking the frequency spacing between the modes is constant and equal to the modulation frequency ω_m, so that $\omega_n - \omega_{n-1} = \omega_m$. Using (7.30) the polarization can then be written as

$$\begin{aligned}
P_M = &- i\Delta\chi'' \varepsilon_0 \psi_0 \sum_n \{ E_n \sin k_n z \exp[i(\omega_n t + \phi_n)] \\
&+ \tfrac{1}{2} E_{n-1} \sin k_{n-1} z \exp[i(\omega_n t + \phi_{n-1})] \\
&+ \tfrac{1}{2} E_{n+1} \sin k_{n+1} z \exp[i(\omega_n t + \phi_{n+1})] \}\ .
\end{aligned} \tag{7.69}$$

The relation between the electric field E and the polarization P can be derived from the Maxwell equations. For a charge-free medium the well-known relation is given by

$$\nabla^2 E - \mu_0 \varepsilon_0 \frac{\partial^2 E}{\partial t^2} - \mu_0 \sigma \frac{\partial E}{\partial t} = \mu_0 \frac{\partial^2 P}{\partial t^2} \ . \tag{7.70}$$

The cavity losses like outcoupling, scattering, etc., are taken into account by the introduction of the conductivity σ. These losses are considered per unit volume. We substitute (7.65, 66, 69) into (7.70) and realize that $dE_n/dz \ll k_n E_n$, leading to

$$\begin{aligned}
\sum_n \psi_0 \{ &- E_n k_n^2 \sin k_n z \exp[i(\omega_n t + \phi_n)] \\
&+ \mu_0 \varepsilon_0 E_n \sin k_n z (\omega_n + \dot{\phi}_n)^2 \exp[i(\omega_n t + \phi_n)] \\
&- i\mu_0 \sigma E_n \sin k_n z (\omega_n + \dot{\phi}_n) \exp[i(\omega_n t + \phi_n)] \} \\
= -i\mu_0 \varepsilon_0 \Delta\chi'' \psi_0 \sum_n \{ &- E_n \sin k_n z (\omega_n + \dot{\phi}_n)^2 \exp[i(\omega_n t + \phi_n)] \\
&- \tfrac{1}{2} E_{n-1} \sin k_{n-1} z (\omega_n + \dot{\phi}_{n-1})^2 \exp[i(\omega_n t + \phi_{n-1})] \\
&- \tfrac{1}{2} E_{n+1} \sin k_{n+1} z (\omega_n + \dot{\phi}_{n+1})^2 \exp[i(\omega_n t + \phi_{n+1})] \} - \mu_0 \varepsilon_0 \psi_0 \\
\sum_n \{ &(\chi_n' - i\chi_n'') E_n \sin k_n z (\omega_n + \dot{\phi}_n)^2 \exp[i(\omega_n t + \phi_n)] \}\ . \tag{7.71}
\end{aligned}$$

Before we continue to solve the above set of equations we shall introduce the following abbreviations. The frequency of the empty-space resonator is

$$\Omega = k_n(\varepsilon_0\mu_0)^{-1/2} \ . \tag{7.72}$$

The single-pass power loss factor S or the effective power reflectivity R of a mirror is related to σ by

$$S = \tfrac{1}{2}\ln(1/R) = \frac{\sigma L'}{c\varepsilon_0} \ , \tag{7.73}$$

where L' is the length of the active medium. The resonances of the empty space resonator are at

$$\Omega_n = \Omega_0 + n\frac{\pi c}{L} \ , \tag{7.74}$$

where L is the length of the resonator, and Ω_0 is a frequency close to the line center. The integer n runs from negative to positive values. The mode frequency is equal to $\omega_n + \dot{\phi}_n$, where we define ω_n as

$$\omega_n = \Omega_0 + n\omega_m \ . \tag{7.75}$$

The frequency difference $\Delta\omega$ between the empty-space mode spacing and the real mode spacing is given by

$$\Delta\omega = \frac{\pi c}{L} - \omega_m \ . \tag{7.76}$$

The single-pass power gain of the nth mode through the active medium is

$$g_n = -\frac{\omega_n\chi_n'' L'}{c} = -k_n L' \chi_n'' \ . \tag{7.77}$$

Note that the single-pass power gain is equal to the round-trip amplitude gain. Similarly the phase retardation of the nth mode during a full round trip through the medium is

$$\psi_n = k_n L' \chi_n' \ . \tag{7.78}$$

With the above abbreviations we solve (7.71) by multiplying both sides by $\psi_0 \sin k_{n_0}z$ and by integrating it over the volume of the resonator. We obtain with $\dot{\phi}_n \ll \omega_n$ for $n = n_0$

$$E_{n+1}e^{i\phi_{n+1}} + E_{n-1}e^{i\phi_{n-1}}$$
$$= E_n e^{i\phi_n}\left[-i\left(\frac{2L}{c\alpha_c}\dot{\phi}_n - \frac{2L}{c\alpha_c}n\Delta\omega + \frac{\psi_n}{\alpha_c}\right) + \frac{g_n}{\alpha_c} - \frac{S}{\alpha_c} - \frac{\alpha_a}{\alpha_c}\right] \ , \tag{7.79}$$

where

$$\alpha_a = \frac{2\omega n}{c} \int_0^a \Delta\chi'' \sin^2 k_n z \, dz \; , \tag{7.80a}$$

$$\alpha_c = \frac{\omega n}{c} \int_0^a \Delta\chi'' \sin k_n z \sin k_{n\pm1} z \, dz \; . \tag{7.80b}$$

In the normal case that the absorption of the modulator has negligible variation over the distance of one wavelength of the optical beam one obtains from (7.80a)

$$\alpha_a = \frac{\omega n}{c} \int_0^a \Delta\chi'' dz \; , \tag{7.81a}$$

which is the maximum single-pass power loss of the modulator. In the above analysis the refractive index n of the modulator has been taken unity; otherwise the calculated value of α_a has to be divided by n. Taking into account that $a \ll L$ we readily find from (7.80b and 81a) that at the position of the mode locker $(z = L)$

$$\alpha_c = -\tfrac{1}{2}\alpha_a \; . \tag{7.81b}$$

Next we look for an analytic solution of (7.79) for the case of a homogeneously broadened Lorentzian line. Since the frequency width of the pulse is small compared with the line width we may expand the gain and the phase retardation of such a profile. We shall see later that the frequencies of the modes are all near the line center. This means that we can restrict the expansion of the gain and phase to, respectively, the second- and the first-order terms to a very good approximation. This can be done in terms of the mode number n. Thus, the gain and the phase retardation for the nth mode will be given by

$$g_n = g_0 + n g_1 + \tfrac{1}{2}n^2 g_2 \; , \tag{7.82a}$$
$$\psi_n = \psi_0 + n\psi_1 \; . \tag{7.82b}$$

After substituting (7.82) into (7.79) it can be argued [7.7] that as long as the term with a quadratic dependence on the mode number is much greater than the linear terms the electrical field amplitudes assume a Gaussian form.

A general Gaussian distribution of the modes (centered at $n = 0$) is then given by

$$E_n e^{i\phi_n} = E_0 \exp(-\alpha n^2 + i\vartheta_1 n) \; , \tag{7.83}$$

where the constant α may be complex, $\alpha = \alpha_1 + i\alpha_2$, and θ_1 must be real, otherwise the mode with $n = 0$ would not have the strongest field. From this we obtain

$$E_{n+1}e^{i\phi_{n+1}} + E_{n-1}e^{i\phi_{n-1}}$$
$$= E_n e^{i\phi_n}[\exp(-2n\alpha - \alpha + i\vartheta_1) + \exp(2n\alpha - \alpha - i\vartheta_1)] \ . \qquad (7.84a)$$

The arguments of the expressions on the right-hand side of (7.84a) are small, as will be seen later. Therefore, we expand the last expression up to second order, and obtain

$$E_{n+1}e^{i\phi_{n+1}} + E_{n-1}e^{i\phi_{n-1}} = E_n e^{i\phi_n}$$
$$\times (2 - 2\alpha + \alpha^2 + 4n^2\alpha^2 - \vartheta_1^2 - 4n\alpha\vartheta_1 i) \ . \qquad (7.84b)$$

By comparing (7.79) and (7.84b) and making use of (7.82) we equate terms of the same order of n for the real and imaginary parts. Then we find the six equations

$$2 - 2\alpha_1 + \alpha_1^2 - \alpha_2^2 - \vartheta_1^2 = (g_0 - S - \alpha_a)/\alpha_c \ , \qquad (7.85a)$$

$$2\alpha_2 - 2\alpha_1\alpha_2 = \frac{(2L\dot{\phi}_n/c + \psi_0)}{\alpha_c} \ , \qquad (7.85b)$$

$$4\alpha_2\vartheta_1 = g_1/\alpha_c \ , \qquad (7.85c)$$

$$-4\alpha_1\vartheta_1 = \frac{(2L\Delta\omega/c - \psi_1)}{\alpha_c} \ , \qquad (7.85d)$$

$$4\alpha_1^2 - 4\alpha_2^2 = \tfrac{1}{2}g_2/\alpha_c \ , \qquad (7.85e)$$

$$8\alpha_1\alpha_2 = 0 \ . \qquad (7.85f)$$

From the last two equations we conclude that $\alpha_2 = 0$ because the right-hand side of (7.85e) is always positive.

By solving (7.85a–e) we shall substitute for the detuning radial frequency $\Delta\omega_t = \Delta\omega - c\psi_1/2L$. We note that $\Delta\omega_t/2\pi = \Delta\nu_t$ is consistent with (7.62) and obtain finally

$$\dot{\phi}_n = \dot{\phi} = -\frac{c}{2L}\psi_0 \ , \qquad (7.86a)$$

$$\alpha_1 = \left(\frac{g_2}{8\alpha_c}\right)^{1/2} \ , \qquad (7.86b)$$

$$\vartheta_1 = -\sqrt{2}\frac{L\Delta\omega_t}{c(\alpha_c g_2)^{1/2}} \ , \qquad (7.86c)$$

$$g_1 = 0 \ , \tag{7.86d}$$

$$g_0 - S = \tfrac{1}{2}\sqrt{2}(g_2\alpha_c)^{1/2} - \frac{2L^2}{c^2}\frac{\Delta\omega_t^2}{g_2} + \tfrac{1}{8}g_2 \ . \tag{7.86e}$$

From this final result it will be clear that the ideal mode-locking frequency is as expected for $\Delta\omega_t = 0$, so that $\Delta\omega_t$ is the frequency-shift from the ideal situation. In that case the right-hand side of (7.86e) can be neglected so that $g_0 = S = (1/2)\ln(1/R)$.

The pulse form can be obtained by summing the fields over all modes. Thus the total field E_{tot} is given by

$$E_{\text{tot}} = \sum_n E_n \exp[\mathrm{i}(\varOmega_0 + n\omega_m + \dot\phi)t + \mathrm{i}\phi_n] \ . \tag{7.87}$$

Making use of (7.83) we get

$$E_{\text{tot}} = E_0[\mathrm{i}(\varOmega_0 + \dot\phi)t] \sum_n \exp(-\alpha n^2 + \mathrm{i}\vartheta_1 n + \mathrm{i}n\omega_m t) \ . \tag{7.88}$$

In order to calculate the radiation intensity we have to take $E_{\text{tot}} \cdot E_{\text{tot}}^*$. The summation over the modes is straightforward. For simplicity, we increase the time ϑ_1/ω_m. In other words, we substitute $t' = t + (\vartheta_1/\omega_m)$ and omit the prime in the final result. We then find with $\alpha = \alpha_1$

$$E_{\text{tot}} \cdot E_{\text{tot}}^* = C^2 \exp\left(-\frac{\omega_m^2 t^2}{2\alpha_1}\right) \ , \qquad \text{where} \tag{7.89}$$

$$C = \left| \sum_n \exp\left[-\alpha_1\left(n - \frac{\mathrm{i}\omega_m t}{2\alpha_1}\right)^2\right] \right| \ . \tag{7.90}$$

For small values of α_1 it is found that C is independent of t. The pulse width, defined as the time between half-intensity points, is then given by

$$\tau_p = 2(2\alpha_1 \ln 2)^{1/2}\omega_m^{-1} \ . \tag{7.91}$$

By means of (7.86b) this becomes

$$\tau_p = 2\left(\frac{g_2}{2\alpha_c}\right)^{1/4}(\ln 2)^{1/2}\omega_m^{-1} \ . \tag{7.92}$$

We note that the delay of the pulse $\Delta\tau$ with respect to the ideal case of minimum absorption of the modulator is thus ϑ_1/ω_m. This corresponds to a phase angle of a modulator from the ideal case, see (7.40), $\theta = \Delta\tau\omega_m/2$ so that $\theta = \vartheta_1/2$. From (7.86c) and also using (7.97) we then calculate for the detuning frequency

$$\Delta\nu_t = -\frac{c^2\theta}{L^2\Delta\omega_N}\sqrt{2g\alpha_a} \ . \tag{7.93}$$

Comparing this with (7.63) we have the same result for small values of θ. It turns out that mode locking occurs only in a small range of $\Delta\nu_t$, with $2L\Delta\nu_t/c$ only a few percent, so that θ is indeed small.

At the line center $g_1 = 0$, so that according to (7.86d) our expansion was made near the line center. The strongest mode does not necessarily coincide with the line center, because the frequency is also determined by cavity resonances. However, it has the resonance frequency closest to the line center, or

$$-\frac{\omega_m}{2} < \xi < \frac{\omega_m}{2} \, , \tag{7.94}$$

where ξ is the shift of the frequency of the dominant mode from the line center. Then, the single-pass intensity gain for the nth mode with frequency $\omega_n + \dot{\phi}$ can for a Lorentzian line be represented by

$$g_n(\omega_n + \dot{\phi}) = g\left[1 + 4\left(\frac{\xi + n\omega_m}{\Delta\omega_N}\right)^2\right]^{-1} \, , \tag{7.95}$$

where $\Delta\omega_N$ is the line width in radians per second, g is the single-pass intensity gain for radiation with frequency that coincides with the central frequency of the line.

If we now compare the expansion of (7.82a) with (7.95) and use the condition that $\Delta\omega_N \gg \omega_m$, we find for the order terms

$$g_0 \simeq g \, , \tag{7.96}$$

$$g_2 \simeq -8\left(\frac{\omega_m}{\Delta\omega_N}\right)^2 g \, . \tag{7.97}$$

Using the latter quantity and (7.81b) we derive from (7.92) the pulse width as

$$\tau_p = \frac{1}{\pi}(2\ln 2)^{1/2}\left(\frac{2g}{\alpha_a}\right)^{1/4}\left(\frac{1}{\Delta\nu_N\nu_m}\right)^{1/2} \, , \tag{7.98}$$

where $\nu_m = \omega_m/2\pi$ and $\Delta\nu_N = \Delta\omega_N/2\pi$.

Comparing (7.98) with (7.52) we see that the analysis in the frequency domain gives the same pulse width as found in the time-domain analysis. The pulse width as a function of the modulation depth α_a is plotted in Fig. 7.8 for $g = \frac{1}{2}\ln(1/R)$ with $R = 0.5$ and a $1:1:3 = CO_2:N_2:He$ gas mixture at one atmosphere. The mode spacing ν_m is equal to 80 MHz. It is seen that the pulse width depends very weakly on the gain and the modulation depth α_a of the modulator. From the pulse-width equation one might expect, at first glance, that the pulse width does not depend on the detuning $\Delta\omega_t$ from the ideal situation. This, however, is not the case, because the phase losses obtained by detuning must be compensated by a higher gain

g and hence the pulse width increases with the detuning. This dependence on $\Delta\omega_t$ turns out to be very weak and the effect of detuning is much more a decrease of pulse power than an increase in pulse width.

The frequency width $\Delta\nu_{\text{pulse}}$ of the light pulse is defined as the frequency between half-power points of the mode spectrum. From (7.81b, 83, 86b, 97) we find

$$\Delta\nu_{\text{pulse}} = (2\ln 2)^{1/2}\left(\frac{\alpha_a}{2g}\right)^{1/4}(\Delta\nu_N\nu_m)^{1/2} , \qquad (7.99)$$

which is the same as (7.53).

We shall now go back to (7.82) and justify the expansions used. Therefore, we shall compare $\Delta\nu_{\text{pulse}}$ with $\Delta\nu_N$ and obtain from (7.99)

$$\frac{\Delta\nu_{\text{pulse}}}{\Delta\nu_N} = (2\ln 2)^{1/2}\left(\frac{\alpha_a}{2g}\right)^{1/4}\left(\frac{\nu_m}{\Delta\nu_N}\right)^{1/2} .$$

In a typical example at atmospheric pressure we calculate a line width of about 4 GHz.

For $\nu_m = 80\,\text{MHz}$ and $(\alpha_a/g)^{1/4}$ about 0.5 we have

$$\frac{\Delta\nu_{\text{pulse}}}{\Delta\nu_N} < \frac{1}{10} .$$

This means that the expansions used in (7.82) do not deviate more than one percent from the exact expressions like (7.95).

7.4 Experimental Investigations of AM Mode Locked Systems

The onset of laser radiation is a random process. Radiation noise is amplified by the medium, and frequency components that are resonant will reach strong field strengths whereas the non-resonant noise radiation remains at a low level due to destructive interference. The build-up of the resonant radiation can be much slower than the gain build-up. This means that, especially when a laser operates at a high pump level, the gain may reach a value far above threshold before gain saturation by the stimulated emission appears. In this unsaturated gain regime the electric fields of the individual modes are independent of each other. Later on during the saturating stage of the fields the strong coupling of the modes in the homogeneously broadened medium will result in the survival of the strongest mode.

In the presence of a mode locker tuned near the axial mode spacing the situation is completely different. The individual modes in the unsaturated regime will now produce side bands that are resonant in the cavity. The phases of the side bands are related to the phase of the modulator. At a resonance frequency side bands produced by various modes interfere with the cavity mode at that frequency. This results in a coupling of modes and the final field consists of a number, depending on the strength of coupling by the modulator, of equally spaced frequency components all oscillating in phase. The output appears in the form of a train of pulses. Using a Ge crystal as an acousto-optic modulator pulse durations of 0.5–1 ns have been reported [7.8].

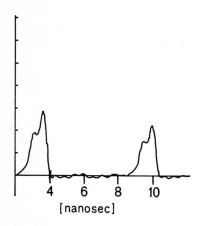

Fig. 7.6. Experimental observations of pulses with substructures in an AM mode locked system. The pulse intensity (in arbitrary units) is plotted as a function of time. The two pulses seperated by about 11 ns are taken from the central part of the mode locked pulse train

[nanosec]

It may happen that during the build-up time more than one set of modes with locked phases appears in the cavity. This gives rise to the appearance of substructures on the individual pulses in the pulse train. These substructures, as for example shown in Fig. 7.6, are random and appear especially when the laser operates at high gain, low modulation depth of the mode locker or when the number of round trips is insufficient due to a short gain period and a long cavity. To circumvent this problem one might lower the pressure of a pulsed CO_2 laser so that a narrower gain profile is obtained. The probability that the laser starts on more than one mode will then be reduced, but the disadvantage is a longer pulse width and a reduced output energy. Another method may be to operate the laser just above threshold. This, however, will also lower the output energy of the generated pulses. Intense and stable pulses are feasible if the field is not initiated by noise but by the injection of a well-defined field into the resonator. This can be single-mode radiation or a short pulse with low intensity but with a well-defined shape. These start signals could be generated internally or, by means of injection, externally. The injection of a short pulse into the laser cavity of an AM mode-locked laser is very difficult because of synchronization of the external pulse-generating device with the active modulator, whereas external injection of a cw laser signal into the cavity is relatively easy [7.9].

Stabilzation without injection is also feasible by applying an extra gain medium with narrow bandwidth into the cavity [7.10]. In the following we shall discuss both stabilization techniques: the addition of a narrow-band gain medium and the injection of continuous monochromatic radiation.

7.4.1 Stabilization of an AM Mode Locked TEA Laser with an Intracavity Low-Pressure CO_2-Laser Amplifier

It has been shown that the addition of a low-pressure gain cell to the TEA laser cavity having a bandwidth of less than the longitudinal mode spacing can alter the mode spectrum of a pulsed TEA CO_2 laser from a multi-longitudinal mode to a single-mode oscillation [7.11].

Two modes of operation of this, so-called, hybrid laser can be distinguished: the gain of the cw section is above threshold, so that the laser already generates single-mode cw radiation before the TEA section is fired, or the cw gain is chosen below threshold. In the first case the oscillation starts from single-mode radiation with a relative high photon density compared to the noise level. As a result, gain saturation of the TEA section will appear at a relatively early stage where the gain has not yet reached its maximum value. Because of this smooth start the giant pulse due to the gain switch will show a longer duration and a lower intensity than what is usually observed with TEA CO_2-laser systems having no additional low-pressure gain. For the low-pressure gain below threshold the pulse starts, relative to the gain rise, later than for the case above threshold and consequently the intensity is higher. For the stabilization of intense mode-locked pulses the low-pressure gain below threshold is therefore more preferable, and will be described [7.9,10].

Due to the additional gain of the low-pressure section near the line center the modes in the center area of the line profile are earlier and more amplified than the others when the gain of the total laser system passes threshold. If the mode spacing is larger than the width of the low-pressure gain profile only the center mode is favored. Therefore, due to mode competition, almost all the energy of the TEA section is directed into this mode.

As a result the output will appear in the form of a gain switched pulse with almost the same intensity and duration as in the case of non-hybrid TEA CO_2 lasers, but now with one single axial mode. When the cavity length is properly tuned so that this mode frequency coincides with the center of the line, the output will be free from any fluctuations due to mode-beating. This in contrast to the output of the usual non-hybrid TEA CO_2 laser. By detuning the cavity so that the oscillating mode does not coincide with the line-center, frequency chirping during the output pulse of the hybrid CO_2 laser can be observed, which is due to the changing dispersion of the TEA section. This effect can be measured by heterodyning the output of the hybrid laser with a stable cw CO_2 laser [7.12].

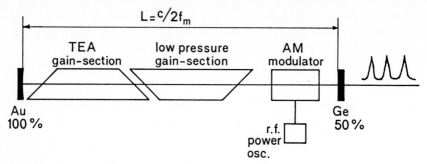

Fig. 7.7. Experimental set-up of a hybrid AM mode locked TEA laser

If we now add the AM mode locker to the hybrid laser, the laser will also start to oscillate on one mode near the line center at the moment the gain passes threshold. But, due to the generation of side bands, adjacent longitudinal modes will be created that are sufficiently stimulated to participate successfully in the mode-competition process, provided that the modulation frequency is equal to the mode spacing. In this way reliable mutual phase locking of all participating modes is possible, because it is ascertained that the laser starts from one longitudinal mode. Figure 7.7 shows the experimental configuration schematically. Besides the observation of very reproducible mode-locking phenomena even at very low modulation depth of the acousto-optic modulator, another interesting technique with the hybrid, AM mode-locked, CO_2 laser is obtained. Namely, the width of the pulses can be easily varied from about 1 to about 4 ns, by just changing the rf power to the modulator. This effect is due to the influence of the small bandwidth of the low-pressure gain section on the pulse width. For low modulation the coupling is weak and during the gain period less modes are forced to oscillate than in the case of stronger modulation. If the gain was present for a longer period, more modes would finally oscillate and the pulse would be narrower. The observed pulses can, therefore, not be considered as the steady state for the coupling process and consequently the steady-state analysis of Sect. 7.2 or 7.3 cannot, in general, predict the pulse width for hybrid systems.

Experimental results of an AM mode-locked system as a function of the modulation depth α, are shown for various conditions in Fig. 7.8. The cavity consists of one total reflector with a curvature of 3000 mm and a flat outcoupling mirror with 50 % reflectivity. The modulation frequency of the modulator is 40 MHz, so that the mode spacing is 80 MHz. The TEA system of the type described in Chap. 6 has a discharge volume of $0.5 \times 0.5 \times 30 \, \text{cm}^3$ and a $1:1:3 = CO_2 : N_2 : He$ gas mixture. The low-pressure section operating below threshold has a discharge length of 55 cm. Both discharges are sealed with ZnSe Brewster windows. For all experiments the pulse with the

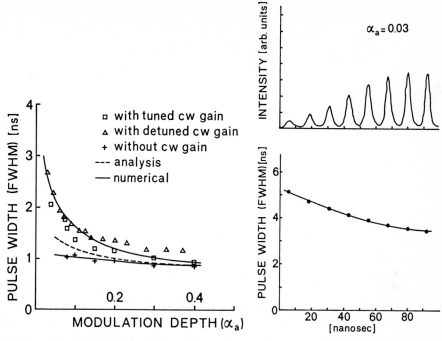

Fig. 7.8. The pulse width is plotted versus the modulation depth for both a hybrid and a normal AM mode-locked system. The width of the most intense pulse in the train is plotted. The straight line represents numerical results. The broken line is calculated according to the analysis (7.98)

Fig. 7.9. Experimental observation of the pulse compression during the pulse train of an AM mode locked hybrid CO_2 laser

highest intensity in the pulse train was selected. The modulation depth as a function of high-frequency power was measured by observing the response on a smooth 200 ns long CO_2-laser pulse from a single-mode hybrid TEA laser.

It is seen in Fig. 7.8 that in the absence of the narrow-band gain medium the pulses are narrower. During the transient of the pulse train after a discharge pulse there is pulse compression that can be observed. In Fig. 7.9 this pulse compression is shown for a modulation depth $\alpha_a = 0.03$. The measurements were done with a photondrag detector and a Tektronix transient digitizer. It is obvious that laser parameters such as gas composition, low-pressure gain, and mirror reflectivity have a direct influence on the efficiency of the mode locking. The measured maximum pulses in Fig. 7.8 are not necessarily band-width limited because the modulation process of the pulse may not have reached the steady state. Apparently without a narrow-band medium the pulse with maximum energy has been more effectively modulated than in the case of an active narrow-band gain medium. This phenomenon will be further treated in Sect. 7.4.3.

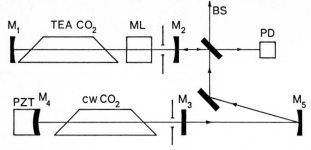

Fig. 7.10. Experimental configuration of a stabilized mode locked system by means of injected cw radiation. M_1 and M_4 are total reflectors with curvatures of 5000 and 3000 mm, respectively; ML is a modulator; M_2 is an outcoupling mirror of 50 % with radius of 2500 mm; M_3 is a plane-parallel outcoupling mirror with 80 % reflectivity. The tuning is obtained by the piezo-electric transducer (PZT). The detection is via a photon-drag detector (PD) [7.9]

7.4.2 Stabilization of an AM Mode-Locked TEA Laser with Injection of Continuous Radiation

The experimental set-up for injection of cw radiation [7.9] is shown schematically in Fig. 7.10. The curvatures of the mirrors of the mode-locked system are 5000 mm for the total reflector and 2500 mm for the outcoupling mirror with 50 % reflection. The modulator and the TEA laser are described in the previous section.

The low-pressure, sealed-off cw CO_2 laser runs on one single axial mode, with an output power of about 1 W. The frequency of this radiation could be tuned by changing the resonator length with a piezo electric transducer. Using mode-matching optics and a 50 % beam splitter the cw radiation was injected into the TEA laser cavity through the outcoupling mirror. The output of the TEA laser consists of a train of nanosecond pulses that were measured with a photon drag detector and a Tektronix transient digitizer. The total system rise-time was estimated to be about 700 ps.

With the mode locker switched off the cw laser was tuned to achieve single longitudinal mode operation of the TEA laser. This was shown by a smooth output pulse that was free from any mode-beating phenomena. Typical output pulses on a 200 ns/div time scale are shown in Fig. 7.11. Note the shorter build-up time of the gain switched pulse as the cw intensity is increased from zero to a few watts. The oscilloscope was triggered externally by the light produced by the spark gap of the TEA discharge.

With the mode locker switched on and the cw laser properly tuned, as described above, stable short pulses were observed even at low values of the modulation depth. In the upper part of Fig. 7.12 two pulses of the train are illustrated for four successive shots in the absence of cw injection. The lower part of Fig. 7.12 displays the same observation under the same conditions, but with cw injection.

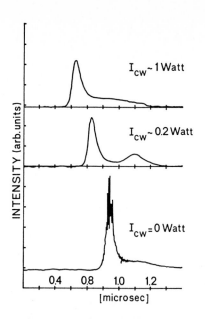

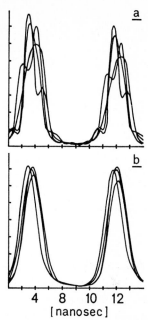

Fig. 7.11. Typical gain switched output pulse of an injection locked TEA laser without modulator [7.9]

Fig. 7.12. Output pulses of an AM mode-locked system without (a) and with (b) cw injection. The modulation depth is 0.07. The time scale is 2 ns per division [7.9]

It was observed that the laser generated stable pulses with high shot-to-shot reproducibility during a large number of shots before the frequency of the cw laser needed to be readjusted due to thermal instabilities of the two cavity lengths. It is expected that this can be further improved by using standard cavity stabilization techniques like Invar bars. Furthermore, both the intensity and the detuning range of the cw laser were found not to be critical for obtaining reliable pulses. An important feature of this stabilization technique is that at high values of the modulation depth no extra pulse broadening occurs due to the additional cw injection. Indeed, for high values of the modulation depth the same relationship between the pulse width and the modulation depth is observed in the absence of injection. This is shown in Fig. 7.13. The solid line represents the analytic determination according to (7.98). Again, the width of the most intense pulse of the train is plotted, although further pulse compression is observed at later events in the train. This is particularly the case for low modulation depths.

For $\alpha_a > 0.15$ the prediction (7.98) agrees very well with the experimental observations, which indicates that for these values of the modulation depth an almost steady-state situation has been reached. For $\alpha_a < 0.15$, however, a large deviation from the prediction is observed. This can be explained as follows. Keeping the medium gain and band width constant, the

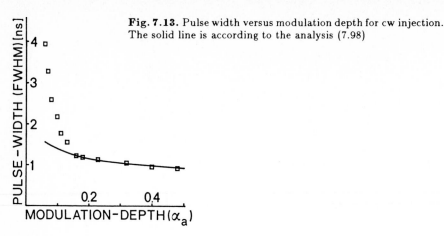

Fig. 7.13. Pulse width versus modulation depth for cw injection. The solid line is according to the analysis (7.98)

pulse compression per round trip due to AM mode locking decreases with decreasing α_a. For lower modulation depths more round trips are required to reach the steady state. The observed discrepancy between the experiment and analysis is therefore caused by the lack of round trips necessary to reach a steady state.

By lowering the cw intensity, pulse narrowing at low modulation depth can be observed, but at the cost of reduced shot-to-shot reproducibility. In conclusion, we may say that with respect to the shot-to-shot reproducibility this technique can be compared with the mode-lock technique having an additional narrow-band gain described in the previous subsection. The extra advantage of this technique is that for strong modulation the high energy pulses in the train are now band-width limited.

7.4.3 The Transient Evolution
of an AM Mode Locked TEA Laser

Although the instabilities of pulse shape and amplitude can be eliminated by applying an intra-cavity low-pressure section or by injecting monochromatic cw radiation there still remains the observation that the durations of the most intense pulses are longer than what is predicted by the steady-state model. This discrepancy increases with pressure because for shorter pulses the mode locking requires more time whereas the gain period decreases with pressure due to increased relaxation rates. In this section we shall discuss the transient behavior of the mode locking process in CO_2 systems and investigate the time taken to reach the steady state [7.13]. The described experiments will confirm the suggestion in previous subsections that the modulation has been insufficient in the short-gain period and the pulses are therefore not band-width limited.

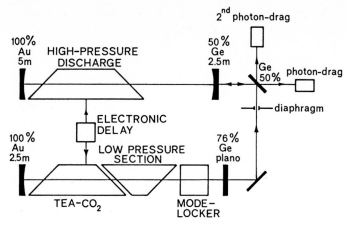

Fig. 7.14. Experimental set up for studying the time evolution of an AM mode locked TEA laser [7.13]

The applied experimental technique is essentially injection mode locking [7.14]. The schematic representation is shown in Fig. 7.14. The mode locking process to be studied occurs in a cavity containing a single discharge TEA CO_2 laser, a low-pressure cw CO_2 discharge operating below threshold, and a modulator. The TEA CO_2 discharge has a volume of $20 \times 1 \times 1 \, cm^3$ sealed with ZnSe Brewster windows and is operating with a $1:1:3 = CO_2:N_2:He$ mixture at 1 atmosphere. The modulator is an acousto-optic germanium crystal with Brewster angles and driven by a 40 MHz rf power oscillator. The modulation depth as a function of rf power was measured by observing the response on a smooth, 200 ns long CO_2 laser pulse from a single-mode, hybrid TEA CO_2 laser.

The low-pressure section is a 55 cm long sealed off tube filled with a $1:2:5 = CO_2:N_2:He$ mixture at 20 torr and has a diameter of 15 mm. The pulse train coming from this AM mode-locked master oscillator is injected in a slave oscillator through a germanium outcoupling mirror. The discharge of this slave oscillator is a uv preionized multi-atmosphere TE CO_2 system with an active volume of $0.8 \times 0.8 \times 30 \, cm^3$ operating with a $1:1:10 = CO_2:N_2:He$ mixture at 5.5 atmospheres [7.15]. The purpose of this high-pressure system is to increase the band width of the slave oscillator in order to minimize the effect of broadening on the injected pulses by the dispersive gain medium [7.16]. Because the full train of pulses is injected into the slave oscillator the cavity lengths of the two oscillators must be equal. An optimal matching was achieved by tuning the cavities for minimum pulse width from the slave oscillator. The amplification of an injected pulse by the slave oscillator is sufficiently large to avoid distortion of the selected pulse by the following injected pulses. To avoid additional pulses in the slave oscillator that originate from reflection on the outcoupling mirror of the master oscillator the slave oscillator was provided with a 2.5 m radius curved outcoupling mirror with a flat surface on the outside. Reflected pulses from this outcoupler are strongly diverging and are practically blocked by an iris in front of the beam splitter. The slave oscillator produces a train of pulses which all, because of the large band width of the active medium, have the same width as the injected one. The energies of the detectable pulses in this train do not depend on the energy of the injected one so that the widths of all pulses in the mode-locked TEA laser, independent of the pulse energies, can be observed. The pulses are detected with two photon-drag detectors and a Tektronix transient digitizer. The response time of the detection system is estimated to be approximately 700 ps. The delay between the two discharges can be varied electronically. The delay is measured by observing the light coming from the last spark gap of the Marx generators of the two discharges. These light signals are transported by two optical fibers to photodiodes.

In Fig. 7.15 some examples are shown of measurements of parts of the pulse train coming from the slave oscillator. Generally the observed pulses are broad at a short delay time and narrow at large delays between the two discharges. As expected, the pulse widths depend on the modulation depth and whether the low-pressure section in the master cavity was switched on or off. In Figs. 7.16–18 results are presented of measurements of the width of pulses from the slave oscillator as a function of the delay between the two discharges. Figures 7.16 and 17 show the measurements for the modulation depth of $\alpha_a = 0.03$ and 0.15, respectively, with the low-pressure

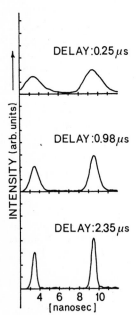

Fig. 7.15. Typical pulses during the mode-locking process taken at various instants

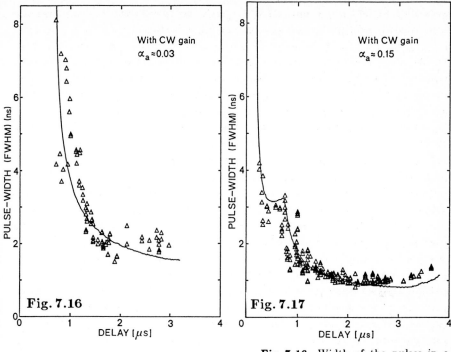

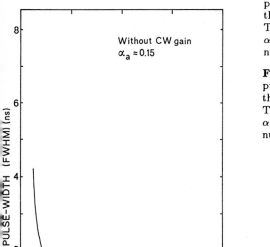

Fig. 7.16. Width of the pulses in a pulse train as a function of time after the gain passes threshold for a hybrid TEA laser with modulation depth $\alpha_a = 0.03$. The solid line represents numerical calculations

Fig. 7.17. Width of the pulses in a pulse train as a function of time after the gain passes threshold for a hybrid TEA laser with modulation depth $\alpha_a = 0.15$. The solid line represents numerical calculations

Fig. 7.18. Width of the pulses in a pulse train as a function of time after the gain passes threshold for a normal TEA laser with $\alpha_a = 0.15$. The solid line represents numerical calculations

section switched on. Figure 7.18 displays the results in the case that the low-pressure section is switched off and $\alpha_a = 0.15$.

From these measurements, it is seen that with the low-pressure section active, the pulse evolves from a large value of the pulse width to a steady state in about $2\,\mu$s. When the low-pressure section was switched off it was not possible to measure the pulse width for low values of the delay with acceptable accuracy because in that area the observed pulses are not stable, i.e., pulses with substructures and double pulses are observed. The pulses observed at a large delay, $>2\,\mu$s, on the other hand, were found to be very stable. With the low-pressure section on, longer pulses are observed. For larger delays, say after $3\,\mu$s, the pulses broaden again. This is because the pulses in the tail of the train are too weak to saturate the low-pressure gain. The pulses are then broadened by the low-pressure section.

In Fig. 7.17 there seems to be a fast evolution to a width of about 3 ns in less than $1\,\mu$s followed by a slower evolution of 1 ns in about $2\,\mu$s. This can be explained as follows. At the beginning the oscillations from the gain of the continuous section together with the weak gain of the pulsed section are just above threshold and are forced by the strong modulation to produce pulses that are mainly determined by the continuous section with small band width. Later on the gain of the pulsed discharge dominates and further compression occurs. For a smaller modulation depth (Fig. 7.16) the pulse-forming mechanism is slower and there is not enough time to form quasi-stationary pulses in the continuous discharge section. As mentioned, the weak pulses during the build-up time and also after the gain switch are

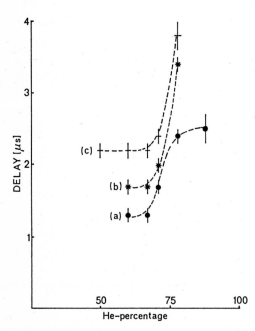

Fig. 7.19. The delay between the firing of the last gap of the Marx generator and the first strong pulse in a train versus He percentage for a hybrid system: (a) cw discharge 4 mA, $\alpha_a = 0.12$; (b) cw discharge 4 mA, $\alpha_a = 0.64$; and (c) without cw discharge $\alpha_a = 0.16$ [7.17]

not easily observable directly from the mode-locked TEA laser. The delay between the firing of the last gap of the Marx system of the discharge and the appearance of the first strong pulse in the gain switch indicates the approach to the steady state of the pulse-forming mechanism. It is observed that the intense pulses during the gain switch are shorter for longer delays [7.17]. In Fig. 7.19 this delay is shown as a function of the He percentage for mixtures of equal CO_2 and N_2 percentages. It is seen that without the cw discharge (c) the delay is longer than for the hybrid systems (a, b) having a cw current of 4 mA. As expected, the delay increases with He concentration because of the lower gain so that more round trips have to be made before reaching saturation and gain switching. Therefore, leaner mixtures have more time to reach steady state and produce narrower pulses.

7.4.4 Numerical Results of the AM Mode Locking

The dynamic processes of pulse formation in a mode-locked TEA laser can also be described numerically [7.13]. We start with the analysis of the energy transfer from the discharge current to the various degrees of freedom as presented in Sect. 6.5, where it has been assumed that each individual degree of freedom is in thermal equilibrium and can be described with its own temperature. Equations (6.53–72) can be used directly. The excitation parameters of the vibrational modes are given by

$$N_e(t)N_{CO_2}h\nu_1\chi_1 = W_1 , \tag{7.100a}$$

$$2N_e(t)N_{CO_2}h\nu_2\chi_2 = W_2 , \tag{7.100b}$$

$$N_e(t)N_{CO_2}h\nu_3\chi_3 = W_3 , \tag{7.100c}$$

$$N_e(t)N_{N_2}h\nu_4\chi_4 = W_4 . \tag{7.100d}$$

In order to calculate the pulse formation we shall now proceed somewhat differently than in Sect. 6.5. The problem is to find the excitation parameters from the experimental conditions. Considering the experimental condition of the TEA discharge in the $1:1:3 = CO_2:N_2:$ He mixture mentioned in the previous subsection we estimate from the cross sections for the energy transfer to the vibrational modes that

$$W_1 = W_2 = 0.7W_3 = 0.14W_4 = W_0 t \exp(-t/\tau_p) , \tag{7.101}$$

where the time dependence is given by the current pulse that is approximated according to the above relation with $\tau_p = 50$ ns. The excitation parameters are considered to follow the current pulse that is indicated in (7.101). The parameter W_0 will now be chosen to reproduce the observed small-signal gain of 3 % cm^{-1} for the P(20) in this mixture. Thus by substituting $\alpha = 0.03$ for the maximum gain in (6.57) we determine W_0 numerically. The field to be calculated starts from the noise. The increase of noise intensity is given by (6.66). The phase of the initial field is not important and can be chosen arbitrarily. The calculation is done within a time window given by the round-trip time $T = 2L/c = 12.5\,\text{ns}$. It is assumed, and this can be verified, that the variations of the interacting quantities are very small during such a time interval. The changes of the densities of the vibrational energies during this time interval are therefore calculated according to (6.53, 58–60) for constant radiation intensity. The intensity used is an average value which is related to the electric field by

$$ I = \frac{c^2}{L} \int\limits_{-\infty}^{\infty} |E|^2 dt \;, \tag{7.102} $$

where the field is defined in such a way that $|E|^2$ represents the photon density. The change of the electric field during one round trip is in turn calculated for constant inversion density.

The electric field $E_j(t)$ after the jth round trip is related to $E_{j-1}(t)$ by

$$ E_j(t) = rT_j(t)\mathcal{F}^{-1}[G_{\text{TEA}}(\omega)G_{\text{cw}}(\omega)\mathcal{F}(E_{j-1}(t))] \;, \tag{7.103} $$

where $T_j(t)$ is the time-dependent transmission function of the acousto-optic modulator, r is the amplitude reflection coefficient including all other linear resonator losses, $G_{\text{TEA}}(\omega)$ and $G_{\text{cw}}(\omega)$ are the frequency-dependent gain of the TEA section and the low-pressure section, respectively, and $\mathcal{F}$ is the Fourier transform operator.

The transmission of the acousto-optic modulator for a round trip can be obtained from (7.31) by

$$ T_j(t) = \exp\left(-2\alpha_a \sin^2\left\{\frac{\omega_m}{2}[t - \delta\tau(j-1)]\right\}\right) \;, \tag{7.104} $$

where $\delta\tau = (2\pi/\omega_m) - T$. The active media of both the TEA and the low-pressure section are assumed to be Lorentzian and are given by (7.33), or

$$ G_{\text{TEA}}(\omega) = \exp\left(\frac{g_{\text{TEA}}(j)}{1 + 2\mathrm{i}(\omega - \omega_L)/\Delta\omega_N}\right) \;, \tag{7.105} $$

$$ G_{\text{cw}}(\omega) = \exp\left(\frac{g_{\text{cw}}(j)}{1 + 2\mathrm{i}(\omega - \omega_L)/\Delta\omega_{\text{cw}}}\right) \;, \tag{7.106} $$

where g_{TEA} and g_{cw} are the gain at the center frequency of the TEA and the low-pressure section, respectively, and $\Delta\omega_{\text{cw}}$ is the line width of the low-pressure discharge.

230

The gain $g_{\mathrm{TEA}} = \alpha$ is obtained for each calculation step from the inversion according to (6.57). For the continuous low-pressure section we obtain

$$g_{\mathrm{cw}}(j) = \frac{g_{\mathrm{cw}}^0}{1 + [I(j)/I_{\mathrm{s}}]} \; , \tag{7.107}$$

where $g_{\mathrm{cw}}^0 = 0.25\,\mathrm{cm}^{-1}$ is the small-signal gain, and $I_{\mathrm{s}} = 5\,\mathrm{W/cm^2}$ is the saturation intensity. The results of the calculations for ideal mode locking ($\delta\tau$ is just equal to the added dispersion or linear delay in the Lorentzian lines) are shown in Figs. 7.16–18 by the solid lines. When the cw section is active we start with a single mode.

The discontinuity of the decreasing pulse in Fig. 7.17 which was explained in the previous subsection is verified with the present calculations. The model also predicts the increase in pulse width after the pulsed gain has disappeared. In the absence of the continuous section the signal starts from noise. It turns out that the results are practically independent of the shape and size of this noise signal. It is seen that the evolution of the pulse width reaches a minimum which can be considered as the steady-state width. This steady state is, as expected, independent of the start pulse. The numerical values of these steady-state pulses are also plotted in Fig. 7.8. It is seen that the analytic calculations are in good agreement with these numerical values.

8. FM Mode Locking of TEA Lasers

The problem of mode locking and pulse forming by means of intracavity FM modulation is, to a large extent, similar to that of AM mode locking described in the previous chapter. Again the problem can be treated either in the time domain or in the frequency domain. The main difference between the results obtained with AM and FM is that for detuning the mode-locking behavior with FM changes asymmetrically when the modulation frequency is detuned. This asymmetry is due to the dispersion of the active medium. In the present chapter we shall consider small detuning, i.e. comparable with the line width, so that the laser remains mode locked. For larger detuning the output signal of the laser changes to an FM laser type of signal which is beside the scope of the present treatment.

8.1 Electro-Optic Phase Modulation

Electro-optic phase modulation can be obtained by the propagation of radiation in a crystal in the presence of an applied periodic electric field. In certain types of crystals the applied electric field causes a change in index of refraction that is proportional to the field. This is the, so-called, electro-optic effect.

The propagation characteristics in a crystal are fully described by means of the index ellipsoid. In the presence of an electric field the equation of the index ellipsoid can be written as [8.1]

$$\frac{1}{n_1^2}x^2 + \frac{1}{n_2^2}y^2 + \frac{1}{n_3^2}z^2 + 2\frac{1}{n_4^2}yz + 2\frac{1}{n_5^2}xz + 2\frac{1}{n_6^2}xy = 1 \ . \tag{8.1}$$

If x, y and z are chosen parallel to the principal dielectric axes of the crystal, then with zero applied field the above equation reduces to

$$\frac{1}{n_x^2}x^2 + \frac{1}{n_y^2}y^2 + \frac{1}{n_z^2}z^2 = 1 \ . \tag{8.2}$$

In the presence of an electric field $\boldsymbol{E} = (E_x, E_y, E_z)$ the linear changes in the coefficients $1/n_i^2$ $(i = 1 - 6)$ will be

$$\Delta \frac{1}{n_i^2} = \sum_{j=1}^{3} r_{ij} E_j \ , \tag{8.3}$$

where the summation over j is $1 = x$, $2 = y$, $3 = z$. The 6×3 matrix with elements r_{ij}, is called the electro-optic tensor. For the modulation of CO_2-laser radiation we can use GaAs and CdTe since these materials are transparent in the $10 \, \mu m$ region. These crystals belong to the $\bar{4}3\,m$ symmetry group [8.2]. They are cubic and have axes of fourfold symmetry along the cube edges and threefold axes of symmetry along the cube diagonals. The only nonvanishing electro-optic tensor elements are $r_{41}, r_{52} = r_{41}, r_{63} = r_{41}$. Further, $n_x = n_y = n_z = n$. The index ellipsoid in the presence of an applied electric field can then be expressed as

$$\frac{x^2 + y^2 + z^2}{n^2} + 2r_{41}(yzE_x + xzE_y + xyE_z) = 1 \ . \tag{8.4}$$

We now apply the field $\boldsymbol{E}$ perpendicular to the $(1\bar{1}0)$ plane of the crystal, i.e. $E_x = -E_y = \frac{1}{2}\sqrt{2}E$, and $E_z = 0$. The index ellipsoid becomes

$$\frac{x^2 + y^2 + z^2}{n^2} + p(yz - xz) = 1 \ , \tag{8.5}$$

where $p = \sqrt{2}r_{41}E$. Next, we look for the new directions x', y', and z' of the principle axes of the ellipsoid. These axes are given by

$$z' = -\frac{1}{\sqrt{2}}x - \frac{1}{\sqrt{2}}y \ , \tag{8.6a}$$

$$y' = \frac{1}{\sqrt{2}}z - \frac{1}{2}x + \frac{1}{2}y \ , \tag{8.6b}$$

$$x' = -\frac{1}{\sqrt{2}}z - \frac{1}{2}x + \frac{1}{2}y \ . \tag{8.6c}$$

Therefore,

$$x = -\frac{1}{2}x' - \frac{1}{2}y' - \frac{1}{\sqrt{2}}z' \ , \tag{8.7a}$$

$$y = \frac{1}{2}x' + \frac{1}{2}y' - \frac{1}{\sqrt{2}}z' \ , \tag{8.7b}$$

$$z = \frac{1}{\sqrt{2}}y' - \frac{1}{\sqrt{2}}x' \ . \tag{8.7c}$$

Substituting (8.7) into (8.5) we obtain

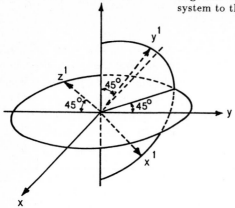

$$x'^2 \left(\frac{1}{n^2} - r_{41}E \right) + y'^2 \left(\frac{1}{n^2} + r_{41}E \right) + z'^2 \frac{1}{n^2} = 1 \ . \tag{8.8}$$

The new axes are drawn in Fig. 8.1.

The applied electric field and the propagation of the optical beam with respect to the principle axes x, y, and z of a crystal belonging to the $\bar{4}3$ m symmetry group is shown in Fig. 8.2. In this figure we also indicated the new principle axes x', y', and z' in the presence of the applied field $E = E_0 \cos \omega_m t$. For frequency modulation of the optical beam, the beam propagates along the z' axis and the direction of polarization coincides either with the x' or y' axis.

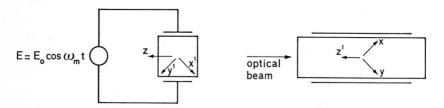

Fig. 8.2. Schematic drawing of the principle axis (x, y, z) and the transformed axis (x', y', z') with respect to the optical beam and applied electric field E

The principle indices of refraction can be deduced from (8.8). In practice, $r_{41}E \ll 1/n^2$ so that the principle indices become

$$n_{x'} = n + \tfrac{1}{2}n^3 r_{41} E_0 \cos \omega_m t \ , \tag{8.9a}$$

$$n_{y'} = n - \tfrac{1}{2}n^3 r_{41} E_0 \cos \omega_m t \ , \tag{8.9b}$$

$$n_z' = n \ . \tag{8.9c}$$

In the case the optical polarization is along the x'-axis the oscillating part of the susceptibility, as derived from (8.9a), is

$$\Delta\chi'_F(t) = \Delta\chi'_F \cos \omega_m t , \quad \text{where} \tag{8.10a}$$

$$\Delta\chi'_F = n^4 r_{41} E_0 . \tag{8.10b}$$

In the case of CdTe the material constants are:

$$n = 2.6 \quad \text{and} \quad r_{41} = 6.8 \times 10^{-12} \, \text{mV}^{-1} .$$

The phase modulator introduces a sinusoidally varying phase perturbation $\delta(t)$ such that the transmission through the modulator is

$$\exp[-i\delta(t)] = \exp(-i\delta \cos \omega_m t) , \tag{8.11}$$

where $\delta = \frac{k_1 a \Delta\chi'_F}{2n^2}$, a is the length of the modulator, and k_1 the wave number in the modulator.

In order to obtain sufficient phase modulation for mode locking the applied field must be high and in the order of $(2\text{–}4) \times 10^5 \, \text{Vm}^{-1}$. As an example the modulator consists of a CdTe crystal with the dimensions $40 \times 5 \times 5 \, \text{mm}^3$. For a cavity length of $123.1 \, \text{cm}$ a modulation frequency $(\nu_m = c/2L)$ of $121.8 \, \text{MHz}$ is required. The output voltage of available oscillators at this frequency may be in the order of $500 \, \text{V}$, which is too low. It is difficult to transform voltages of these high frequencies in conventional ways because of high dielectric and radiation losses.

A relatively simple way to transform an output voltage of $500 \, \text{V}$ and $121.8 \, \text{MHz}$ to higher values of the voltage is to use a, so-called, Lecher system. It consists of two pipes placed parallel to each other at a certain distance. The two ends of the pipes are short circuited, so that an electromagnetic (EM) wave is able to build up a standing EM wave between them, with the condition that the length L of the pipes is matched to the frequency of the EM wave $(f = c/2L)$. Figure 8.3 shows the construction of such a Lecher system and the standing wave which can be generated with this system. This Lecher system is energized by the high-frequency oscillator. The pipes are directly connected to the transmitter at a distance L_1 from one of the short-circuited ends. For proper operation we have to satisfy the following three conditions.

1. The Lecher system must be resonant for the applied frequency $(f = \nu_m)$. In this case a standing wave of half a wavelength is built up with an oscillating frequency equal to the modulation frequency.
2. The position where the oscillator is connected to the pipe system, i.e. the length L_1, must be chosen in such a way that the energy transfer is maximum. This can be obtained by making the input impedance of

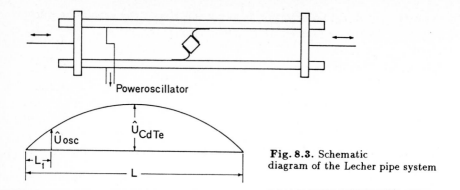

Poweroscillator

$\hat{U}_{osc}$

$\hat{U}_{CdTe}$

L_1

L

Fig. 8.3. Schematic
diagram of the Lecher pipe system

the Lecher system equal to the output impedance of the transmitter.
This condition determines the length L_1.

3. In order to prevent conduction losses in the pipes it is necessary to
gild the pipes. This will strongly increase the quality of the resonator
system. Furthermore, the pipe system must be packed in a metal
box which is silvered on the inside. This silver-coated box decreases
radiation losses considerably and will therefore increase the quality Q
of the resonator system.

If these three conditions are satisfied, we obtain a system with a maximum
output voltage at the middle of the pipe system, being the place where the
CdTe crystal is mounted. The maximum voltage over the CdTe crystal is

$$\hat{U}_{CdTe} = \frac{\hat{U}_{osc}}{\sin\left(\pi L_1/L\right)} \ .$$

In our example we use the following values: $\hat{U}_{osc} = 500\,\text{V}$, $L = 120\,\text{cm}$, and
$L_1 = 10\,\text{cm}$. Hence the maximum voltage over the crystal will be about
$2000\,\text{V}$, which is sufficient for mode locking. With this applied voltage we
obtain $E_0 = 4 \times 10^5\,\text{Vm}^{-1}$ and $\Delta\chi'_F = 1.2 \times 10^{-4}$.

8.2 Time-Domain Analysis of FM Mode Locking

The analysis of the pulse forming by FM mode locking in the time domain
[8.3] for a homogeneously broadened system starts again with a Gaussian
pulse shape. As in the case of AM modulation a self-consistent equation
for the pulse after a round trip gives the pulse parameters. Following the
sequence of the pulses as indicated in Fig. 7.5 we find for $E_3(t)$ after passing
the modulator twice

$$E_3(t) = E_2(t) \exp[-2i\delta(t)] , \qquad (8.12)$$

where $\delta(t)$ is given by (8.11).

The phase variation reaches a maximum for $\omega_m t = 0$ and a minimum for $\omega_m t = \pi$. For the ideal mode locking, i.e. the modulator does not perturb the pulse which is assumed short compared to the modulation period, the pulse passes through the modulator at one or the other extremum of the phase variation. For the maximum phase variation we substitute $\cos \omega t = 1 - 2 \sin^2(\omega t/2)$ and at the minimum $\cos \omega t = -1 + 2 \cos^2(\omega t/2)$. Since the pulse is short the modulation during transmission can then be approximated by

$$\exp[-2i\delta(t)] = \exp(\mp 2i\delta \pm i\delta\omega_m^2 t^2) . \qquad (8.13)$$

In the more general case when the pulse goes through the modulator at a phase angle θ from the ideal case the transmission can be written as

$$\exp[-2i\delta(t)]$$
$$= \exp[\mp 2i\delta \cos \theta \pm 2i\delta \sin \theta(\omega_m t) \pm i\delta \cos \theta(\omega t)^2] . \qquad (8.14)$$

For a positive value of θ the pulse lags behind the modulation signal. The first term in (8.14) is the additional phase shift that changes the optical length of the cavity, so that the length is now

$$L' = L \pm (\delta/\pi)\lambda \cos \theta . \qquad (8.15)$$

The second term is a Doppler frequency shift and the third term gives a linear frequency chirp to the pulse.

In the case of ideal mode locking the procedure is the same as that in Sect. 7.2.1 except that we replace α_a by $\mp 2i\delta$. The self-consistency requirement gives

$$\gamma = \mp \frac{1}{2}i\delta\omega_m^2 \pm \frac{1}{2}\left(-\delta^2\omega_m^4 \mp \frac{1}{4}\frac{i\delta\omega_m^2 \Delta\omega_N^2}{g}\right)^{1/2} . \qquad (8.16)$$

Again we have the conditions $\omega_m \ll \omega_N$ and the second term under the square root sign is much larger than the first. We now may approximate γ by

$$\gamma = \frac{(\omega_m \Delta\omega_N)}{4}\sqrt{\frac{\mp i\delta}{g}} , \qquad (8.17)$$

where the real part of γ has been taken positive because of the Gaussian pulse shape. Comparing (7.36) with (8.17) we have

$$\gamma = \alpha - i\beta = (1 \mp i)(\omega_m \Delta\omega_N/4)\sqrt{\delta/2g} .$$

We now obtain for the pulse width

$$\tau_p = \frac{1}{\pi}(2\ln 2)^{1/2}\left(\frac{2g}{\delta}\right)^{1/4}\left(\frac{1}{\nu_m \Delta \nu_N}\right)^{1/2}, \tag{8.18}$$

for the pulse bandwidth

$$\Delta \nu_{\text{pulse}} = (2\ln 2)^{1/2}\left(\frac{2\delta}{g}\right)^{1/4}(\nu_m \Delta \nu_N)^{1/2}, \tag{8.19}$$

and for the modulation frequency

$$\nu_m = \left(\frac{2L}{c} \pm 2\frac{\delta\lambda}{\pi c} + \frac{2g}{\Delta \omega_N}\right)^{-1}. \tag{8.20}$$

For the gain we find, as a good approximation,

$$g = -\frac{1}{2}\ln R. \tag{8.21}$$

So far the pulse parameters are given for ideal mode locking.

8.2.1 Self-Consistency of the Circulating Pulse with Detuning

Detuning the modulation frequency means that the pulse passes the modulator at some angle θ from one or the other extremum of the phase variation. Then the pulse experiences a Doppler shift and in consecutive passes through the modulator the pulse-center frequency is shifted until some equilibrium is reached where the Doppler shift of the modulator is cancelled by an equal and opposite frequency shift from the active medium. The frequency shift normalized to the line width is again $\eta = (\omega_0 - \omega_L)/\Delta\omega_N$. When the pulse passes through the active medium we find similarly to (7.44)

$$E_2(t) = \frac{E_0}{2\sqrt{\gamma A'}}\exp\left(g' - \frac{(t-B')^2}{4A'}\right)\exp(i\omega_0 t), \tag{8.22}$$

where A', B' and g' are given by (7.57). One important difference from the case without detuning is that B' is now complex. Consider the term $(t-B')^2/4A'$ and split B' into its real and imaginary part; then

$$\begin{aligned}
(t-B')^2/4A' = &\left(t - \frac{2g(1-4\eta^2)}{\Delta\omega_N(1+4\eta^2)^2}\right)\frac{1}{4A'} \\
&+ \frac{4ig\eta}{\Delta\omega_N(1+4\eta^2)^2 A'}\left(t - \frac{2g(1-4\eta^2)}{\Delta\omega_N(1+4\eta^2)^2}\right) \\
&- \frac{16g^2\eta^2}{\Delta\omega_N(1+4\eta^2)^4 A'}.
\end{aligned} \tag{8.23}$$

Now consider the expansion

$$\frac{4ig\eta}{\Delta\omega_N(1+4\eta^2)^2 A'} = \frac{K_1}{4A'} + iK_2 \tag{8.24}$$

with K_1 and K_2 real. The values of K_1 and K_2 can be calculated as

$$K_2 = \frac{g\eta\,\Delta\omega_N(\alpha^2+\beta^2)(1+4\eta^2)}{g(1-12\eta^2)(\alpha^2+\beta^2)+\alpha(1+4\eta^2)^3\Delta\omega_N^2}\,, \tag{8.25a}$$

$$K_1 = K_2\left(\frac{\beta}{\alpha^2+\beta^2} - \frac{16g(6\eta-8\eta^3)}{\Delta\omega_N^2(1+4\eta^2)^3}\right). \tag{8.25b}$$

Substituting (8.24) back into (8.23) and using the result for calculating $E_2(t)$ we obtain

$$E_2(t) = \frac{E_0}{2\sqrt{\gamma A'}}\exp\left[g' - \left(t - \frac{2g(1-4\eta^2)}{\Delta\omega_N(1+4\eta^2)^2} + \frac{K_1}{2}\right)^2\frac{1}{4A'} + i(\omega_0 - K_2)t \right.$$

$$\left. + \frac{K_1^2}{16A'} + iK_2\frac{2g(1-4\eta^2)}{\Delta\omega_N(1+4\eta^2)^2} + \frac{16g^2\eta^2}{\Delta\omega_N(1+4\eta^2)^4 A'}\right]. \tag{8.26}$$

The last expression reveals some interesting effects of the active medium on the pulses:

1. The delay of the pulse envelope by the active medium has changed from $2g/\Delta\omega_N$ for ideal mode locking to

$$\frac{2g(1-4\eta^2)}{\Delta\omega_N(1+4\eta^2)^2} + \frac{K_1}{2}$$

and this will compensate exactly for the detuning.
2. The frequency shift K_2 in (8.26) will cancel out the Doppler shift from the modulator.
3. The last three exponentials show additional attenuation and phase shift.

Using the transmission through the modulator, as given by (8.14), the self-consistency equation reveals, similarly to (7.59), the following conditions:

$$T_m = \frac{2L}{c} \pm 2\frac{\delta\lambda}{\pi c}\cos\theta + \frac{2g(1-4\eta^2)}{\Delta\omega_N(1+4\eta^2)^2} + \frac{K_1}{2}\,, \tag{8.27a}$$

$$K_2 = \pm 2\delta\omega_m\sin\theta\,, \tag{8.27b}$$

$$\gamma = \frac{1}{4A'}\mp i\delta\omega_m^2\cos\theta\,, \tag{8.27c}$$

$$\left| \frac{r}{2\sqrt{\gamma A'}} \exp\left(g' + \frac{K_1^2}{16A'} + \frac{16g^2\eta^2}{\Delta\omega_N(1 + 4\eta^2)^4 A'}\right) \right| = 1 \ . \qquad (8.27\text{d})$$

In the last condition only the magnitude is considered because the phase angle is of no consequence.

From (8.27c) the complex constant γ can be solved. We find with the same approximation as before

$$\gamma \simeq \frac{\omega_m \Delta\omega_N}{4\sqrt{g}} \left[\mp i\delta \cos\theta\left(1 + 2i\eta\right)^3\right]^{1/2}$$

$$= \pm \frac{\omega_m \Delta\omega_N}{4\sqrt{g}} \sqrt{\delta} \cos\theta (1 + 4\eta^2)^{3/4}(\cos\psi + i\sin\psi) \ , \qquad (8.28)$$

where $\psi = \mp\frac{\pi}{4} + \frac{3}{2}\tan^{-1}(2\eta)$. Since the real part of $\gamma = \alpha - i\beta$ must be positive we have the $+$ sign of (8.28) for positive $\cos\psi$ and the $-$ sign for negative $\cos\psi$. We now obtain for the pulse width

$$\tau_p = \frac{1}{\pi}(2\ln 2)^{1/2}\left(\frac{g}{\delta\cos\theta}\right)^{1/4}(1 + 4\eta^2)^{-3/8}\left(\frac{1}{\nu_m\Delta\nu_N|\cos\psi|}\right)^{1/2} \qquad (8.29)$$

and for the pulse bandwidth

$$\Delta\nu_{\text{pulse}} = (2\ln 2)^{1/2}\left(\frac{\delta\cos\theta}{g}\right)^{1/4}(1 + 4\eta^2)^{3/8}\left(\frac{\nu_m\Delta\nu_N}{|\cos\psi|}\right)^{1/2} \ . \qquad (8.30)$$

As before, the detuning can be calculated from (8.27a). Similarly to (7.62) the detuning is defined as

$$\Delta\nu_t = \frac{c}{2L} - \frac{c^2}{4L^2}\left(\frac{2g}{\Delta\omega_N} \pm \frac{2\delta\lambda}{\pi c}\right) - \nu_m \ ,$$

where $\nu_m = 1/T_m$. Using (8.27a) we find

$$\Delta\nu_t = -\frac{c^2}{4L^2}\left[\frac{2g}{\Delta\omega_N}\left(1 - \frac{(1 - 4\eta^2)}{(1 + 4\eta^2)^2}\right) \pm \frac{2\delta\lambda}{\pi c}(1 - \cos\theta) - \frac{K_1}{2}\right] \ . \ (8.31)$$

The dependence of η, θ, and g on the detuning for a given laser system are, in principle, given by (8.25a, 27b, d and 31). However, the best procedure for obtaining these relations is to use η as the independent parameter. After a few iterations θ, g and $\Delta\nu_t$ are obtained numerically.

8.3 Frequency-Domain Analysis of FM Mode Locking

The analysis in the frequency domain [8.4] is similar to the treatment in Sect. 7.3. Again we start with the field distribution of axial modes and the asumption that the distribution is Gaussian over the mode frequencies. The electric field polarizes the medium and the resulting polarization is given by (7.66). The polarization of the phase modulator is

$$P_F = \varepsilon_0 \sum_n \psi_0 \Delta\chi'_F(t) E_n \sin k_n z \exp[i(\omega_m t + \phi_n)] \quad , \tag{8.32}$$

where $\Delta\chi'_F(t)$ is given by (8.10). Since the mode-frequency spacing is equal to ω_m we can evaluate P_F as

$$P_F = \tfrac{1}{2}\varepsilon_0 \Delta\chi'_F \psi_0 \sum_n \{\exp[i(\omega_n t + \phi_{n-1})] E_{n-1} \sin k_{n-1} z$$
$$+ \exp[i(\omega_n t + \phi_{n+1})] E_{n+1} \sin k_{n+1} z\} \quad . \tag{8.33}$$

Substituting the electric fields and the polarizations into the Maxwell equations and using the abreviations mentioned in Sect. 7.3 we finally obtain similarly to (7.79) the following coupling between adjacent mode fields

$$E_{n+1} e^{i\phi_{n+1}} + E_{n-1} e^{i\phi_{n-1}}$$
$$= \frac{1}{\delta} E_n e^{i\phi_n} \left[-\left(\frac{2L}{c}\dot{\phi}_n - \frac{2L}{c} n\Delta\omega + \psi_n \right) - ig_n + iS \right] \quad , \tag{8.34}$$

where the modulator is located near the mirror at $z = L$. The parameters ψ_n, g_n, and S are given by (7.78, 77, 73), respectively, and δ is

$$\delta = \frac{\omega_n}{c} \int_0^a \Delta\chi'_F \sin k_n z \sin k_{n\pm 1} z \, dz \quad . \tag{8.35}$$

Taking into account that $a \ll L$ we obtain for δ

$$\delta = -\frac{\omega_n}{2c} \int_0^a \Delta\chi'_F dz \quad ,$$

where $\Delta\chi'_F$ is given by (8.10b). In the above analysis the refractive index n of the modulator has been taken unity; otherwise the calculated value of δ has to be divided by n.

Assuming a Gaussian distribution of the mode fields the mutual relation within such a distribution is, to a good approximation, expressed by (7.84b). In the following we look for an analytic solution of (8.34) for the case of a homogeneously broadened line. Due to the interactions of the radiation field with the inverted medium the gain profile changes over its full frequency range. In general, it turns out that the frequency width of the pulse is small as compared to the line width. This is illustrated in Fig. 8.4, where the gain and the mode amplitudes are plotted against the frequency. This means that in the frequency region of interest we may, to a good approximation, expand the gain profile and the phase retardation of such a line profile to the second order. This can be done in terms of the mode number n. Thus, the gain and the phase retardation for the nth mode will be

$$g_n = g_0 + n g_1 + \tfrac{1}{2} n^2 g_2 \ , \tag{8.36}$$

$$\psi_n = \psi_0 + n \psi_1 + \tfrac{1}{2} n^2 \psi_2 \ . \tag{8.37}$$

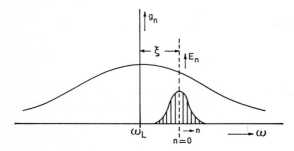

Fig. 8.4. Schematical presentation of the mode spectrum with respect to the optical line. The center pulse frequency $(n = 0)$ has a frequency shift ξ with respect to the line center

By comparing (7.84b) and (8.34) and making use of (8.36,37) we equate terms of the same order of n for the real and imaginary part, and obtain the following six equations

$$2 - 2\alpha_1 + \alpha_1^2 - \alpha_2^2 - \vartheta_1^2 = -\frac{2d}{\delta c} \dot{\phi}_n - \frac{\psi_0}{\delta} \ , \tag{8.38a}$$

$$-2\alpha_2 + 2\alpha_1 \alpha_2 = \frac{S}{\delta} - \frac{g_0}{\delta} \ , \tag{8.38b}$$

$$4\vartheta_1 \alpha_2 = -\frac{\psi_1}{\delta} + \frac{2L}{c\delta} \Delta\omega \ , \tag{8.38c}$$

$$-4\vartheta_1 \alpha_1 = -\frac{g_1}{\delta} \ , \tag{8.38d}$$

$$4\alpha_1^2 - 4\alpha_2^2 = -\frac{1}{2\delta} \psi_2 \ , \tag{8.38e}$$

$$8\alpha_1 \alpha_2 = -\frac{1}{2\delta} g_2 \ . \tag{8.38f}$$

By solving this set of equations we have to consider two different situations,

242

which are described by the sign of the modulation parameters δ. For a positive value of δ we are dealing with a pulse that passes the modulator when its phase perturbation is at or close to its maximum value. The other situation, given by a negative value of δ, corresponds to the passage of the pulse at or close to the minimum of the phase perturbation. The two solutions given different values for α_1 and α_2. It turns out that the frequency shift ξ of the center pulse frequency changes sign by changing the sign of δ. The values of the pulse width and pulse bandwidth as a function of the detuning frequency do not depend on the sign of δ. Because the two solutions are similar we shall consider only a positive value of δ.

From (8.38e, f) we find

$$\alpha_1 = \frac{1}{4\sqrt{\delta}} [-\psi_2 + (\psi_2^2 + g_2^2)^{1/2}]^{1/2} \quad \text{and} \tag{8.39}$$

$$\alpha_2 = \mp \frac{1}{4\sqrt{\delta}} [\psi_2 + (\psi_2^2 + g_2^2)^{1/2}]^{1/2} . \tag{8.40}$$

The sign of α_2 depends on the sign of g_2. In practical situations with a mode-center frequency close to the line center where g_2 is negative we have to use the positive sign in (8.40), otherwise the negative sign. For values of $\alpha_2 \neq 0$ we are dealing with a frequency chirp in the light pulse, i.e. in the time domain there is a linear frequency chirp during the pulse. Using (8.39, 40) we obtain from (8.38c, d)

$$\frac{2L}{c} \Delta\omega_t = \mp g_1 \frac{[\psi_2 + (\psi_2^2 + g_2^2)^{1/2}]^{1/2}}{[-\psi_2 + (\psi_2^2 + g_2^2)^{1/2}]^{1/2}} \quad , \quad \text{where} \tag{8.41}$$

$$\Delta\omega_t = \Delta\omega - \frac{c}{2L} \psi_1 . \tag{8.42}$$

The negative and positive signs refer again to the positive and negative value of g_2, respectively.

From this equation it is readily seen that for $g_2 = 0$ at the lower frequency side of the profile, where ψ_2 is positive, the detuning $\Delta\omega_t$ goes to infinity. This means that by detuning the system for positive values of $\Delta\omega_t$ the center frequency of the pulse will never be smaller than the frequency for which g_2 at the lower frequency side of the profile becomes zero. For negative values of $\Delta\omega_t$ the center pulse frequency moves on the higher frequency part of the profile. It is also seen from (8.41) that for $\Delta\omega_t = 0$ the value of g_1 must be zero, i.e. the pulse-center frequency coincides with the line center. The frequency shift from the ideal situation is not given by $\Delta\omega_t$, because by detuning the system the pulse-center frequency moves away from the line center, so that the gain at the line center has to increase in order to balance the cavity losses, and accordingly ψ_1 will increase. The detuning from the ideal situation $\Delta\omega_{\text{det}}$ is then expressed by

$$\Delta\omega_{\text{det}} = \Delta\omega_t + \Delta\frac{c}{2L}\psi_1 , \tag{8.43}$$

where $\Delta c\psi_1/2L$ is the change of $c\psi_1/2L$ with respect to the ideal situation. It turns out that as $\Delta\omega_{\text{det}}$ decreases from the zero value, the center frequency of the pulse shifts continuously to higher frequencies on the high-frequency side of the line profile. Furthermore, there is only a small frequency range for negative detuning, as compared to positive detuning, provided there is sufficient gain in the medium. The gain of the system for the pulse can be obtained from (8.38b). Substituting (8.39, 40) we find

$$g_0 = S \mp [\frac{1}{4}\delta\psi_2 + \frac{1}{4}\delta(\psi_2^2 + g_2^2)^{1/2}]^{1/2} + \frac{g_2}{8} . \tag{8.44}$$

The pulse form can be obtained by summing the field over all modes. Thus the total field is

$$E_{\text{tot}} = \sum_n E_n \exp[i(\Omega_0 + n\omega_m + \dot\phi)t + i\phi_n] . \tag{8.45}$$

Making use of (7.83) we get

$$E_{\text{tot}} = E_0[i(\Omega_0 + \dot\phi)t] \sum_n \exp(-\alpha n^2 + i\vartheta_1 n + in\omega_m t) . \tag{8.46}$$

In order to calculate the radiation intensity we have to take $E_{\text{tot}} \cdot E_{\text{tot}}^*$. The summation over the modes is straightforward. For simplicity, we increase the time by ϑ_1/ω_m. In other words, we substitute

$$t' = t + \vartheta_1/\omega_m \tag{8.47}$$

and omit the prime in the final result. We then find

$$E_{\text{tot}} \cdot E_{\text{tot}}^* = C^2 \left| \exp\left(-\frac{\omega_m^2 t^2}{2\alpha}\right) \right| , \quad \text{where} \tag{8.48}$$

$$C = \left| \sum_n \exp\left[-\alpha\left(n - i\frac{\omega_m t}{2\alpha}\right)^2\right] \right| . \tag{8.49}$$

Because the time duration of the pulse is small it turns out that C can be considered as time independent.

The pulse width, defined as the time interval between half intensity points, is according to (8.48)

$$\tau_p = \frac{2}{\omega_m}(\ln 2)^{12}\left(\frac{2\alpha_1^2 + 2\alpha_2^2}{\alpha_1}\right)^{1/2} . \tag{8.50}$$

Using (8.39, 40) we obtain

$$\tau_p = \frac{2}{\omega_m} (\ln 2)^{1/2} \delta^{-1/4} \frac{(\psi_2^2 + g_2^2)^{1/4}}{[-\psi_2 + (\psi_2^2 + g_2^2)^{1/2}]^{1/4}} \ . \tag{8.51}$$

From (8.47) we find that the pulse passes through the modulator at an instant $\Delta t = -\vartheta_1/\omega_m$ away from the peak of the modulation cycle. So we find that

$$\Delta\phi_{\text{pulse}} = -\vartheta_1 \tag{8.52}$$

is the phase difference between the modulator and the pulse. (For positive values of $\Delta\phi_{\text{pulse}}$ the pulse is lagging.)

The frequency bandwidth $\Delta\nu_{\text{pulse}}$ of the laser pulse is defined as the frequency between half-power points of the mode spectrum. From (7.83) and (8.39) we get

$$\Delta\nu_{\text{pulse}} = \frac{2\nu_m (2 \ln 2)^{1/2} \delta^{1/4}}{[-\psi_2 + (\psi_2^2 + g_2^2)^{1/2}]^{1/4}} \ , \tag{8.53}$$

where $\nu_m = \omega_m/2\pi$. The line profile being known, the pulse parameters τ_p, $\Delta\nu_{\text{pulse}}$, $\Delta\phi_{\text{pulse}}$, g_0 and ξ, being the frequency difference between the pulse center frequency and line center frequency, can be calculated as a function of $\Delta\omega_{\text{det}}$ by means of (8.50, 53, 52, 44, 41 and 43).

For a Lorentzian line the single-pass intensity gain for the nth mode with frequency $\omega_n + \dot{\phi}$ is given by (7.95). The phase retardation of the nth mode for a round trip through such a medium is

$$\psi_n = \frac{2(\xi + n\omega_m)}{\Delta\omega_N} g \left[1 + 4 \left(\frac{\xi + n\omega_m}{\Delta\omega_N} \right)^2 \right]^{-1} \ . \tag{8.54}$$

Comparing (8.36) with (7.95) we find

$$g_0 = g(1 + p^2)^{-1} \ , \tag{8.55}$$

$$g_1 = -2gpq(1 + p^2)^{-2} \ , \tag{8.56}$$

$$g_2 = 2gq^2(3p^2 - 1)(1 + p^2)^{-3} \ , \quad \text{where} \tag{8.57}$$

$$\frac{2\xi}{\Delta\omega_N} = p \quad \text{and} \quad \frac{2\omega_m}{\Delta\omega_N} = q \ .$$

similarly we obtain by comparing (8.37) with (8.54)

$$\psi_0 = gp(1 + p^2)^{-1} \ , \tag{8.58}$$

245

$$\psi_1 = gq(1 - p^2)(1 + p^2)^{-2} , \tag{8.59}$$

$$\psi_2 = 2gpq^2(p^2 - 3)(1 + p^2)^{-3} . \tag{8.60}$$

By means of (8.55–60) the pulse parameters are calculated for the experimental parameters, $S = 0.3$, $\nu_m/\Delta\nu_N = 0.04$ and different values of the modulation parameters δ as a function of the frequency detuning from the ideal situation. The results are plotted in Figs. 8.5–9. It is found that the frequency shift ξ is practically independent of the modulation depth parameter δ. It should be noted that for the ideal situation, $\Delta\omega_{det} = 0$, where $p = 0$, the pulse parameters are easily derived. Substituting (8.57) for $p = 0$ into (8.51, 53) we find for ideal mode locking

$$\tau_p = \frac{1}{\pi}(2\ln 2)^{1/2}\left(\frac{2g}{\delta}\right)^{1/4}\frac{1}{(\nu_m\Delta\nu_N)^{1/2}} \quad \text{and} \tag{8.61}$$

$$\Delta\nu_{pulse} = (2\ln 2)^{1/2}\left(\frac{2\delta}{g}\right)^{1/4}(\nu_m\Delta\nu_N)^{1/2} . \tag{8.62}$$

These values are the same as those found for a time-domain solution of an optical pulse in a resonator, see (8.18, 19).

It is seen that the pulse parameters change asymmetrically with the detuning frequency and that the negative detuning range is small compared to the positive detuning, which seems to be unlimited provided there is sufficient gain. The asymmetry of the pulse behavior evolves from the asymmetry of the dispersion of the active medium.

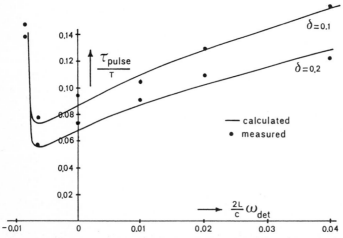

Fig. 8.5. The ratio of τ_{pulse} to the round-trip time T of the optical pulse is plotted as a function of the relative frequency detuning for $S = 0.3$, $\nu_m/\Delta\nu_N = 0.04$ and two different values of δ. The dots represent experimental results

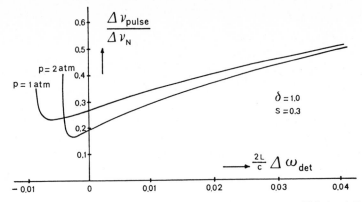

Fig. 8.6. The ratio of the pulse band width to the line width is plotted as a function of the detuning frequency for $S = 0.3$, $\nu_m/\Delta\nu_N = 0.04$ and for 1 and 2 atmosphere $CO_2 : N_2 : He = 1 : 2 : 4$ mixtures

The condition to be fulfilled for pulse forming by means of mode locking is to find a way of obtaining a constant axial frequency separation equal to the modulation frequency. In any laser system the axial mode frequency spacing is not constant for all modes due to the dispersion of the active medium. Furthermore, in the case of pulse forming the gain is not constant for all modes due to the gain profile, so that in order to maintain over a period a certain mode amplitude distribution there must be a redistribution of the amplitudes by passing through the modulator. Both the constancy of mode frequency and the redistribution of the mode amplitudes is achieved by the nonlinear phase spacing between the axial modes, given by α_2. By this mechanism the modes with higher and lower frequencies will pass the modulator effectively at different times.

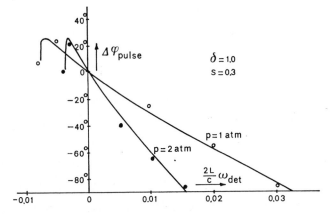

Fig. 8.7. The phase difference between the modulator and the pulse is plotted as a function of the detuning frequency for $S = 0.3$, $\nu_m/\Delta\nu_N = 0.04$ and for 1 and 2 atmosphere $CO_2 : N_2 : He = 1 : 2 : 4$ mixtures

247

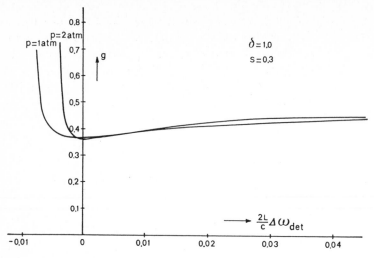

Fig. 8.8. The single-pass power gain at the line center as a function of the frequency detuning for $S = 0.3$, $\nu_m/\Delta\nu_N = 0.04$ and for 1 and 2 atmosphere $CO_2 : N_2 : He = 1:2:4$ mixtures

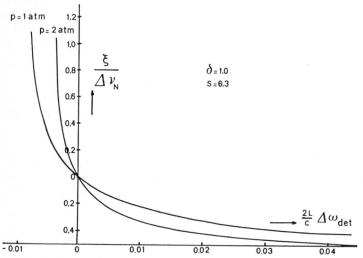

Fig. 8.9. The ratio of the center pulse frequency shift to the line width is plotted versus the relative frequency detuning for $S = 0.3$ and $\nu_m/\Delta\nu_N = 0.04$ for 1 and 2 atmosphere CO_2 mixtures. It turns out that this shift is practically independent of the modulation depth parameter

In the case of detuning the "natural" mode spacing is not equal to the modulation frequency. The pulse will now not pass the modulator at the extremum of the phase variation of the modulator, i.e. $\Delta\omega_{pulse} \neq 0$. The modulator will then give a Doppler shift to the mode spectrum and therefore also to the axial mode separation. The phase difference between modulator and pulse would yield a further change in frequency each time the pulse

248

passes the modulator. This, however, is not the case, because the position of the mode spectrum with respect to the line profile will be such that after a passage through the medium the mode spectrum has suffered an opposite frequency change compared with that caused by the modulator. The modes closer to the line center suffer more gain than those further away from the line center. This means that the pulse-center frequency, i.e. the frequency of the mode having largest amplitude, has changed effectively towards the line center by passing the medium.

8.4 Experiments with FM Mode-Locked TEA Lasers

The experimental system consists basically of a TEA CO_2 laser and a CdTe crystal for electro-optic phase modulation [8.4]. The TEA laser $(1 : 2 : 4 = CO_2 : N_2 : He$ mixture) employs two brass uniform-field electrodes of 30 cm length and a separation distance of 5 mm, shaped according to *Chang* [8.4], with $k = 0.01$. The discharge is fed from a Blumlein circuit with a total capacitance of 6 nF and a charge voltage of 20 kV. The cavity is formed by one totally reflecting mirror with a radius of curvature of 2.5 m and a germanium flat with a reflection coefficient of 70 %.

The mode locking occurs by means of active mode locking with an electro-optic device in order to produce the required periodic phase modulation. The modulator (Sect. 8.2) consists of a CdTe crystal with both ends antireflection coated and dimensions $40 \times 5 \times 5 \, \text{mm}^3$. The modulator is driven by an oscillator that delivers a maximum voltage of 2000 V to the crystal. The long-term frequency stability of the oscillator was within 10 kHz. The modulation frequency at 120.42 MHz was controlled by a quartz crystal. The detuning of the system was performed simply by changing the distance between the mirrors while the modulator frequency remained fixed at this stabilized frequency. The center mirror separation distance for ideal mode-locking is the distance for which the observed beat frequency of the modes of the system with the modulation power switched off is equal to the modulator frequency.

Although the above analyses have been done for steady-state conditions, and the experiments were done under pulsed-discharge conditions, we notice that for the mixture used, the inversion duration time is long enough to obtain a quasi-stationary condition. The observed pulse train consisted of about 20–30 pulses with a duration of about 250 ns. The pulse width at the beginning of the train and at the end of the train were practically equal. By detuning the laser system for fixed parameters of electrical input energy and modulation depth the pulse form disappears abruptly for a small negative detuning. However, for positive detuning the pulse form does not change, but the height of the pulses decreases very slightly. In contrast to AM, the output does not disappear for positive detuning of the system. In

Fig. 8.5 we have also plotted the observed pulse duration as a function of the detuning for two different values of δ. In order to be able to measure the pulse duration with a hot-hole detector with a rise time of 200 ps one has to use small values for the modulation parameter δ; otherwise the measured pulse width would be detector limited. From the figure we see that the detuning is asymmetric and that there is an excellent agreement between the calculated and measured values of τ_{pulse}.

The phase shift of the pulse with respect to the modulation signal was obtained by putting both the unchanged modulation signal and the pulse signal on a dual-beam oscilloscope. The phase shift in which we are interested is then simply obtained by subtracting from the observed phase shift on the oscilloscope, its value for the ideal situation with no detuning. The observed values for gas mixtures of, respectively, 1 and 2 atmospheres are plotted in Fig. 8.7.

9. Passive Mode Locking

The technical requirements for active mode locking, as described in the previous two chapters, are not simply fulfilled. The external source for driving the modulator must be accurately frequency stabilized at the pulse repetition frequency. An alternative for obtaining short pulses, called passive mode locking, does not require these complications. The synchronization of the radiation modulation in the cavity with the pulse oscillation is done by the pulse itself. The mode locking is simply obtained by placing a saturable absorber in the cavity. Passive mode locking for obtaining short pulses turns out to be very practical and easy to perform. In this chapter the physics and performances of passive mode locked systems are described.

9.1 Basic Principles

Since the oscillating radiation in a cavity has a constant period $(2L/c)$ the laser cavity itself can provide a periodic modulation of the radiation field. This principle is used by passive mode locking where no external source for driving the modulator is used. If, for instance, an absorber is placed within the cavity the circulating radiation is periodically modulated. Using a saturable absorber and provided the amplification of the active medium is sufficient to compensate for the absorption, mode locking may be obtained. This can be understood as follows.

Consider the interaction of two light pulses with a saturable absorber. The absorption versus intensity is shown in Fig. 9.1. At low incident power the absorption A is constant, independent of the incident power. At increasing incident power the absorption begins to decrease because the radiation depletes the ground state of the absorber (saturation of the absorber). Next we consider the absorption of two light pulses close behind each other with intensities I_1 and I_2, respectively. The transmitted pulses will differ from the incident pulses in two ways. First, the intensity ratio of the transmitted pulses I_1^1/I_2^1 is smaller than that of the incident pulses, because the weaker pulse is absorbed more than the stronger pulse. Secondly the transmitted pulses are shortened because the wings are also absorbed more than the peaks. We have assumed that the absorber responds to the instantaneous

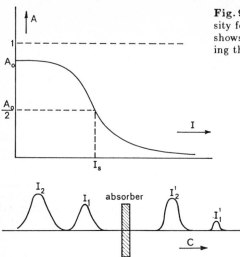

Fig. 9.1. Transmission versus incident intensity for a saturable absorber. The lower part shows the effect of two successive pulses passing through the saturable absorber

pulse intensity, i.e. short relaxation time. This means that the excited state lifetime of the absorber is short compared to the pulse duration. If this condition is not satisfied, the trailing part of the pulse interacts with a medium already bleached by the pulse front and the absorber is ineffective in shortening the end of the pulse. Only the pulse front is shortened by the absorber and the pulse becomes asymmetric.

Next we consider the circulating pulse in a resonator containing an active medium and a saturable absorber. At the onset of laser action the phases of the cavity modes are distributed in a random way and the radiation field in the cavity contains many irregular fluctuations. As the power builds up, the peak intensity of the biggest fluctuation will at some time become comparable to the saturation intensity of the absorber. This peak intensity will start bleaching the absorber. Thus the biggest peak will be absorbed less than the rest of the fluctuations. Assuming that the amplification of the active medium is the same for all fluctuations, it follows that the peak will grow at the fastest rate. This growth rate increases because bleaching becomes more effective at higher intensities. One finally ends up with one big circulating pulse, having extracted most of the energy of the active medium. The output consists of a pulse train. The laser gain is then saturated by this pulse before the smaller fluctuations have grown to a comparable level. The gain of the medium minus the absorption will become negative for the other small fluctuations and they will disappear. Thus the initially chaotic energy distribution is transformed by the saturable absorber into a simple circulating pulse. This process of selectively amplifying the peak fluctuations and discriminating against all weaker spikes corresponds in the frequency domain to locking of the phases of the cavity modes similar to what we have seen in the previous chapters with active mode locking.

9.2 Mode Locking with Fast Saturable Absorption

In the previous chapters analytic treatments of forced mode locking were given. Those analyses were facilitated by the assumption of Gaussian pulses, which turn out to provide very good approximations. Such an approach is not possible in the case of passive mode locking. Nevertheless, passive mode locking has received a great deal of theoretical attention. Most of the theoretical work is based on the statistical process of the build up of short pulses from noise [9.1–4]. In particular, the work of *Letokhov* [9.1] and *Fleck* [9.2] was a breakthrough in the theoretical understanding of this type of mode locking. They presented some computer solutions for a sample in the pulse train after the laser and the saturable absorber are well within the saturation regime, thereby showing that the pulse shapes may be treated deterministically. This result stimulated efforts to develop a nonstatistical treatment, or quasi-steady-state analysis [9.5].

An analytic treatment of steady-state saturable absorber mode locking of a homogeneously broadened laser with a closed-form solution for the pulse shape was then given by *Haus* [9.5]. This analysis will be followed. It contains the following assumptions:

a)   the relaxation time of the absorber is short compared to the pulse width.

b)   The dependence of the absorption of the saturable absorber upon power can be expanded to first order in power.

c)   The relaxation time of the laser medium is relatively long so that the laser gain is approximately time independent.

d)   The mode-locked pulse changes only slightly during one passage through the system.

e)   The pulse spectrum contains only a narrow part of the laser line width.

Although assumption b) imposes the most severe practical constraints it is shown later on that the qualitative conclusions still apply in the case of strong saturation of the absorber. If we also assume that the gain period is sufficiently long for the build up of the pulse the analysis is useful to the understanding of cw passive mode locking of CO_2 TE systems by means of a germanium saturable absorber having a fast relaxation time.

9.2.1 The Fundamental Equation

The mode locking will be studied in the time domain. Similarly to Sect. 7.2 the action on the mode-locked pulse within the laser cavity by each component of the sytem is analyzed and self-consistency after a round trip will be required. We also use the approximation that the pulse band width is narrow compared with the total laser line width. Consider the situation illustrated in Fig. 9.2, which is similar to Fig. 7.5 if we replace the modulator

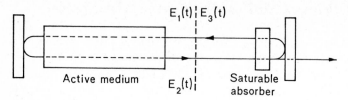

Fig. 9.2. Schematic round trip of a laser pulse through the cavity

by the saturable absorber. The length of the absorber is a small fraction of the total length of the cavity. We also assume as in Sect. 7.2 a uniform intensity profile over the optical beam cross section.

Starting from $E_1(t)$ we obtain for the Fourier transform of the pulse after a round trip through the active medium

$$E_2(\omega) = G(\omega)E_1(\omega) \ , \tag{9.1}$$

where $G(\omega)$ according to (7.34) for $\eta = 0$ is

$$G(\omega) = \exp\left\{g\left[1 - 2\mathrm{i}\left(\frac{\omega - \omega_0}{\Delta\omega_N}\right) - 4\left(\frac{\omega - \omega_0}{\Delta\omega_N}\right)^2\right]\right\} \ . \tag{9.2}$$

It should be noted that the center frequency of the pulse ω_0 coincides practically with the center frequency ω_L of the line because in this way the highest gain is obtained. Because the gain contains power of $\omega - \omega_0$ it is convenient to write the electric field $E(t)$ in terms of a slowly time-varying envelope $v(t)$ and the exponential $\exp(\mathrm{i}\omega_0 t)$:

$$E(t) = v(t)\exp(\mathrm{i}\omega_0 t) \ . \tag{9.3}$$

The Fourier transform of $E(t)$ becomes

$$E(\omega) = v(\omega - \omega_0) \ . \tag{9.4}$$

Going back to (9.1) we obtain for $E_2(t)$

$$E_2(t) = \mathcal{F}^{-1}\{G(\omega)v_1(\omega - \omega_0)\} \ , \tag{9.5}$$

where $\mathcal{F}^{-1}$ means the inverse Fourier transform. After passage of the pulse twice through the saturable absorber, the amplitude of the pulse apart from a phase shift is obtained by multiplying it by $\exp[-2L(t)]$. All cavity losses (including outcoupling) can be taken into account by multiplying the field by $\exp(-\omega_0 T_R/2Q)$ which accounts for the exponential decay of the pulse during T_R as determined by the quality Q of the cavity for the axial modes. The field $E_3(t)$ is then

$$E_3(t) = E_2(t) \exp\left[-2L(t) - \frac{\omega_0 T_R}{2Q}\right] .$$ (9.6)

Further, when the pulse returns to the reference plane the delay time is equal to the cavity round-trip time T_R or

$$E_4 = E_3(t - T_R) .$$ (9.7)

As mentioned above, the pulse width is small compared with the line width. In practical cases for systems of about 60 cm length and an outcoupling mirror with an amplitude reflectivity of about 90 % the gain g is about 0.1 so that the exponentials in (9.2) can be expanded. Taking the inverse Fourier transform it should be noted that the nth power of $i\omega$ becomes d^n/dt^n in the time domain. Each multiplication in the frequency domain by $i(\omega - \omega_0)$ corresponds to $(d/dt) - i\omega_0$ in the time domain, which by using (9.3) becomes $\exp(i\omega_0 t)[dv(t)/dt]$. Substituting (9.5,6) into (9.7) and expanding the exponentials we finally obtain

$$v_4(t) = \left[1 - 2L(t) - \frac{\omega_0 T_R}{2Q}\right.$$
$$\left. + g\left(1 - \frac{2}{\Delta\omega_N}\frac{d}{dt} + \frac{4}{\Delta\omega_N^2}\frac{d^2}{dt^2}\right)\right]v_1(t - T_R) .$$ (9.8)

It is convenient to simplify the notation by introducing

$$\frac{2gQ}{\omega_0 T_R} = G ,$$ (9.9)

which means that the gain is normalized to the loss. In the absence of the saturable absorber $G = 1$. We also introduce

$$Q_A^{(t)} = \frac{\omega_0 T_R}{4L(t)} ,$$ (9.10)

which may be interpreted as the time-dependent Q produced by the saturable absorber. We then have for (9.8)

$$v_4(t) = v_1(t - T_R) - \frac{\omega_0 T_R}{2Q}$$
$$\left[1 + \frac{Q}{Q_A^{(t)}} - G\left(1 - \frac{2}{\Delta\omega_N}\frac{d}{dt} + \frac{4}{\Delta\omega_N^2}\frac{d^2}{dt^2}\right)\right]v_1(t - T_R) .$$ (9.11)

The first term in the square brackets is the effect of the linear cavity loss, the second term represents the modulation by the saturable absorber, the term with d/dt expresses the effect of the gain and dispersion of the active medium, and the last term with d^2/dt^2 describes diffusion of the pulse along the time coordinate. Self-consistency requires an additional small time delay or advance after one period, because of the change of the pulse envelope with respect to the center frequency. The pulse period T_P is not necessarily equal to the round-trip time T_R. So we substitute

$$v_4(t) = v_1(t - T_R + \delta T) \; . \tag{9.12}$$

The time δT is small so that the last expression can be expanded with respect to δT. Taking into account only the first order of δT, by substituting (9.12) into (9.11), we find the desired equation for the steady-state pulse envelope $v(t)$

$$\left[1 + \frac{Q}{Q_A^{(t)}} - G\left(1 + \frac{4}{\Delta\omega_N^2} \frac{d^2}{dt^2} \right) + \frac{2G + \delta}{\Delta\omega_N} \frac{d}{dt} \right] v(t) = 0 \; , \tag{9.13}$$

where

$$\delta = \frac{2Q\Delta\omega_N \delta T}{\omega_0 T_R} \; . \tag{9.14}$$

9.2.2 Saturable Absorber and Laser Gain

If the relaxation time of the absorber is fast compared with the rate of change of the intensity the population difference is an instantaneous function of the intensity. This is for instance applicable to Ge saturable absorbers. For homogeneous broadening we then have

$$n = \frac{n_0}{1 + (P/P_A)} \; , \tag{9.15}$$

where n_0 is the equilibrium population density of the absorbing level, P is the radiation power equal to $A_A |v(t)|^2$, where A_A is the cross section of the optical beam in the absorber, and P_A is the saturation power of the absorber. For inhomogeneous broadening we obtain similarly

$$n = \frac{n_0}{\sqrt{1 + (P/P_A)}} \; . \tag{9.16}$$

We assume that P/P_A is small so that n can be expanded to first order in power; i.e.,

$$n = n_0 \left(1 - \frac{A_A |v(t)|^2}{a P_A} \right) \; , \tag{9.17}$$

where a is 1 or 2 for the homogeneous and inhomogeneous absorption line, respectively. The amplitude absorption after single passage through the absorber is given by

$$L(t) = \frac{1}{2}\sigma_A \Theta_A n_0 \left(1 - \frac{A_A |v(t)|^2}{aP_A}\right) , \qquad (9.18)$$

where σ_A is the optical cross section of the absorbing particles and Θ_A is the length of the absorber. Using (9.18, 10) we obtain

$$\frac{Q}{Q_A(t)} = q\left(1 - \frac{A_A |v(t)|^2}{aP_A}\right) , \quad \text{where} \qquad (9.19)$$

$$q = \frac{2Q\sigma_A \Theta_A n_0}{\omega_0 T_R} . \qquad (9.20)$$

For the fast absorber the differential equation for the pulse envelope is obtained by substituting (9.19) into (9.13). The intensity field $v(t)$ depends not only on the absorption and amplification but also on the beam geometry. This is especially true for a focused beam through the absorber in order to improve the saturation. By introducing the power field $u(t) = v(t)\sqrt{A_A}$ we can avoid this complication. We then obtain for the pulse envelope

$$\left[1 + q + \frac{2G + \delta}{\Delta\omega_N}\frac{d}{dt} - G\left(1 + \frac{4}{\Delta\omega_N^2}\frac{d^2}{dt^2}\right)\right]u(t) = q\frac{|u(t)|^2}{aP_A}u(t) . \qquad (9.21)$$

It should be noted that for a homogeneously broadened line the saturation power is given by

$$P_A = \frac{\hbar\omega_0 A_A}{\sigma_A \tau_A} , \qquad (9.22)$$

where τ_A is the relaxation time of the absorbing medium. If the absorber behaves like a two-level system with equal relaxation times of the upper and lower levels, σ_A has to be replaced by $2\sigma_A$.

The relaxation time of the homogeneously broadened CO_2-laser medium is slow compared with the pulse-repetition time. This means that G does not have any appreciable time dependence. Then G is given by

$$G = \frac{G_0}{1 + (P/P_L)} , \qquad (9.23a)$$

where G_0 is the small-signal value of G and P_L is the saturation power

$$P_L = \frac{\hbar\omega_0 A_L}{\sigma_L \tau_L} , \qquad (9.23b)$$

where σ_L is the cross section of stimulated emission of the laser transition (Sect. 3.2), A_L the cross section of the laser beam in the active medium and τ_L the relaxation time of the active medium.

9.2.3 Pulse Form

We shall now look for the solution of (9.21) corresponding to mode-locked pulses. The time shift parameter δ is an adjustable parameter to be determined by the boundary conditions of the problem. We are looking for periodic solutions. Equation (9.21) can be considered as the motion of a particle with displacement $u(t)$ in a potential well

$$-\frac{1}{2}(1 + q - G)u^2 + \frac{1}{4}\frac{qu^4}{aP_A} \ . \tag{9.24}$$

The motion term containing d/dt can be considered as a dissipative force. It is clear that no periodic solutions exist when this term is present and thus we may state that the pulses have their repetition period if $2G + \delta = 0$. The remaining equation becomes

$$\frac{4G}{\Delta\omega_N^2}\frac{d^2u(t)}{dt^2} = (1 + q - G)u(t) - \frac{q|u(t)|^2}{aP_A}u(t) \ . \tag{9.25}$$

Considering the last equation again as the motion of a particle with mass $4G/\Delta\omega_N^2$ and displacement $u(t)$ it is seen that if the particle is launched in the potential well at the displacement

$$u_0 = \left(\frac{2aP_A}{q}(1 + q - G)\right)^{1/2} \tag{9.26}$$

with zero velocity, it moves to the origin and stops there. The solution is pulse-like and symmetric in time and has the time dependence

$$u(t) = \frac{u_0}{\cosh(t/\tau_p)} \ , \quad \text{where} \tag{9.27}$$

$$\tau_p = \frac{2}{\Delta\omega_N}\left(\frac{G}{1 + q - G}\right)^{1/2} \ . \tag{9.28}$$

This solution is an isolated pulse. A succession of periodic pulses of any desired period T_p is obtained by launching the particle at a lower height. However, requiring $\tau_p \ll T_p$, which is in experimental agreement with well mode-locked pulses, the single pulse according to (9.27) is an excellent approximation to one period of the periodic pulse train.

It is seen that the pulse has exponential tails, which should be contrasted with the Gaussian time dependence of active mode locking, treated in Chaps. 7 and 8. Further, we note that G is less than $1 + q$, which means that in the steady state the laser is below threshold with respect to the linear loss in the absence of laser power. This is possible because the laser power bleaches the saturable absorber and thus reduces the loss below the linear loss. In fact, the condition $1 + q - G > 0$ is a necessary requirement for the stability of a train of isolated pulses, because for $1 + q - G < 0$ and the long dead time between pulses, relatively slow noise perturbations could grow between pulses and the solution would not be stable.

In practice, the parameters that are specified by the system are the normalized small-signal gain G_0, the saturation power of the laser P_L, the normalized small-signal absorption parameter q, and the saturation power of the absorber P_A. The pulse width τ_p and the laser power can be described in terms of these parameters. The pulse energy is given by

$$\int_{-\infty}^{\infty} \frac{u_0^2 dt}{\cosh^2(t/\tau_p)} = 2u_0^2 \tau_p \ . \tag{9.29}$$

The repetition frequency is $1/T_p$ so that the average laser power is given by

$$P = \frac{2u_0^2 \tau_p}{T_p} \ . \tag{9.30}$$

Introducing the last expression into (9.28) and using (9.26) one obtains the relation

$$qK \frac{P}{P_L} = \frac{2G}{\Delta \omega_N \tau_p} \ , \quad \text{where} \tag{9.31}$$

$$K = \frac{1}{8} \frac{P_L}{a P_A} T_p \Delta \omega_N \ . \tag{9.32}$$

The coefficient K is a measure of the laser saturation power normalized to the saturation absorption power. Equations (9.31, 32) supplemented by the dependence of G upon the laser power P yield three equations for the unknowns P/P_L, G, and $\tau_p \Delta \omega_N$ in terms of q, K, and G_0. Using (9.28) to eliminate G we obtain from (9.31, 23a) one equation for $\tau_p \Delta \omega_N$ and one for P/P_L, i.e.,

$$\frac{G_0}{1+q} \left(\frac{1}{2} \tau_p \Delta \omega_N + \frac{2}{\tau_p \Delta \omega_N} \right)^2$$
$$= \frac{1}{2} \tau_p \Delta \omega_N \left(\frac{1}{2} \tau_p \Delta \omega_N + \frac{2}{\tau_p \Delta \omega_N} + \frac{1+q}{qK} \right) \ , \tag{9.33}$$

$$\frac{qK}{1+q} \frac{P}{P_L} = \left(\frac{1}{2} \tau_p \Delta \omega_N + \frac{2}{\tau_p \Delta \omega_N} \right)^{-1} \ . \tag{9.34}$$

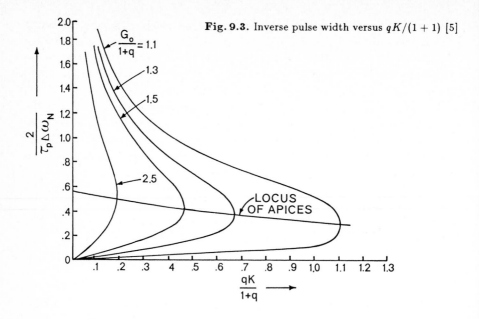

Fig. 9.3. Inverse pulse width versus $qK/(1+1)$ [5]

The solution of $2/\tau_p \Delta\omega_N$ versus $qK/(1+q)$ obtained from (9.33) is plotted in Fig. 9.3. The excess gain $G_0/(1+q)$ is used as a parameter. It is seen from the figure that only under certain conditions mode locking is obtained. No mode locking is obtained if for a fixed excess gain parameter $G_0/(1+q)$ the value of $qK/(1+q)$ becomes too large, i.e. the signal absorption of the saturable absorber becomes too large or its saturation power P_A becomes too small. The figure also indicates that for each appropriate set of parameters, in general, two solutions corresponding to different pulse widths are obtained. It can be shown that only the solution with the larger pulse width τ_p is stable [9.5]. Only the domain bounded by the locus of apices and the abscissa of Fig. 9.3 contains stable solutions. For small values of $2/\tau_p \Delta\omega_N$ in this domain one obtains from (9.33)

$$\frac{2}{\tau_p \Delta\omega_N} = \left(\frac{G_0 - 1 - q}{1 + q}\right) \frac{qK}{1 + q} \,. \tag{9.35}$$

Thus the stable branches of the curves in Fig. 9.3 are approximately straight lines.

The powers of the mode locked pulses are obtained from (9.34) and plotted in Fig. 9.4. The two branches of solutions having different pulse width for the same $qK/(1+q)$ as obtained from (9.33) show also different powers. Only the lower branches corresponding to the larger pulse widths are of interest.

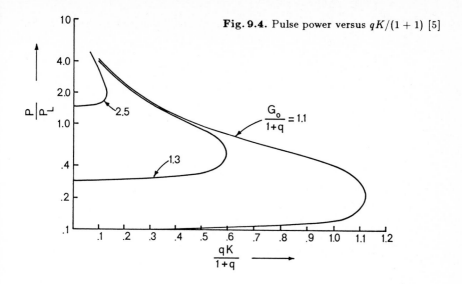

Fig. 9.4. Pulse power versus $qK/(1+1)$ [5]

9.2.4 General Mode-Locking Criteria

So far the mode-locking solution, pulse form, and the stability domain of the system parameters have been described. Next, we want to discuss how to optimize the parameters to achieve shorter pulses, maximum pulse energy, or both. For this purpose it is useful to study Fig. 9.5 where the locus of apices of Fig. 9.3 is plotted as the stability boundary in the $G_0/(1+q) - qK/(1-q)$ plane. The range of parameter values lying outside this boundary corresponds to unstable solutions. The loci of constant $1/\tau_p \Delta\omega_N$ are also shown in this figure.

The choice of the minimum value of τ_p depends on the line width $\Delta\omega_N$. The lower limit of $1/\tau_p \Delta\omega_N$ may be set by the value $\tau_p \Delta\omega_N = 40$ which is drawn in Fig. 9.5. In order to get single-pulse mode locking it is also necessary that the round-trip time is larger than twice the half-power width of the pulse; otherwise there will be undesired pulse overlap. This means that $T_R > 3.5 \tau_p$.

With respect to the laser and saturable absorber parameters we continue as follows. A tangent along the curve of maximum pulse width is drawn. This line intersects the abscissa at K and the ordinate at G_0. In our example both values are equal to 1.44. The value of K, as given by (9.32), depends on the ratio of the saturation powers of the active medium and saturable absorber. The saturation power P_A depends on absorbing species and cross section but not on the density or material thickness. The q value of the absorber is determined by its thickness (Θ) and its concentration (n_0). If one varies the q value for the fixed values of G_0 and K one moves in Fig. 9.5 along a straight line. The tangent mentioned above is such a line. It has at the abscissa the value $q = \infty$ and at the ordinate $q = 0$. The point for

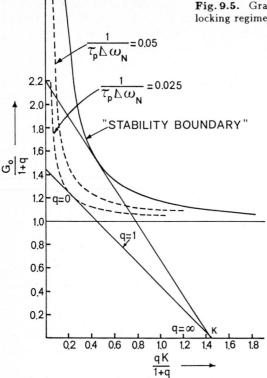

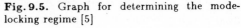
Fig. 9.5. Graph for determining the mode-locking regime [5]

which the small signal absorption loss is equal to the cavity loss $(q = 1)$ is the half-point on the line. It is seen from the figure that a small-signal gain G_0 in excess of 1.44 requires a smaller saturation power of the absorber so that K becomes larger than 1.44 to obtain well-separated pulses. Further, we see that for the given K value of 1.44 the maximum small-signal gain G_0 is equal to 2.2.

9.2.5 Mode Locking the CO_2 System

The theory presented so far used the assumption that the pulse intensity is small compared to the saturation intensity. This, however, is not a necessary condition for obtaining good mode locking. The assumption was only used for mathematical reasons. If this assumption is dropped one does not obtain a closed-form solution, yet the resulting pulse is symmetric and has exponential tails. This follows from the fact that the differential equation for $u(t)$ is second order, and in the tails, the saturation of the absorber can be disregarded, making the equation linear in this limit.

In the preceding analysis it was also assumed that the laser gain is time independent, that the passage of a single steady-state pulse produces a negligible temporal variation of the gain, and that the saturation of the gain to its steady-state value was accomplished by the passage of many round trips. This means that the relaxation time of the active medium τ_L is much greater than the pulse width. This, however, is not the case for a CO_2 system. As we know, the CO_2 system has a fast rotational relaxation time and a much slower vibrational relaxation time or vibrational energy transfer time between N_2 and CO_2. The gain is thus time dependent. We must reconsider the mode locking conditions.

If the parameters of the system are such that the mode-locked pulses are longer than the rotational relaxation time, but shorter compared with the vibrational relaxation time, the saturation of the gain can then be described by the expression

$$G' = \frac{G_0}{[1 + (P/P_L)][1 + (|u|^2/P_{rot})]} \quad , \tag{9.36}$$

where P_L is the total saturation power of the CO_2 system and P_{rot} is the saturation energy of a rotational level. It should be noted that P is the time average laser power, whereas $|u|^2$ is the instantaneous pulse power. Expansion of G' for $|u|^2/P_{rot} < 1$ gives

$$G' = G\left(1 - \frac{|u|^2}{P_{rot}}\right) \quad . \tag{9.37}$$

We may now study the additional conditions necessary to achieve mode locking in a CO_2 system. We have to replace G in (9.25) by the field-dependent G' given by (9.37). This gives us the extra term $G|u|^2/P_{rot}$. In the left-hand side of (9.25) we may still use only the time-independent part of G because the term $(4/\Delta\omega_N^2)d^2/dt^2$ describes the dispersive effect which is relatively small. (The Fourier spectrum of the pulse is narrow compared with the laser line width and hence small changes in the coefficient multiplying this operator may be neglected). The potential that is now obtained for the motion of the particle with the displacement $u(t)$ has the form

$$-\frac{1}{2}(1 + q - G)u^2 + \frac{1}{4}\left(\frac{q}{aP_A} - \frac{G}{P_{rot}}\right)u^4 \quad . \tag{9.38}$$

The particle with displacement u can only move from a position $u > 0$ with zero potential and with zero velocity to the origin, also having zero potential if $q/aP_A > G/P_{rot}$. Thus we have the additional condition

$$G\frac{aP_A}{P_{rot}} < q \quad . \tag{9.39}$$

This means that the rotational saturation power which determines the instantaneous response of the medium must not be too small. The saturation power is proportional to the beam area. In general, the ratio of the beam area at the amplifier to that at the absorber is very critical and will be chosen much larger than one in order to fulfill the last condition.

9.3 Mode Locking by P-Type Germanium

Experimental studies with passive mode-locking techniques of CO_2 lasers have been performed using saturable gases such as SF_6 [9.6], BCl [9.7], and hot CO_2 [9.8] and bleachable semiconductors like Ge [9.9]. Investigations by *Gibson* and co-workers [9.9] have shown that properly doped Ge can serve as a fast relaxing passive mode-locking element. It can be simply made as a plane-parallel flat that is even simultaneously used as the outcoupling reflector. However, experimental work has shown that stable and clean mode locking is obtained by inserting a plane-parallel flat with anti-reflection coatings inside the cavity as close as possible to one of the mirrors [9.10]. One may use p-type Ga-doped or In-doped single-crystal germanium of about 1-Ω cm resistivity. The small-signal absorption coefficient for Ga-doped material is about $1\,\mathrm{cm}^{-1}$ and for In-doped Ge about $6\,\mathrm{cm}^{-1}$. Using the latter material for a 1 mm thick element, nearly band-width limited pulses of 400 ps in a 600 torr CO_2 laser have been produced [9.10]. The gain medium of that system was 60 cm long and had a small-signal gain of 3.8 % per cm. The cavity was 2.53 m long and formed by one total reflector with a radius of curvature of 3 m and an outcoupling flat of 85 % reflectivity. Axial-mode operation was assured by means of a diaphragm. For the reliable generation of shorter pulses the most critical parameters were found to be the position of the saturable absorber and the excess gain of the system. Decreasing the separation between the absorbing plate and the output mirror reduces the pulse duration. The pulse broadening by the increased separation between absorber and mirror is due to the interference effects between the part of the pulse traveling towards the mirror and the reflected portion of the pulse. If twice this separation distance is comparable to the pulse rise time, then the peak of the pulse may overlap with its tail. In such a situation the pulse tail in the saturable absorber is enhanced relative to the pulse peak, which leads to pulse broadening. Decreasing the excess gain increases the stability of the pulses. This can be expected from Fig. 9.5 because it shows that we are further away from the stability boundary. The shortest pulses of 400 ps for the above-described system were obtained by operating the system near threshold with a small-signal gain of 1.86 % per cm. The cavity design with the saturable absorber near the outcoupling mirror is such that the ratio of the beam size in the laser medium to that in the saturable absorber is 9. This ratio is observed to be very critical for operation.

The saturable Ge absorber can be described by a two-level model with a relaxation time of a few picoseconds which is considerably less than the shortest pulses available from high-pressure CO_2 lasers. The absorber can therefore respond in a steady-state fashion over the incident pulse duration and the absorption can be considered as an instantaneous function of the intensity.

Further, experimental [9.11,12] and theoretical [9.13] studies on the saturation behavior of Ge have revealed that the behavior can be described as an inhomogeneously broadened absorption, as given by (9.16). In the expansion of the absorption, as expressed by (9.17), we have $a = 2$. The saturation power density which depends only on the type of absorbing species and not on the doping concentration was found in the range of $3\,MW/cm^2$ [9.14]. This saturation power is, in general, larger than that of the laser. The inequality (9.39) which is difficult to satisfy will therefore require a large beam ratio. This may be the reason why mode locking is very critical on this beam ratio and that values close to 10 are required for obtaining good mode locking. This requirement is less critical with the use of gases as the absorber because of their lower saturation powers.

The advantage of a saturable semiconductor compared to a gas absorption cell is not only its simple use but much more its small thickness and close position to the mirror. Reproducable pulse trains (less than 10 % variation in peak intensity) with a high contrast ratio of the largest pulse to the second largest pulse (larger than 100) in the window of a round trip are obtained.

For a TEA system the shortest pulses using p-type Ge are about 0.5 ns [9.9]. Since the pulse width depends on the line width one might expect to obtain much shorter pulses for laser systems operating at multi-atmosphere pressures. For instance, at 10 atm pressure the pressure-broadened line width is about 50 GHz so that as far as the line width is concerned the pulse duration may be an order of magnitude smaller. However, since this increase in line width results in a considerable overlap of the adjacent rotational lines and the total line width of such a system will be effectively about 600 GHz, pulses of the order of a few picoseconds should be expected. This large pulse compression is not realized in a TE system. Using a uv-preionized system operating at 12 atm pressure, pulses of about 150 ps have been generated [9.15]. There are two reasons for this limited compression. First there is the reduced gain pulse. Due to higher pressure the inversion relaxes faster. The time constant is inversely proportional to the pressure. Second, the shorter the pulse the more round trips are required to take full profit of the large line width.

9.3.1 Active and Passive Mode Locking

Although with passive mode locking, in general, shorter pulses can be generated than with active mode locking the pulse structure and reliability of the performance is less. This can be significantly improved with simultaneous active and passive mode locking. In addition to higher energy stability, less jitter and reliable signal pulse operation in the time window $2L/c$ other properties of the laser can be improved. The main attractive feature of the combination of passive and active mode locking is that the minimum pulse width is determined by the passive mode locking mechanism whereas the operating parameters of the laser are not critical and can be varied over quite a wide range, permitting certain properties of interest to be selectively optimized.

9.3.2 Colliding Pulse and Locking

A sharp improvement of passive mode locking can also be obtained by using a resonator configuration in which two mode-locked pulses coming from opposite directions collide in a saturable absorber. This technique was first demonstrated with a dye laser [9.16]. One method is to place the saturable absorber exactly in the middle between the cavity mirrors. In such a cavity two pulses circulating in opposite directions will be generated. These pulses simultaneously traverse the absorber. Another method is a ring laser with pulses circulating in both directions. For both methods the interaction of the counterpropagating pulses creates a transient grating in the population of absorbing molecules, which synchronizes, stabilizes and shortens the pulses. The synchronization occurs because minimum energy is lost in the saturable absorber when the two pulses meet in the absorber.

A third method is the use of a cyclic interferometer containing the saturable absorber as an end mirror (Fig. 9.6) [9.17]. This can be understood as follows. If the beam splitter has an equal power division for the two beams, then any signal coming into this interferometer from the laser will be divided into two parts which will travel around the ring in opposite directions and recombine so that all the signal will return to the laser along the same axis. This happens for any signal irrespective of frequency or amplitude distributions, provided the beam splitter remains 50/50 at all frequencies. The collision of the two pulses in the saturable absorber is thus obtained by locating the absorber exactly at the midpoint as illustrated in Fig. 9.6. In this configuration there is again only one circulating pulse that passes the gain medium twice before outcoupling.

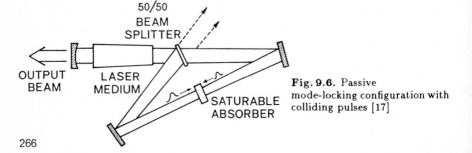

Fig. 9.6. Passive mode-locking configuration with colliding pulses [17]

10. Short-Pulse Amplification

In order to understand the behavior of short-pulse amplification in CO_2 systems, two physical processes have to be considered in detail. First it is necessary to understand the interaction of the radiation field with the inverted medium. This interaction becomes complicated when we are dealing with high-power amplification where we want to extract all the available energy from the inverted medium. The complication results from the fact that the leading edge of the propagating pulse experiences the highest gain, whereas later on the amplification becomes much less due to saturation. This affects not only the pulse shape but also the gain distribution under the pulse profile while propagating. In the case of a single transition where the population densities of the upper and lower level are only disturbed by the interaction of a one-dimensional optical pulse (plane wave) this problem can be treated analytically. Thereby it is assumed that the gain is not frequency dependent over the spectrum range of the pulse. The treatment of this problem by *Frantz* and *Nodvik* [10.1] gives a useful insight into the amplification of an intense short pulse.

The second process that has to be taken into account in pulse amplification is the molecular relaxation. As we discussed in Chap. 2, there are many rotational sublevels connected to a vibrational transition. Those sublevels are strongly coupled by molecular collisions. This means that if a rotational upper sublevel is depleted by the pulse interaction this sublevel will be refilled from the non-lasing sublevels belonging to the same vibrational band. A similar effect occurs with the lower sublevel. The density of the lower level increases and consequently relaxation to other sublevels will follow. As well as this rotational relaxation there may also be vibrational relaxations during the amplification process.

The pulse interaction also disturbs the energy distribution within the molecular vibrations associated with the transitions. This means that intramode relaxations occur. Furthermore there may also be relaxations between the molecular vibrations. As we discussed in Chap. 2, there is a strong coupling between the ν_1 and ν_2 vibrations due to the near resonances so that an increase of energy of the ν_1 and ν_2 vibration by the laser process will cause an energy transfer to the ν_2 and ν_1 vibration, respectively.

All these relaxation phenomena act, in fact, as pumping sources for the inverted lasing transition, the strengths of which depend on the distortion of the lasing levels from their local equilibrium distribution. It will be clear that this relaxation pumping during the pulse propagation is only of interest if the characteristic time of a particular relaxation process is comparable to or smaller than the pulse duration. In the case of nanosecond pulses in TEA CO_2 amplifiers only the rotational relaxation and the intramode relaxation turn out to be important.

In this chapter we shall first consider the simple two-level system and after that we shall treat numerically the more complicated situation that includes relaxation too. It will be demonstrated that the efficiency of energy extraction by short pulses increases considerably, depending on pulse duration and gas density, if multi-frequency pulses are chosen that interact with several transitions simultaneously.

10.1 Pulse Propagation in a Two-Level System

We consider the interaction of a radiation field of intensity I with a two-level system. A one-dimensional beam propagating in the x direction is incident on an inverted medium occupying the region $0 \leq x \leq L$. The number densities of the upper and lower states are n_2 and n_1, respectively. The inversion density is $\delta = n_2 - n_1$. Before the beam enters the medium $\delta = \delta_0$, independent on x. The change of intensity of the radiation field is given by

$$\frac{\partial I}{\partial t} + c\frac{\partial I}{\partial x} = c\sigma\delta I , \tag{10.1}$$

where σ is the stimulated emission cross section which will be considered constant, independent of the pulse spectrum. The inversion density changes according to

$$\frac{\partial \delta}{\partial t} = -\frac{2\sigma}{h\nu}\delta I , \tag{10.2}$$

where it is assumed that during the short pulse the inversion is only influenced by the radiation field. The coherent effect of the induced polarization of the medium is negligible because it is also assumed that the dephasing time of the polarization by collisions is short compared to the pulse duration time. After some mathematical manipulations the solution of (10.1,2) for I and δ can be written as [10.1]

$$I(x,t) = \frac{I_0(t - x/c)}{1 - [1 - \exp(-\sigma\delta_0 x)]\exp\left[-\frac{2\sigma}{h\nu}\int\limits_{-\infty}^{t-x/c} I_0(t')dt'\right]} , \tag{10.3}$$

$$\delta(x,t) = \frac{\delta_0 \exp(-\sigma\delta_0 x)}{\exp\left[\frac{2\sigma}{h\nu}\int\limits_{-\infty}^{t-x/c} I_0(t')dt'\right] + \exp(-\sigma\delta_0 x) - 1} , \tag{10.4}$$

where $I_0(t)$ is the intensity of the incoming pulse at $x = 0$. The pulse intensity that enters the amplifier at time t leaves it at time $t + L/c$. Consequently for an input intensity $I_0(t)$ the corresponding output intensity is obtained by

$$I_L(t) = I(L, t + L/c) . \tag{10.5}$$

Substituting (10.5) into (10.3) the intensity distribution of the outgoing pulse can be obtained as a function of the intensity distribution of the incoming pulse. Integrating the calculated intensity the amplified pulse energy is obtained. Thus both pulse amplification and pulse shape are given. In the case of pulse amplification only we use for the outgoing pulse

$$dE_L = I(L, t + L/c)dt . \tag{10.6}$$

For the incoming pulse we use

$$dE_0 = I_0(t)dt \tag{10.7}$$

and for the part of the pulse that has passed $x = 0$

$$E_0(t) = \int\limits_{-\infty}^{t} I_0(t')dt' . \tag{10.8}$$

The outgoing pulse is now

$$dE_L = \frac{dE_0}{1 - [1 - \exp(-\sigma\delta_0 L)]\exp[-E_0(t)/E_s]} , \tag{10.9}$$

where E_s is the saturation energy

$$E_s = \frac{h\nu}{2\sigma} . \tag{10.10}$$

Equation (10.9) can also be written as [10.2]

$$dE_L = E_s\frac{\partial}{\partial t}\ln(1 + \exp(\alpha_0 L)\{\exp[E_0(t)/E_s] - 1\}) , \tag{10.11}$$

where $\alpha_0 = \sigma\delta_0$ is the small-signal gain. After integration we obtain

$$\frac{E_L}{E_s} = \ln\{1 + \exp(\alpha_0 L)[\exp(E_0/E_s) - 1]\} , \tag{10.12}$$

where E_0 is the total input energy of the pulse. If $E_0 \exp(\alpha_0 L)/E_s \ll 1$, we obtain

$$E_L = E_0 \exp(\alpha_0 L) \tag{10.13}$$

which is just the small-signal gain equation. Under conditions of large saturation, $E_0/E_s \gg 1$, we find

$$E_L = E_0 + \alpha_0 L E_s = E_0 + \frac{h\nu\delta_0 L}{2} \tag{10.14}$$

which states that the increase in the radiation energy is one half the energy stored in the inverted medium. (The factor 2 arises from the fact that of each photon produced one upper state is changed to a lower state which means that the inversion is decreased by two particles.) The term $h\nu\delta_0 L/2$ is thus the maximum available energy from the amplifier in the case of a two level system.

10.2 Pulse Propagation with Rotational Relaxation

In the previous section we assumed that the changes of upper and lower levels were only caused by the interacting radiation field. This approach is only realistic for CO_2 amplifiers if the pulse duration is so short that all relaxation processes that contribute to the inversion are negligible. This, however, is not the case if the product of total gas pressure times the pulse time is comparable to or larger than 5×10^{-10} bar $\cdot$s. For these pulses the rotational relaxation affects the inversion and the available energy may become much more than that predicted by the two-level model. Nevertheless a nanosecond CO_2-laser pulse consisting of one single vibrational-rotational transition may not extract at one atmosphere all the available energy from the vibrational band. In order to extract more energy the incident pulse must consist of several vibrational-rotational lines. Even more dramatic improvements are possible if the pulse contains vibrational-rotational transitions in both the (00^01-I) and (00^01-II) bands.

10.2.1 Rotational Relaxation

The distorted population density of the rotational sublevel interacting with the field is collisionally coupled to all other rotational levels. During collisions there will be a net change of rotational states to counterbalance the distortion. One might then argue the effect of the rotational energy change on the collisional transfer propability. It is expected that the larger the rotational change Δj the smaller the transfer probability. Experiments with low pressure sytems have indeed revealed that there is some dependence on Δj. However, this effect is very small [10.3, 4]. Therefore in the following we neglect this effect on Δj and consider all collisions with equal probability, i.e. we use the reservoir model [10.5].

Let n_j be the population density of the jth rotational sublevel of a vibrational state, see (2.20), and $k_{jj'}$ the collision frequency with which a molecule in the sublevel j is transferred to the sublevel j'. The rate of change of the number density n_j within a vibrational level is then

$$\frac{dn_j}{dt} = \sum_{j'} (n_{j'} k_{j'j} - n_j k_{jj'}) \ . \tag{10.15}$$

We replace $k_{j'j}$ by k_j because it is independent on j'. We now write for (10.15)

$$\frac{dn_j}{dt} = k_j \sum_{j'} n_{j'} - n_j \sum_{j'} k_{j'} \ . \tag{10.16}$$

At equilibrium where $n_j = \bar{n}_j$ and $dn_j/dt = 0$ we must have the condition

$$\frac{k_j}{\sum\limits_{j'} k_{j'}} = \frac{\bar{n}_j}{\sum\limits_{j'} n_{j'}} = P(j) \ , \tag{10.17}$$

where according to (2.20)

$$P(j) = \left(\frac{2hcB}{kT}\right)(2j+1)\exp\left[-F(j)\frac{hc}{kT}\right] \quad \text{and} \tag{10.18a}$$

$$\sum_{j'} n_{j'} = N \tag{10.18b}$$

is the population density of the vibrational band considered. This result means that the probability of a collision involving a transfer or exchange of rotational energy forming a molecule in the jth state is directly proportional to the probability distribution of that state within the vibrational level.

The rotational relaxation time τ_r of a molecule, defined as the time for changing the rotational energy, is given by

$$\tau_r = \left(\sum_{j'} k_{j'}\right)^{-1} \tag{10.19}$$

which is about 0.15–0.2 ns in typical CO_2-laser mixtures at 1 atm. It should be noted that τ_r is the lifetime of a rotational state before it changes to *any*

271

other state which is *not* equal to the time during which all rotational states have been effectively changed into one particular state, i.e. the considered j state. The fraction of the manifold of rotational states of a vibrational level that is transferred into the j state is equal to $P(j)$. This means that the time needed to transfer all molecules with this vibrational state into its j state is equal to $\tau_r/P(j)$. In the case where the j level is the lasing sublevel it is evident that $\tau_r/P(j)$ is the minimum time the pulse needs to draw on the total rotational population.

If n_{vj} is the lower-laser-level density we may now write for (10.16) by using (10.17, 19)

$$\frac{dn_{vj}}{dt} = \frac{1}{\tau_r}[P(j)N_v - n_{vj}] \ . \tag{10.20}$$

Assuming the same value for the rotational relaxation time of the upper level we have similarly

$$\frac{dn_{v'j'}}{dt} = \frac{1}{\tau_r}[P(j')N_{v'} - n_{v'j'}] \ . \tag{10.21}$$

10.2.2. Short Pulse Multiline Energy Extraction with Rotational Relaxation

Energy extraction by a short pulse can be considerably increased if the pulse contains several transitions of a vibrational band [10.6]. The treatment will then be a summation of several two-level systems, as discussed in Sect. 10.1. This will be further extended by including the rotational relaxations of the non-lasing rotational sublevels. For efficient extraction it is not necessary to include both P and R transitions because they interact with the same upperlevels. Since the gain of a P transition is higher we consider in the following only P transitions.

Similar to (10.1) we obtain for the radiation field

$$\frac{\partial I}{\partial t} + c\frac{\partial I}{\partial x} = c\sum_j \sigma_j \delta_j I_j \ , \tag{10.22}$$

where the intensity I_j is the part of the field that contains the frequency of the P transition with lower level j and inversion density δ_j, $I = \sum_j I_j$, δ_j is according to (3.31) for a P transition

$$\delta_j = n_{v'j'} - \frac{2j-1}{2j+1}n_{vj} \tag{10.23}$$

and $\sigma_j = \sigma_{v'j'\to vj}$, as given in (3.15). For the population density of the upper laser level we have with relaxation

272

$$\frac{dn_{v'j-1}}{dt} = \frac{1}{\tau_r}[P(j-1)N_{v'} - n_{v'j-1}] - \frac{\sigma_j}{h\nu}\delta_j I_j \ . \tag{10.24a}$$

Similarly we find for the lower level

$$\frac{dn_{vj}}{dt} = \frac{1}{\tau_r}[P(j)N_v - n_{vj}] + \frac{\sigma_j}{h\nu}\delta_j I_j \ . \tag{10.24b}$$

Multiplying both sides of (10.24b) by $(2j-1)/(2j+1)$ and subtracting this from (10.24a) readily yields

$$\frac{d\delta_j}{dt} = \frac{1}{\tau_r}(\bar{\delta}_j - \delta_j) - \frac{\sigma_j}{h\nu}\delta_j I_j\left(1 + \frac{2j-1}{2j+1}\right) \tag{10.25}$$

with $\bar{\delta}_j$ as the "quasi-equilibrium" population

$$\bar{\delta}_j = P(j-1)N_{v'} - \frac{2j-1}{2j+1}P(j)N_v \ . \tag{10.26}$$

The rate of change of the total population density of a vibrational level is obtained by taking the sum of all rotational sublevels, i.e. the sum of (10.24a) for all rotational levels. For the total population of a vibrational level the rotational relaxation plays no role because the total density does not change. We obtain

$$\frac{dN_{v'}}{dt} = -\sum_j \frac{\sigma_j}{h\nu}\delta_j I_j \tag{10.27}$$

and similarly for the lower level

$$\frac{dN_v}{dt} = \sum_j \frac{\sigma_j}{h\nu}\delta_j I_j \ . \tag{10.28}$$

The value of σ_j for a P transition at the line center can be obtained from (3.15, 27). Equations (10.25–28) have been solved numerically by *Feldman* [10.7] with the approximation $P(j-1){\simeq}P(j)$ and $2j-1{\simeq}2j+1$. These results are plotted in Fig. 10.1 where the extracted energy versus incident pulse energy for a Gaussian pulse consisting of only the $P(20)$ transition of the (00^01-I) vibrational band is calculated for various values of the pulse width over rotational relaxation time (τ_p/τ_r). The propagation distance was 1 m.

The maximum energy that is available from the vibrational inversion, E_{avbl}, is obtained when $\bar{\delta}_j = 0$ which means that the total vibrational inversion is used for the pulse amplification. Suppose that $N_{v'}^0$ and N_v^0 are the vibrational densities just before the arrival of the pulse; the available

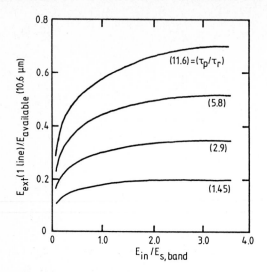

Fig. 10.1. Computed energy extraction versus incident pulse energy for an incident Gaussian pulse consisting of the $P(20)$ transition of the $(00^01\text{-}I)$ band, plotted for various values of pulsewidth over rotational thermalization time (τ_p/τ_r). The propagation distance is 1 m [10.7]

energy is then

$$E_{\text{avbl}} = h\nu L \Delta N_{v'} \ , \tag{10.29}$$

where L is the length of the amplifier, and $\Delta N_{v'}$ is the available density of the upper level

$$\Delta N_{v'} = N_{v'}^0 - N_{v'} \tag{10.30}$$

with the conditions $\bar{\delta}_j = 0$ and the conservation of particles $N_{v'}^0 + N_v^0 = N_{v'} + N_v$. From these conditions we readily obtain

$$\Delta N_{v'} = N_{v'}^0 - (N_{v'}^0 + N_v^0)\left(1 + \frac{P(j-1)}{P(j)}\frac{2j+1}{2j-1}\right)^{-1} \tag{10.31}$$

which for large j becomes

$$\Delta N_{v'} \sim \tfrac{1}{2}(N_{v'}^0 - N_v^0) \ . \tag{10.32}$$

The E_{avbl} is also, according to (10.14), equal to $\alpha_0 L E_{\text{s,band}}$ where $\alpha_0 = \delta_j^0 \sigma_j$ and $E_{\text{s,band}}$ the saturation energy of the lasing vibrational band. The initial inversion density δ_j^0 is

$$\delta_j^0 = P(j-1)N_{v'}^0 - \frac{2j-1}{2j+1}P(j)N_v^0 \ . \tag{10.33}$$

The saturation energy is then obtained by comparing the last expression for E_{avbl} with (10.29). We obtain

$$E_{s,band} = h\nu \frac{\Delta N_{v'}}{\delta_j^0 \sigma_j} \quad . \tag{10.34}$$

Substituting (10.31, 33) into (10.34) we finally find

$$E_{s,band} = \frac{h\nu}{\sigma_j \left[P(j-1) + \frac{2j-1}{2j+1} P(j) \right]} \quad . \tag{10.35}$$

The available energy is thus given by

$$E_{avbl} = \sigma_j \delta_j^0 L E_{s,band} \quad . \tag{10.36}$$

In the case of the $P(20)$ of the (00^01-I) band with $A_p = 0.173$ (Table 3.1) and $\Delta\nu_p = 4\,\text{GHz}$ we can calculate σ_j according to (3.15, 27) and with (10.35) we find $E_{s,band} = 110\,\text{mJ/cm}^2$. It should be noted that in the absence of rotational relaxation, i.e., $\tau_p \ll \tau_r$, the saturation energy was according to (10.10) simply given by $E_s = h\nu/2\sigma_j$ and we would obtain $E_s = 7\,\text{mJ/cm}^2$.

For large j values (10.35) may be approximated by

$$E_{s,band} \sim \frac{h\nu}{2\sigma_j P(j)} \quad . \tag{10.37}$$

The value of $E_{s,band}$ used in Fig. 10.1 is the approximation given by (10.37). The salient feature in Fig. 10.1 is that the ratio of the extracted energy to the available energy is independent of the small-signal gain and of the initial inversion density, and that it decreases markedly as τ_p/τ_r decreases. The value $\tau_p/\tau_r = 5.8$ corresponds to a 1 ns pulse in a 1 atm. amplifier. It is seen that for this value only 50 % of the available energy is extracted.

It should be noted that in the absence of rotational relaxation $(\tau_p \ll \tau_r)$ the saturation energy is equal to $h\nu/2\sigma_j$ whereas for long pulses $(\tau_p \gg \tau_r)$ it is given by (10.35). For intermediate values of τ_p the saturation energy is not independent of the energy. This can be understood from the fact that for increasing energy and constant τ_p of a Gaussian pulse the pulse saturates the lasing transition and thus the period of "relaxation pumping" increases; hence leading to an energy dependence of the saturation energy.

This can be circumvented by considering a rectangular pulse in the time domain so that the time of relaxation pumping remains constant. Since the time needed to transfer all molecules of a vibrational upper level into the $(j-1)$ state is given by $\tau_r/P(j-1)$ we may write for the saturation energy $E_s(\tau_p)$ as a function of the pulse time τ_p

$$E_s(\tau_p) = E_{s,band}\{1 - \exp[-\tau_p P(j-1)/\tau_r]\} \\ + E_s \exp[-\tau_p P(j-1)/\tau_r] \quad . \tag{10.38}$$

For long pulses we again obtain $E_s(\tau_p) = E_{s,band}$.

275

10.2.3 The Effect of Rotational Relaxation on the Inversion and Pulse Form

The previous analysis may be used to calculate the transient behavior of the medium and the pulse. Let us consider a Gaussian pulse of 1 ns (FWHM) at the entrance of a 60 cm long amplifier with a small-signal gain of 2.4 % cm^{-1} for the $P(20)$ transition of the (00^01-I) band. At the time $t = 0$ the maximum of this pulse enters the amplifier and the corresponding radiation leaves the amplifier at $t = 2$ ns. The change of the population inversion density of the $P(20)$ transition near the end of this amplifier is calculated as a function of time by means of (10.25–28). For comparison, this is done with and without rotational relaxation. The results are depicted in Fig. 10.2 for various input energies E_{in}. It is clearly seen that the relaxation time of the considered inversion density is much more than τ_r and that even a strong 1 ns pulse of $E_{in} = 213\,\text{mJ/cm}^2$ does not extract the available energy of the vibrational band. Figure 10.3 shows the vibrational population densities of the lasing levels. The extracted energy increases considerably if the pulse contains more transitions. In Fig. 10.4 the inversion density of the $P(20)$ transition is depicted for 1 ns pulses of $106\,\text{mJ/cm}^2$ input energy consisting of one to five rotational lines. Figure 10.5 shows again the vibrational densities. Comparing Figs. 10.4 and 5 with, respectively, Figs 10.2 and 3 the

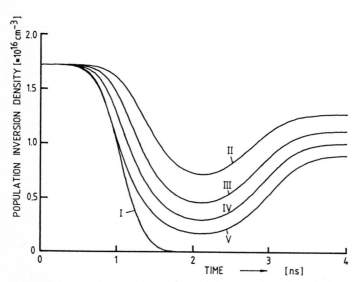

Fig. 10.2. Time evolution of the population inversion density of the $P(20)$ transition at the end of an amplifier with a length of 60 cm having a small-signal gain of 2.4 % cm^{-1} for various input energies of a Gaussian pulse of 1 ns.
Curve I: $E_{in} = 106\,\text{mJ/cm}^2$ and no rotational relaxation
Curve II: $\tau_r = 0.16\,\text{ns}$, $E_{in} = 21\,\text{mJ/cm}^2$
Curve III: $\tau_r = 0.16\,\text{ns}$, $E_{in} = 53\,\text{mJ/cm}^2$
Curve IV: $\tau_r = 0.16\,\text{ns}$, $E_{in} = 106\,\text{mJ/cm}^2$
Curve V: $\tau_r = 0.16\,\text{ns}$, $E_{in} = 213\,\text{mJ/cm}^2$

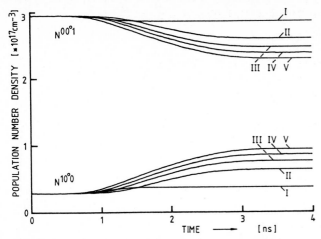

Fig. 10.3. Time evolution of the vibrational population number density at the end of an amplifier with a length of 60 cm having a small-signal gain of $2.4\,\%\,\mathrm{cm}^{-1}$ for various input energies of a Gaussian pulse of 1 ns.
Curve I: no rotational relaxation, $E_{\mathrm{in}} = 106\,\mathrm{mJ/cm^2}$
Curve II: $\tau_{\mathrm{r}} = 0.16\,\mathrm{ns}$, $E_{\mathrm{in}} = 21\,\mathrm{mJ/cm^2}$
Curve III: $\tau_{\mathrm{r}} = 0.16\,\mathrm{ns}$, $E_{\mathrm{in}} = 53\,\mathrm{mJ/cm^2}$
Curve IV: $\tau_{\mathrm{r}} = 0.16\,\mathrm{ns}$, $E_{\mathrm{in}} = 106\,\mathrm{mJ/cm^2}$
Curve V: $\tau_{\mathrm{r}} = 0.16\,\mathrm{ns}$, $E_{\mathrm{in}} = 213\,\mathrm{mJ/cm^2}$

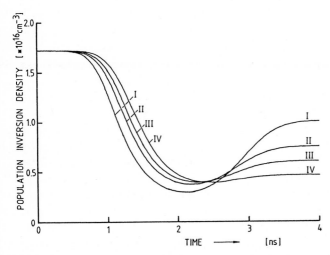

Fig. 10.4. Time evolution of the population inversion density of the $P(20)$ transition at the end of an amplifier with a length of 60 cm. $E_{\mathrm{in}} = 106\,\mathrm{mJ/cm^2}$, $\alpha_0 = 2.4\,\%\,\mathrm{cm}^{-1}$. The spectra of the Gaussian pulse of 1 ns are:
Curve I: $P(20)$,
Curve II: $P(18) + P(20)$,
Curve III: $P(18) + P(20) + P(22)$,
Curve IV: $P(16) + P(18) + P(20) + P(22) + P(24)$

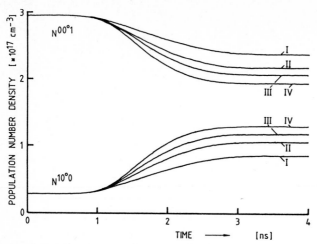

Fig. 10.5. Time evolution of the population number density of the upper and lower vibrational level at the end of an amplifier with a length of 60 cm. $E_{in} = 106 \, \text{mJ/cm}^2$, $\alpha_0 = 2.4 \, \% \, \text{cm}^{-1}$. The spectra of the Gaussian pulses of 1 ns are:
Curve I: $P(20)$,
Curve II: $P(18) + P(20)$,
Curve III: $P(18) + P(20) + P(22)$,
Curve IV: $P(16) + P(18) + P(20) + P(22) + P(24)$

effect of multiline amplification is substantial. A pulse having five lines may empty almost the whole band, as is indicated in Fig. 10.4. The effect of more lines on the relaxation pumping is physically equivalent to enlarging the pulse length. For m lines the relaxation pumping is m times faster than for one line so that m lines in a 1 ns pulse extract about the same energy as one line in a m ns pulse, provided the pulse energies are equal.

The saturation also effects the pulse form. The leading edge experiences the largest gain and will be more amplified than the trailing edge. The resulting distortion of the pulse shape is illustrated in Fig. 10.6. In the absence of rotational relaxation only the front of the pulse is somewhat amplified, whereas with rotational relaxation all parts of the pulse are amplified and the pulse symmetry is less distorted, although the leading edge is more amplified. The pulse is also broadened by the relaxation pumping because the pumping occurs mainly near the pulse tail. Moreover, the point of highest intensity shifts forwards as the pulse propagates. The pulse form depends on the input energy, small-signal gain, pulse time, and line spectrum of the pulse. Figure 10.7 shows the calculated pulse widths after amplification over 40 cm for an incoming Gaussian pulse of 1 ns containing the $P(20)$ line only. The small-signal gain is 3.3 % per cm. The time difference of the half maximum of the leading and trailing edge to the pulse center is also plotted as a function of $E_{in}/E_{s,band}$. It is clear that the pulse form remains unchanged for both weak and strong pulses. Weak pulses will be undistorted because all parts of the pulse experience the same gain. Intense pulses add a rel-

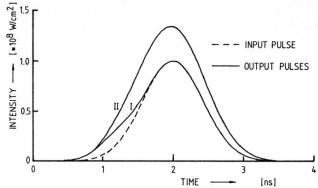

Fig. 10.6. Input and output pulse shape for two different cases. $E_{in} = 106\,\text{mJ/cm}^2$. *(Dotted curve)* Input pulse shape. *(Curve I)* Two-level situtation (no rotational relaxation). *(Curve II)* Normal rotational relaxation, $\tau_r = 0.16\,\text{ns}$

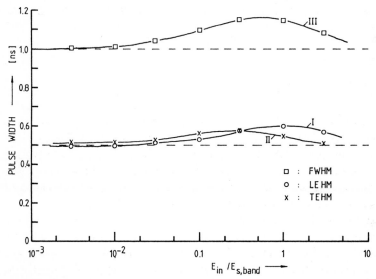

Fig. 10.7. Calculated pulse width of an incoming Gaussian pulse of 1 ns containing $P(20)$ line only. The amplifier is 40 cm long and has a small signal gain of $3.3\,\%\,\text{cm}^{-1}$. *(Curve I)* Leading edge half maximum. *(Curve II)* Trailing edge half maximum, *(Curve III)* Full width half maximum

atively negligible amount of energy to the pulse energy so that the pulse shape remains practically unchanged.

10.2.4 Multiband Energy Extraction

As discussed previously, the maximum energy density extractable utilizing multi-line pulses within the (00^01-I) transition is, using (10.29, 32), approximately given by

279

$$E_{\text{avbl}} \simeq \tfrac{1}{2}(N_{v'}^0 - N_1^0)h\nu L \qquad\qquad (10.39)$$

which indicates that at most $1/2$ of the inversion in the band can be converted into radiation. This limitation arises from the buildup of the lower level as the upper level is depleted. Since the relaxation time of the lower level is larger than 20 ns at 1 atm, this level acts as a bottleneck, limiting depletion of the upper level. This limitation can be circumvented by taking advantage of the inversion in the $(00^01$-II) band.

If the incident pulse contains simultaneously lines from both vibrational bands, a larger portion of the upper level density can be converted into radiation. An additional advantage of the $(00^01$-II) band is the fact that the photons at 9.4 μm contain 10 % more energy than those from the $(00^01$-I) band so that 10 % more energy is converted from the upper-state particles that radiate into the $(00^01$-II) band. Dividing the transfer equally over these bands the rate of depletion of the upper vibrational level is twice the rate of filling each of the lower vibrational levels. This offers one the possibility of depleting approximately $2/3$ of the initial vibrational inversion density, assuming the population densities of the two lower vibrational levels remain equal.

A still bigger part of the upper-level density can be converted into radiation if the vibrational reservoir of the lower level is further extended to include also the degenerate levels (02^20) and $(02^{-2}0)$; see Sect. 2.2. These levels are strongly collisionally coupled to the (II) level, being a mixture of (10^00) and (02^00), as analysed in Sect. 2.4. The three levels of the ν_2 vibration are closely spaced in the energy diagram, as shown in Fig. 2.3. The intramode relaxation, i.e. the thermalization of a vibrational degree of freedom, is very fast and comparable to rotational relaxation. This means that during pulse amplification one may expect that the (02^20) and $(02^{-2}0)$ levels are also populated. Even for nanosecond pulses at 1 atm a substantial part of this additional lower level reservoir is used. In the case where intramode relaxation is faster than the pulse duration so that the population densities of these three levels are approximately equal, the highest energy extraction will then be obtained for a transfer ratio of $1:3$ for the $(00^01$-I) and $(00^01$-II) vibrational mode, respectively. In this way one obtains the possibility of depleting approximately 80 % of the initial inversion density between two lasing vibrational levels.

Depending on pulse duration and gas pressure, the intramode relaxation as a pumping source is, of course, not limited to the above-mentioned group of levels. The upper level is collisionally coupled to the vibrational ladder of the ν_3 vibration so that depletion of the (00^01) level generates a pumping source within the thermalization process of the ν_3 vibration. A similar situation arises with respect to the thermalization of the ν_1 and ν_2 vibrations. These effects will be further discussed in Sect. 10.7.

10.3 Experimental Technique for Generating Multi-Line Short Pulses

As indicated earlier, the number of lines that have to be stimulated for an efficient process depends on the pulse duration and gas pressure. It is most advantageous to choose primarily the strongest transitions, not only because in that case the main part of the stored energy is stimulated, but also because relaxation with large Δj occurs with substantial probability [10.3,4]. The relaxation rates of the non-stimulated transitions increase in proportion to the number of emptied states, i.e. stimulated states. Therefore, multi-line stimulation markedly increases the energy that is extracted.

Besides the efficiency, the multi-line content of the interacting short pulse is also important for prevention of pulse broadening during amplification, because relaxation pumping by the non-stimulated transitions amplifies mainly the pulse tail. The relaxation process broadens the pulse to a time that depends on the period it takes to extract all the stored energy. This is in contrast to pulse sharpening which can be expected if the main part of the stored energy is extracted by stimulation without relaxation pumping [10.7].

In the past, several techniques have been described for short-pulse multi-line oscillation. One method is to employ transition-selective reflectors as etalons [10.8,9], or absorbers [10.10] such as SF_6. These techniques have in common that the losses of some oscillations with high gain have been relatively increased, so that competition between the transitions changes in favor of the weaker ones. The gain-to-loss ratio is then more or less equal and there is no "survival of the strongest". However, these techniques do not permit independent control of the lines and therefore there is a restricted number of lines in the spectrum. The main problem in obtaining multi-line oscillation is the strong coupling between the lines by thermal collisions. At the onset several lines start to oscillate, but as a consequence of this coupling strong competition will occur between them, and as is the case in a normal system, after some time ($< 10^{-7}$ s), the strongest line, $P(20)$, will survive at the expense of the others. This troublesome competition can be avoided by spatial separation of the lines in the gain medium. Therefore, in the past several resonator configurations have been investigated, in which the various transitions have their own specific region of interaction with the medium while they coincide before leaving the cavity. This aim can be achieved with a double-grating configuration, containing two similar parallel gratings [10.11].

Such a system is not however applicable to short-pulse generation by amplitude mode locking, as the round-trip time of all transitions is different. This problem can be avoided by using a configuration with only one grating, as shown schematically in Fig. 10.8 [10.12]. This configuration has the additional advantage of a simpler alignment procedure. The system has

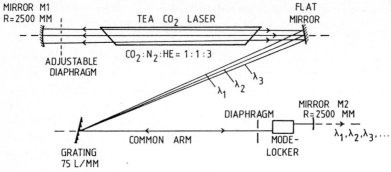

Fig. 10.8. Mode-locked multi-line CO_2 laser system

two sections: the common arm, containing the mode locker, and the arm with angular dispersion. The mirror at the latter section has its center of curvature on the grating, so that the optical path lengths of the system is wavelength independent, provided that the dispersive behavior of the medium is neglected. Further, it is also necessary to phase lock the transitions. This is possible with active AM mode locking since the pulse of each line will always pass the modulator near its minimum loss. Thus, near the outcoupling mirror the pulses will coincide in space and in time. Moreover, one must fulfill the condition that the pulse trains of the oscillating lines have sufficient overlap in time for selecting a single multi-line pulse by means of a fast shutter. This is especially difficult with fast pulsed discharges due to gain variations of the lines. The resulting spread of buildup times can be reduced by making the gain-to-loss ratio of the distinct lines more or less equal.

For the fundamental mode the beam width of an individual line as determined by an aperture in the common arm is about 6 mm. If, for example, ten lines are selected, then in the dispersive arm, the total manifold of lines must be spread over about 50 mm in order to have sufficient separation. These separated lines must also experience minimum disturbance from the active medium because in the common arm all lines must coincide. This requires a homogeneously excited system with an aperture of at least 50 mm. Fortunately, homogeneous single discharges with large aperture as discussed in Chap. 6 have proven to be reliable and applicable to this technique.

10.3.1 Experimental Set-up

The experimental configuration, as shown in Fig. 10.8, has an optical length between the mirrors M_1 and M_2 of 375 cm, as dependent on the modulation frequency of 40.028 MHz of the mode locker. As a result, two pulses travel through the cavity independently of one another. Between the mirrors the grating with 75 l/mm and a blaze angle of 23.86° is placed at the center of curvature of M_1 in such a way that maximum angular dispersion (first order) is obtained. Since the experimental performances are influenced by any slight disturbance, especially mechanical vibrations caused by the pulsed discharge, the experimental setup is mounted on a rigid optical table. In the chosen configuration the TEA section, provided with Brewster NaCl windows, cannot disturb the optical alignment. The principle of the TEA laser has been described in Chap. 6. It employs two identical uniform field electrodes at a separation distance of 50 mm. The length of the electrodes is 400 mm. The mode locker is placed close to the outcoupling mirror. Amplitude modulation is accomplished by means of periodic acoustic waves in a germanium crystal. A diaphragm is placed near the mode locker to select the fundamental mode.

10.3.2 Multi-Line Experiments

If this laser operates without additional diffraction losses the oscillation of eight adjacent rotational transitions of the (00^01-I) band is observed: $P(14) - P(28)$. Each line has an overlap of about 10 % with its neighbors. The output-pulse shape for this case is shown in Fig. 10.9 (solid line). It is apparent from this figure that the distinct oscillations do not reach their peak intensity at the same time. The pulse train length of an individual line is observed to be about 200 ns. This is due to the spread in buildup times as a consequence of differences in the gain of the various rotational lines. The time delays between the lines were measured by means of two photon-drag detectors (Fig. 10.10). The first detector observes the leading edge of the total pulse, which is used as a trigger signal. The total beam is further analysed by the monochromator, so that all lines can be successively read out by a transient digitizer. The results are averaged; they are shown in Column A of Table 10.1.

It is concluded that only two or three lines oscillate simultaneously. These differences in buildup times can be reduced by introducing additional diffraction losses for the various lines. Each line will then be independently

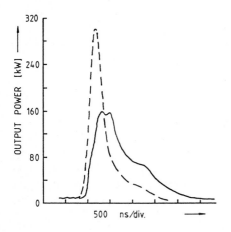

Fig. 10.9. Output pulse shape of a mode-locked multi-line CO_2 TEA laser (the photon-drag detector signal has passed a 20 MHz low-pass filter). *(Solid line)* no additional diffraction losses in the cavity, 790 ns (FWHM). *(Dashed line)* additional diffraction losses in the cavity, 330 ns (FWHM)

Table 10.1. Time difference between the spectral lines. *(A)* Pulse shape as in Fig. 10.9 *(solid line)*. *(B)* Pulse shape as in Fig. 10.9 *(dashed line)*

Spectral line	Time difference [ns]	
	A	B
$P(14)$	180	16
$P(16)$	130	24
$P(18)$	60	16
$P(20)$	0	0
$P(22)$	390	28
$P(24)$	400	20
$P(26)$	720	630
$P(28)$	1090	1050

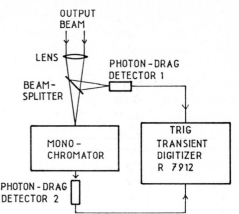

Fig. 10.10. Set-up for determining the time delays between spectral lines

controlled. The results are shown in Column B of Table 10.1. Again, the values are averages and taken relative to $P(20)$. The total jitter is 80 ns. In this way six lines are oscillating simultaneously. The overall pulse shape corresponding to this configuration is also shown in Fig. 10.9 (dashed line). The overall pulse length has to do with the mentioned jitter of line oscillations. If the "slow" lines $P(26)$ and $P(28)$ are eliminated from the output pulse, its FWHM is approximately 320 ns. The line spectrum as obseved by the monochromator is shown in Fig. 10.11. This spectrum is obtained by continuously pulsing the system and turning the monochromator. Each line is resolved by thirteen shots. Apart from the reduced time delays, the additional advantage is that the line energies are now much more equal. The fact that the spectrum of Fig. 10.11 is not the result of a single shot or an averaged result indicates that the laser system has good reproducibility.

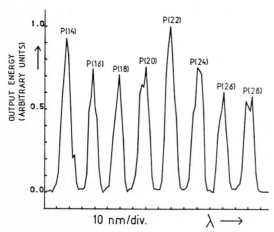

Fig. 10.11.
Energy spectrum of the output pulse shown in Fig. 10.9 by the dashed line

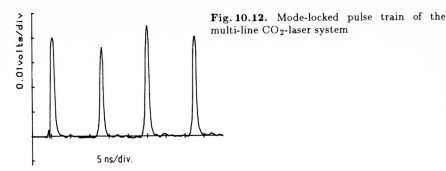

Fig. 10.12. Mode-locked pulse train of the multi-line CO_2-laser system

PART OF OSCILLATOR PULSE TRAIN

The output consists of a train of pulses of which one single output pulse has a width of 1.15 ns and an energy of approximately 3.7 mJ. Part of this pulse train is exhibited in Fig. 10.12. This measured pulse width of a pulse containing six transitions has no significant difference from the pulse width consisting of only one transition. In principle, resonance dispersion will modify the group velocity of the pulses differently for each of the rotational lines. It depends on the gain. If steady-state conditions are applicable, the gain is equal to the losses, which vary only little for the various lines. The differences in group velocity result in a time delay between the pulses of the rotational lines. For steady state these time delays can be calculated as in Chap. 7 and are found to be of the order of a few tenths of a nanosecond, which is negligible.

In principle, the maximum number of oscillating lines depends on the width of the TEA section, the dispersion of the grating, and the distance between this grating and the TEA section. To avoid strong line competiton the lines do not need to be fully separated. It is observed that in the case of about 50 % overlap all the adjacent lines will oscillate, although their intensity distribution is quite different from that observed for fully separated lines; especially around the strong $P(20)$, the intensities of the neighboring lines are relatively small.

10.4 Single Pulse Selection

From the train of optical pulses generated by the mode-locked oscillator, a single pulse can be selected by means of a fast shutter. A convenient shutter can be obtained by using the electro-optical effect of a crystal, for example a CdTe crystal. The optical properties of this crystal are described in Sect. 8.1.

Such a pulse selection apparatus, usually called switch-out, is shown in Fig. 10.13. The output pulse train of a mode-locked system with polarization

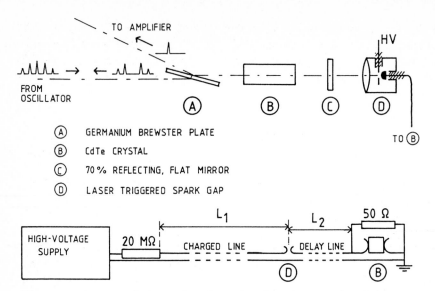

Fig. 10.13. Optical and electrical configuration of an electro-optic switchout system

in the plane of the drawing is incident on a Brewster plate. After passing this Brewster plate, the CdTe crystal, and a reflector of about 70 % the transmitted part of the pulse train is focussed on a high-voltage switch that triggers the high-voltage pulse for the electro-optic CdTe crystal. As soon as the high voltage is applied to the crystal it behaves like a quarter wavelength plate for the pulses that pass the crystal. The direction of polarization is then changed by 90° so that the reflected pulse incident on the Brewster Ge plate will experience a reflection of 87 %. For selecting one particular pulse the timing and duration of the applied voltage must be properly chosen. The high-voltage circuit is also displayed in Fig. 10.13. The duration of the high-voltage pulse and the moment of arrival at the crystal are determined independently by the lengths L_1 and L_2, respectively. For a mode-locked TEA system with nanosecond pulses at 12 ns separation, good results are obtained with $L_1 = 1$ m and $L_2 = 3.5$ m resulting in a high-voltage pulse of 10 ns duration arriving at the crystal 17.5 ns after firing the laser-triggered spark gap [10.6]. The breakdown between the two electrodes of the laser-triggered spark gap is adjusted by gas pressure and electric field strength between the electrodes so that for the triggering, a strong pulse near the center of the train can be chosen. Within one nanosecond the switch is closed. A coaxial line with a characteristic impedance of 50 Ω is used and thus the resistance across the crystal must also be 50 Ω in order to avoid pulse reflections.

In this arrangement the crystal is used as a double-pass electro-optic Pockels cell gate and is optically equivalent to a quarter wavelength plate

[10.13]. The principle can be described as follows. When the applied field E of the high-voltage pulse is perpendicular to the $(1\bar{1}0)$ plane of the CdTe crystal the indices of refraction for the two polarization directions of the incident beam along the electrically induced birefringent axes are according to (8.9) given by

$$n_{x'} = n + \tfrac{1}{2}n^3 r_{41} E \;, \tag{10.40a}$$

$$n_{y'} = n - \tfrac{1}{2}n^3 r_{41} E \;, \tag{10.40b}$$

where the direction of propagation is along z' axis. The x', y', z' directions with respect to the principle axes x, y, and z are given by (8.6a–c).

If the polarization direction of the incident beam makes an angle of $45°$ with respect to the electrically induced birefringent axes (x' and y') then the component along the x' axis (if E is positive) is retarded with respect to the component along the y' axis. The phase difference Γ between the two components can then be simply calculated by means of (10.40). We obtain

$$\Gamma = \frac{2\pi}{\lambda} L n^3 r_{41} E \;, \tag{10.41}$$

where λ is the wavelength of the incident radiation and L the length of the crystal. The field is

$$E = \frac{V}{d} \;, \tag{10.42}$$

where V is the voltage across the crystal, and d the distance between the electrodes. Since the pulse passes the crystal twice, the outgoing radiation is for $\Gamma = \frac{\pi}{2}$ again linearly polarized with a directional change of $90°$. The required "quarter-wave" voltage $V_{\lambda/4}$ is

$$V_{\lambda/4} = \frac{\lambda d}{4 L n^3 r_{41}} \;. \tag{10.43}$$

Using the values $n^3 r_{41} = 1.2 \times 10^{-10}\,\text{m/V}$, $L = 34\,\text{mm}$, and $d = 10\,\text{mm}$ one finds $V_{\lambda/4} = 6.5\,\text{kV}$. For obtaining this voltage across the crystal the line must be charged at twice this value or $13\,\text{kV}$.

The pulse with this polarization change of $90°$ and again incident on the Ge plate at the Brewster angle will be $87\,\%$ reflected. More generally, for an arbitrary voltage, the energy of the switched-out pulse, E_{out}, is

$$E_{\text{out}} = 0.87 R E_{\text{in}} \sin^2\left(\frac{\pi}{2}\frac{V}{V_{\lambda/4}}\right) \;, \tag{10.44}$$

where E_{in} is the energy of a laser pulse from the mode-locked train and R is the reflectivity of the mirror C. (It is assumed that absorption and reflection losses by the crystal are negligible.)

Finally we notice that the technique described here can also be used to cut a short pulse out of a laser pulse from a system that is not mode locked. The radiation pulse of a system that also contains within the cavity a continuous low-pressure CO_2 discharge with small band width will be smooth and without the irregularities that are caused by mode beating phenomena because here one single longitudinal mode is favored. If such a long pulse is incident on the crystal and if for instance $L_1 = 10$ cm, a one nanosecond pulse will be switched out.

10.5 Prepulse, Retropulse and Parasitic Radiation Protection

The technique of single-pulse selection from a mode-locked pulse train is often a prerequirement for short-pulse amplification. The transmitted energy of the other non-selected pulses must be as low as possible to avoid undesirable loss of inversion in the amplifiers and additional pulses. The discrimination against undesired radiation will be an important feature in an amplification chain.

A major source of feed-through energy of the gate is the small amount of permanent birefringence in the electro-optic crystal. Another point of concern is the reflections of an amplified pulse passing various optical components. Even weak reflected radiation near the end of a chain may be amplified to pulses of high intensity that reach the mode-locked system. Such reflected backward travelling pulses may seriously damage the optical components of the switch-out and mode-locked system. The residual gain that is used by the backward travelling waves can be, in the case of nanosecond pulses in atmospheric amplifiers, as high as 50 to 80 % of the original gain. These retropulses must be kept below about 1 J/cm^2 which is roughly the damage threshold of antireflection coatings and NaCl windows. For this reason the energy density of the selected amplified pulse will be decreased, if necessary, at several places of the amplification chain, by increasing the beam size. Prepulse protection can be achieved by the use of saturable absorbers, i.e. absorbers with low transmission for weak pulses and high transmission for pulse energies above the saturation energy of the absorber. For instance one might think of p-type Ge absorbers which have a broad band but unfortunately they have a low damage threshold. Gaseous absorbers in cells with windows of high damage threshold have the advantage of easily adjustable absorption by changing the gas pressure [10.14]. The use of saturable absorbers has the additional advantage of suppressing parasitic radiation that is produced by amplified spontaneous emission. Without absorbers a long amplifier may generate amplified spontaneous emission to an intolerable level. It is a rule of practice that the product

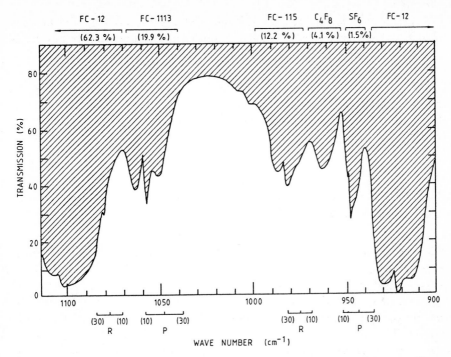

Fig. 10.14. Small-signal transmission of the gas Mix 804 vs wave number $p = 9.07$ torr, $L = 10$ cm, $T = 25.5°$ C. The P and R branches of the $(00°1$-II$)$ and $(00°1$-I$)$ bands are shown on the wave number scale; the legend at the top of the figure identifies the components in the gas, the percentage composition and the region of useful absorption [10.15]

$\alpha_0 L$ of the small-signal gain and the amplifier length must be smaller than 5 or 6 in order to avoid these parasitic oscillations. Since the absorption of a single gas component will not cover the whole CO_2 laser spectrum the gaseous absorber will consist of a mixture of several components. Figure 10.14 shows the small signal transmission of a broad band gas mixture that was originally developed at Los Alamos National Laboratories with the name Mix 804 [10.15]. The saturated transmission of this mixture is depicted in Fig. 10.15. The absorbers will prevent weak prepulses, retropulses and parasitic oscillations from reaching an unacceptable energy level.

Optical damage by strong retropulses can be successfully eliminated by a method based on optical breakdown as shown in Fig. 10.16. It is well known that gases have a "limiting radiation flux" beyond which the absorbtion increases drastically. The limiting flux is $28 \, \text{J}/\text{cm}^2$ in air at atmospheric pressure. Using for instance a diaphragm of approximately $2 \times 10^{-3} \, \text{cm}^2$, as shown in the configuration of Fig. 10.16, a maximum energy of 56 mJ can be transmitted in the atmosphere. In this case the maximum energy of the transmitted beam with a diameter of 5 mm will be about $74 \, \text{mJ}/\text{cm}^2$. If such

289

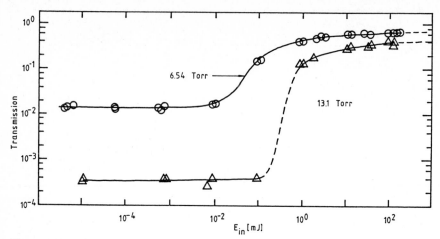

Fig. 10.15. Saturated transmission properties for gas Mix 804 for two different gas cell pressures. The wavelength and absorption cell geometries are $\lambda = 10.6\,\mu m$, $P(20)$, $L = 77,5\,cm$, $\tau_p = 1.2\,ns$, area $1/e^2 \simeq 1.7\,cm^2$ [10.15]

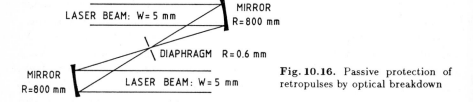

Fig. 10.16. Passive protection of retropulses by optical breakdown

a configuration is installed just after the switch-out, the retropulses will not damage the optical components of the switch-out [10.6].

10.6 Experimental Studies of Short Pulse Multi-Line Energy Extraction

The experimental arrangement for multi-line energy extraction is shown in Fig. 10.17. A single nanosecond pulse is selected from the output pulse train of the oscillator by the double-pass CdTe Pockels cell switch-out system described in Sect. 10.4.

The selected pulse is amplified by a preamplifier and fed into the final amplifier. The amplifiers are separated by a 5 cm long saturable absorber cell containing 70 torr of Mix 804, in order to prevent parasitic oscillations [10.14]. To obtain a well-defined Gaussian energy distribution, the pulse is spatially filtered before entering the final amplifier. Both amplifiers are TEA laser modules of the type described in Sect. 6.10, filled with a $CO_2 : N_2 : He = 1 : 1 : 4$ laser mixture. Part of the input pulse is split off using

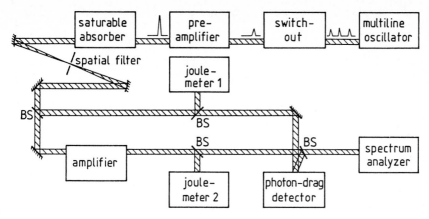

Fig. 10.17. Schematic set-up of multi-line short-pulse energy extraction

a beam splitter to determine the input energy and the input pulse shape. The output pulse is used to determine the output pulse energy, shape, and spectrum. The output pulse instead of the input pulse is used to determine the pulse spectrum because the pulse energy of the latter is too small. However, the output spectrum of the pulse is somewhat different from the input spectrum, due to a different degree of saturation of the various spectral lines. The pulse energies are determined by pyroelectric detectors. Both input and output pulse shapes are determined with a single photon-drag detector connected to a transient digitizer. The pulses reach the detector with a time delay of 6 ns.

With this setup measurements were performed on energy extraction from an amplifying medium by a laser pulse of $\simeq 1$ ns duration with a variable frequency spectrum. The experiments can be divided into four categories [10.16]:

1.   energy extraction at a single line;
2.   energy extraction at two lines with variable spacing;
3.   energy extraction at three lines with variable spacing; and
4.   energy extraction at a maximum number of adjacent lines.

In all these experiments the input pulse shape was monitored and appeared to be nearly Gaussian with a duration between 1.0 and 1.2 ns (FWHM). In the experiments with two and three lines the intensity ratio of the strongest and weakest line is between 2 : 1 and 1 : 1. For more then three lines this ratio could be larger.

Figure 10.18 exhibits the results of energy extraction experiments at one, two, three, and a maximum number of six adjacent lines. To gain an idea of the saturation parameter E_s the data were fitted to the curves according to (10.12). Although the small-signal gain is somewhat different for the

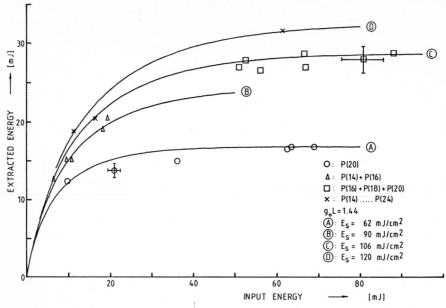

Fig. 10.18. Energy extraction of a variable number of adjacent lines; solid curves are calculated according to (10.12) and the E_s values of the fitted curves are indicated

various lines, these differences can be neglected. For all lines considered the small-signal gain was taken to be $2.4\,\%\,\mathrm{cm}^{-1}$. The length of the amplifier was 60 cm, so that $\alpha_0 L = 1.44$. The parameters E_s that result in the best fit are shown in Fig. 10.18. They are calculated using (10.12) while correcting for the Gaussian nature of the beam in the amplifier which has its $1/e$ intensity point at $r = 1.2\,\mathrm{mm}$ (beam size was determined with the amplifier switched off). Formula (10.12) is therefore integrated over $E_{\mathrm{in}}(r)$.

To determine the effect of a possible Δj dependence in the rotational relaxation processes on the energy extraction of the amplifier, experiments were performed with an incoming pulse consisting of two or three lines with a variable spacing. The results are displayed in Fig. 10.19. The values of E_s that are found in these experiments are listed in Table 10.2. It is

Table 10.2. Saturation energy for two and three lines with variable spacing

Spectrum	E_s [mJ/cm^2]
$P(14) + P(16)$	90 ± 4
$P(14) + P(18)$	90 ± 4
$P(14) + P(20)$	92 ± 4
$P(14) + P(22)$	97 ± 4
$P(16) + P(18) + P(20)$	106 ± 4
$P(14) + P(18) + P(22)$	114 ± 4

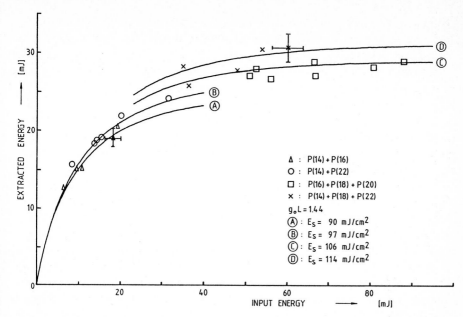

Fig. 10.19. Energy extraction with a fixed number of lines but with variable spacing. Solid curves are calculated according to (10.12)

seen that a small increase in energy extraction is obtained when the lines are more widely separated. The difference, is of the order of 8 %, which is within the accuracy of the measurements. The reservoir model for rotational relaxation that excludes the Δj dependence of multiline amplification is therefore applicable.

10.7 Multiple-Pass Pulse Amplification

In Sect. 10.2.2 it has been shown that the available energy for an initial vibrational inversion density is given by

$$E_{\mathrm{avbl}} = h\nu L \Delta N_{v'} \ . \tag{10.45}$$

This represents the energy of all rotational lines in a given inversion density between two vibrational levels. However, the extracted energy can be increased if we can make use of the intramode relaxation, i.e. the energy of higher vibrational levels. The vibrational energy density of the upper vibrational level N_{00^01} will, after depletion by the laser pulse, return to a new equilibrium value within the vibrational energy distribution of the ν_3 vibration. In other words, due to intramode relaxation the energy of higher states will be transferred to the lasing state thus producing more available energy.

Similarly the increased density of the lower vibrational state can relax to a new equilibrium whereas the laser-induced production to the lower level density is thermalized within its degree of freedom. Further, since there is a close reasonance between the ν_1 and ν_2 vibrations the capacity of the lower level can be increased considerably. Even in the absence of intramode relaxation but with a Fermi-resonance equilibrium between the lower levels (I) and (II) together with the levels $(02^{-2}0)$ and (02^20), approximately 80 % of the original inversion between the upper and lower vibrational state can be converted into radiation instead of the 50 % given by (10.32).

The effect of intramode relaxation and thermal equilibrium between the ν_1 and ν_2 vibrations depends on gas pressure and pulse duration. The intramode relaxation at one atmosphere is of the order of one nanosecond whereas the Fermi relaxation time at that pressure is about 30 ns [10.17], so that for one nanosecond pulses the intramode relaxation cannot be neglected. However in the case of a single-line pulse there is such a narrow channel for the intramode energy transfer that the intramode relaxation can be neglected. Only in the case of multi-line operation the intramode relaxation has to be considered in nanosecond pulses at one atmosphere.

Instead of increasing the pulse length and/or the gas pressure, the maximum available radiation energy from the vibrational energy of a system can also be obtained by multiple-pass amplification. Then the pulse passes the medium several times. During the time between two successive round trips through the medium the above-mentioned relaxation processes occur. To obtain an effective Fermi relaxation process it is necessary to have a round-trip time of at least 30 ns. A schematic drawing of a multiple-pass amplification is depicted in Fig. 10.20 [10.18].

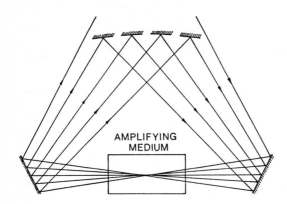

AMPLIFYING MEDIUM

Fig. 10.20. Schematic drawing of a multiple-pass amplifier

The effect of all vibrational energy on the amplification of a laser pulse has been treated by *Harrach* [10.19]. The available energy of all vibrational levels is taken into account under the condition that the vibrational energy exchange between the levels of a vibrational mode is sufficient fast compared with the total interaction time of the optical pulse. This means that for this energy extraction the vibrational energy distribution can be considered as thermal and we may use the temperature model described in Chaps. 3 and 6. We will consider two cases: complete intramode relaxation and complete intramode relaxation together with complete Fermi relaxation between the ν_1 and ν_2 vibrations so that the vibrational temperatures T_1 and T_2 are equal. The application of each model depends on the product $\tau_p p$ of pulse duration and gas pressure of the amplifier or on the total round-trip time of the multiple-pass system.

10.7.1 Intramode Vibrational Relaxation

Intramode thermalization occurs for $\tau_p p > 1$ ns $\cdot$ bar and multi-line operation, or multiple-pass operation with round-trip times larger than the rotational relaxation time.

We start with vibrational energy distributions given with the temperatures T_1, T_2, and T_3 (Sect. 6.5). The energy extraction by the pulse finishes as soon as the inversion between the lasing levels is reduced to zero, which does not mean that at that instant the relevant vibrational temperatures are equal. For reasons of simplicity, we introduce for the ν_1, ν_2, and ν_3 vibrations the parameters q, r, and s, respectively:

$$q = \exp(-h\nu_1/kT_1) \, , \tag{10.46}$$

$$r = \exp(-h\nu_2/kT_2) \, , \tag{10.47}$$

$$s = \exp(-h\nu_3/kT_3) \, . \tag{10.48}$$

For this analysis both rotational and vibrational distributions are thermalized. The vibrational energy per unit volume of the CO_2 molecules is

$$E_{vib}(q,r,s) = N_{CO_2} \left(\frac{h\nu_1 q}{1-q} + \frac{2h\nu_2 r}{1-r} + \frac{h\nu_3 s}{1-s} \right) \, . \tag{10.49}$$

Making in (10.33) the approximation $j - 1 \sim j$ we write for the initial inversion in the $(00^0 1\text{-I})$ band

$$\delta_j^0 = P(j) N_{CO_2} \frac{s_0 - q_0}{Q(q_0, r_0, s_0)} \, , \tag{10.50a}$$

where the partition function Q is

$$Q(q_0, r_0, s_0) = [(1-q_0)(1-r_0)^2(1-s_0)]^{-1} \, . \tag{10.50b}$$

After the passage of the pulse and the thermalization of the relevant vibrations the distribution parameters are q_e, r_0 and s_e. The total vibrational energy is now given by (10.49) with the new values q_e, r_0, and s_e. If all the available energy has been extracted we arrive at the population parameters

$$s_e = q_e \, . \tag{10.51}$$

The extracted laser energy E_{rad} must be equal to the difference of vibrational energy of the initial and final states

$$E_{rad} = E_{vib}(q_0, r_0, s_0) - E_{vib}(q_e, r_0, s_e) \ . \tag{10.52}$$

This energy is also equal to the fraction $(\nu_3 - \nu_1)/\nu_3$ of the ν_3 vibration

$$E_{rad} = h(\nu_3 - \nu_1) N_{CO_2} \left(\frac{s_0}{1 - s_0} - \frac{s_e}{1 - s_e} \right) \ . \tag{10.53}$$

Solving (10.52, 53), and using (10.51) one finds

$$s_e = q_e = \frac{d_0}{2 + d_0} \ , \quad \text{where} \tag{10.54}$$

$$d_0 = \frac{q_0}{1 - q_0} - \frac{s_0}{1 - s_0} \ . \tag{10.55}$$

The maximum amount of energy that can be extracted is now given by (10.53) with s_e given by (10.54). If this theory is compared with E_{avbl} (only complete rotational relaxation) given by (10.39) we find for the ratio $\gamma = E_{rad}/E_{avbl}$

$$\gamma = \frac{2Q(q_0, r_0, s_0)}{s_0 - q_0} \left(\frac{s_0}{1 - s_0} - \frac{s_e}{1 - s_e} \right) \ . \tag{10.56}$$

The increase in output energy by the factor γ can thus be calculated from the initial vibrational temperatures. For $T_1 = T_2 = 400\,\text{K}$ and $T_3 = 1500\,\text{K}$, we find $\gamma = 1.5$.

10.7.2 Fermi Relaxation

In the previous subsection we excluded an energy drain into the second lower laser vibrational state. The two lower vibrational levels are closely coupled by Fermi resonance. For TEA systems the Fermi relaxation time is about 30 ns [10.17]. Let us assume that these two lower levels maintain in equilibrium so that $T_1 = T_2$. The total vibrational energy is then given by

$$E_{vib} = N_{CO_2} \left[h\nu_1 \left(\frac{r}{1 - r} + \frac{r^2}{1 - r^2} \right) + \frac{h\nu_3 s}{1 - s} \right] \ , \tag{10.57}$$

where we have used the relations $\nu_1 = 2\nu_2$ and $q = r^2$ for $T_1 = T_2$. Similar to the previous section we have now

$$s_e = q_e = r_e^2 \ . \tag{10.58}$$

The extracted energy is

$$E_{rad} = E_{vib}(r_0, s_0) - E_{vib}(r_e, s_e) \ . \tag{10.59}$$

Solving (10.53, 58, and 59) now yields

$$r_e = \sqrt{q_e} = \sqrt{s_e} = \{[4(e_0 + 3)e_0 + 1]^{1/2} - 1\}[2(e_0 + 3)]^{-1} \ , \tag{10.60}$$

where

$$e_0 = \frac{s_0}{1 - s_0} + \frac{r_0^2}{1 - r_0^2} + \frac{r_0}{1 - r_0} \ . \tag{10.61}$$

The extracted radiation energy is now calculated by means of (10.53) and using s_e given by (10.60). This will be, in fact, the maximum extractable energy from the CO_2 molecules. The new factor γ is obtained by using (10.56) and s_e given by (10.60). Taking again the example of $T_1 = T_2 = 400\,K$ and $T_3 = 1500\,K$, we obtain $\gamma = 2.6$. Comparing this value of γ with that of previous subsections where Fermi relaxation was excluded we see that the use of the ν_2 vibration gives a considerable increase of the extracted energy.

References

Chapter 1

1.1   C.K.N. Patel: Phys. Rev. Lett., **12**, 588 (1964)
1.2   F. Legay, N. Legay-Sommaire: Compt. Rend., **259**, 99 (1964)
1.3   H. Sugawara, K. Kuwabara, S. Takemori, A. Wada, K. Sasaki: In
      Gas Flow and Chemical Lasers, Sixth Intern. Symp., Jerusalem, 8–12
      Sept. 1986, Springer Proc. Phys. Vol. 15 (Springer, Berlin, Heidelberg
      1987) p.265
1.4   A.E. Hill: Appl. Phys. Lett., **12**, 324 (1968)
1.5   P.B. Corkum: IEEE J. QE-**21**, 216 (1985)
1.6   H. Kogelnik, T. Li: Proc. IEEE **54**, 1312 (1966)
1.7   H.K.V. Lotsch: Optik **26**, 112 (1967)
1.8   H.K.V. Lotsch: Optik **28**, 65, 328, 555 (1968/69) and **29**, 130, 622
      (1969)
      H.K.V. Lotsch: Optik **30**, 1, 181, 217, 563 (1969/70)
1.9   A.E. Siegmann: Appl. Opt. **13**, 353 (1974)
1.10 W.F. Krupke, W.R. Sooy: IEEE J. QE-**5**, 575 (1969)
1.11 A.G. Fox, T. Li: Bell Syst. Techn. J. **40**, 453 (1961)
1.12 A.E. Siegman: Proc. IEEE **53**, 277 (1965)
1.13 L. Bergstein: Appl. Opt. **7**, 495 (1968)
1.14 W. Streifer: IEEE J. QE-**4**, 229 (1968)
1.15 Yu.A. Ananev, G.N. Vinokurov, L.V. Kovalchuk, N.A. Sventsitskaya,
      V.E. Sherstobitov: Sov. Phys. JETP **31**, 420 (1970)

Chapter 2

2.1   D.M. Dennison: Rev. Mod. Phys. **3**, 280 (1931)
2.2   D.M. Dennison: Rev. Mod. Phys. **12**, 175 (1940)
2.3   H. Statz, C.L. Tang, G.F. Koster: J. Appl. Phys. **37**, 4278 (1966)
2.4   L.I. Schiff: *Quantum Mechanics* (McGraw-Hill, New York 1955)
2.5   L. Pauling, E.B. Wilson Jr.: *Introduction to Quantum Mechanics*
      (McGraw-Hill, New York 1937)
2.6   G. Herzberg: *Molecular Spectra and Molecular Structure II: Infrared
      and Raman Spectra of Polyatomic Molecules* (Van Nostrand, Prince-
      ton, NJ 1964)
2.7   E. Fermi: Z. Physik **71**, 250 (1931)

2.8 M. Silver, T.S. Hartwick, M.J. Posakony: J. Appl. Phys. **41**, 4566 (1970)

2.9 R. Beck. W. Englisch, K. Gürs: *Table of Laser Lines in Gases and Vapors*, 3rd ed., Springer Ser. Opt. Sci., Vol. 2 (Springer, Berlin, Heidelberg 1980)

2.10 F.R. Petersen, E.C. Beaty, C.R. Pollack: J. Molec. Spectroscopy **102**, 112 (1983)

2.11 C. Freed, L.C. Bradley, R.G. O'Donnell: IEEE J. **QE-6**, 1195 (1980)
L.C. Bradley, K.L. Soohov, C. Freed: IEEE J. **QE-22**, 234 (1986)

2.12 C. Freed, R.G. O'Donnell: Metrologia **13**, 151 (1977)

Chapter 3

3.1 K. Shimoda: *Introduction to Laser Physics*, 2nd. ed., Springer Ser. Opt. Sci., Vol. 44 (Springer, Berlin, Heidelberg 1986)

3.2 A.R. Edmonds: *Angular Momentum in Quantum Mechanics* (Princeton Uni. Press, Princeton NJ 1957) pp. 65–67

3.3 V.V. Nevdakh: Sov. J. Quantum Electron. **14**, 1091 (1984)

3.4 P.W. Anderson: Phys. Rev. **76**, 647 (1949)
J.H. van Vleck, H. Margenau: Phys. Rev. **76**, 1211 (1949)

3.5 E.T. Gerry, D.A.Leonard: Appl. Phys. Lett. **8**, 227 (1966)

3.6 U.P. Oppenheim, A.D. Devir: J. Opt. Soc. Am. **58**, 585 (1968)

3.7 O.R. Wood: Proc. IEEE **62**, 355 (1974)

3.8 R.L. Abrams: Appl. Phys. Lett. **25**, 609 (1974)

3.9 L.O. Hocker, M.A. Kovacs, C.K. Rhodes, G.W. Flynn, A.Javan: Phys. Rev. Lett. **17**, 233 (1966)

3.10 G.J. Ernst, W.J. Witteman: VIII Intl. Quantum Electronics Conf., San Francisco (1974) Paper S-8

3.11 W.J. Witteman: Philips Res. Repts. **21**, 73 (1966)

3.12 K.R. Manus, H.J. Seguin: J. Appl. Phys. **43**, 5073 (1972)

3.13 F. Kaufman, J.R. Kelso: J. Chem. Phys. **28**, 510 (1958)

3.14 J.E. Morgan, H.I. Schiff: Can. J. Chem. **41**, 903 (1963)

3.15 M.J.W. Boness, G.J. Schulz: Phys. Rev. Lett. **21**, 1031 (1968)

3.16 R.D. Hake, A.V. Phelps: Phys. Rev. **158**, 70 (1967)

3.17 G.J. Schulz: Phys. Rev. **116**, 1141 (1959)

3.18 G.J. Schulz: Phys. Rev. **125**, 229 (1962)

3.19 G.J. Schulz: Phys. Rev. **135**, A988 (1964)

3.20 P.O. Clark, M.R. Smith: Appl. Phys. Lett. **9**, 367 (1966)

3.21 D.C. Tyte, R.W. Sage: Proc. IRE, Conf. on Lasers and Opto-Electronics (1969), Southampton, England

3.22 J. Polman, W.J. Witteman: IEEE J. QE-6, 154 (1970)

3.23 J.B. Moreno: Sandia Laboratory, Report SLA-73-1024 (1974)

3.24 C.B. Moore, R.E. Wood, B.L. Hu, J.T. Yardley: J. Chem. Phys. **46**, 4222 (1967)

3.25 W.J. Witteman: J. Chem. Phys. **35**, 1 (1961)
3.26 R.L. Taylor, S. Bitterman: Rev. Mod. Phys. **41**, 26 (1969)
3.27 K.J. Siemsen, J. Reid, C. Dang: IEEE J. QE-**16**, 668 (1980)
3.28 T.L. Cottrell, J.C. McCoubrey: *Molecular Energy Transfer in Gases* (Butterworths, London 1961)
3.29 P.O. Clark, J.Y. Wada: IEEE J. QE-**4**, 263 (1968)
3.30 P. Bletzinger, A. Garscadden: Appl. Phys. Lett. **12**, 289 (1968)
3.31 M.Z. Novgorodov, A.G. Sviridov, N.N. Sobolev: IEEE J. QE-**7**, 508 (1971)
3.32 V.N. Chirkov, A.V. Yakovleva: Opt. Spectrosc. **28**, 441 (1970)
3.33 W.J. Witteman: J. Chimie Physique **1**, 107 (1967)
3.34 G.M. Schindler: IEEE J. QE-**16**, 546 (1980)

Chapter 4

4.1   M.J. Druyvesteyn, F.M. Penning: Rev. Mod. Phys. **12**, 87 (1940)
4.2   H. Kogelnik, T. Li: Proc. IEEE, **54**, 1312 (1966)
4.3   G.J. Ernst, W.J. Witteman: IEEE J. QE-**9**, 911 (1973)
4.4   H. Kogelnik: Appl. Opt. **4**, 1562 (1965)
4.5   G.J. Ernst: Opt. Commun. **25**, 368 (1978)
4.6   C.P. Christensen, C. Freed, H. Haus: IEEE J. QE-**5**, 276 (1969)
4.7   P.K. Cheo: IEEE J. QE-**3**, 683 (1967)
4.8   W.J. Witteman: IEEE J. QE-**2**, 375 (1966)
4.9   R.J. Carbone: IEEE J. QE-**3**, 373 (1967)
4.10 W.J. Witteman: Appl. Phys. Lett. **11**, 337 (1967)
4.11 R.G. Pike, D. Hubbard: J. Res. Nat. Bur. Stand. **59**, 127 (1957)
4.12 W.J. Witteman, H.W. Werner: Phys. Lett. **26A**, 454 (1968)
4.13 R.R. Reeves Jr., P. Harteck, B.A. Thompson, R.W. Waldron: J. Phys. Chem. **70**, 1637 (1966)
4.14 W.J. Witteman: IEEE J. QE-**5**, 92 (1969)
4.15 W.J. Witteman: IEEE J. QE-**4**, 786 (1968)
4.16 J. Reid, K. Siemsen: Appl. Phys. Lett. **29**, 250 (1976)
4.17 J. Reid, K. Siemsen: J. Appl. Phys. **48**, 2712 (1977)
4.18 G.J. Ernst, W.J. Witteman: IEEE J. QE-**7**, 484 (1971)
4.19 W.J. Witteman, R.J. Carbone: IEEE J. QE-**6**, 462 (1970)
4.20 A. Maitland, M.H. Dunn: *Laser Physics* (North Holland, Amsterdam 1969) p. 187
4.21 A.L.S. Smith, S. Moffat: Opt. Commun. **30**, 213 (1979)
4.22 M.C. Skolnick: IEEE J. QE-**6**, 139 (1970)
4.23 K.M. Abramski, J. Van Spijker, W.J. Witteman: Appl. Phys. B**36**, 149 (1985)
4.24 V.J. Stefanov: J. Phys. E**3**, 1027 (1970)
4.25 H. Jacobs, A.J. Karecman, J. Schumacher: J. Appl. Phys. **38**, 3412 (1967)

Chapter 5

5.1 S.A. Losev: *Gasdynamic Laser*, Springer Ser. Chem. Phys. Vol. 12 (Springer, Berlin, Heidelberg 1981)

5.2 A.E. Hill: Appl. Phys. Lett. **16**, 423 (1970)

5.3 T.F. Deutsch, F.A. Horrigan, R.I. Rudko: Appl. Phys. Lett. **15**, 88, (1969)

5.4 W.B. Tiffany, R. Targ, J.D. Foster: Appl. Phys. Lett. **15**, 11 (1969)

5.5 C.O. Brown: Appl. Phys. Lett. **17**, 388 (1970)

5.6 A.C. Ackbreth, J.W. Davis: IEEE J. QE-**8**, 139 (1972)

5.7 A.E. Hill: Appl. Phys. Lett. **18**, 194 (1971)

5.8 C.J. Buczek, R.J. Wayne, P. Chenausky, R.J. Freiberg: Appl. Phys. Lett. **16**, 321 (1970)

5.9 C.O. Brown, J.W. Davis: Appl. Phys. Lett. **21**, 480 (1972)

5.10 I. Sejima, S. Shigeo: Cleo '82, Technical Digest, ThG2 (1982)

5.11 M. Gastaud, A. Pons, P. Bousselet, G. Hutin: Sixth Intern. Symp. on Gas Flow and Chemical Lasers, Jerusalem, 8–12 Sept. 1986

5.12 H. Sugawara, K. Kuwabara, S. Takemori, A. Wada, K. Sasaki: In *Gas Flow and Chemical Lasers*, Sixth Intern. Symp., Jerusalem, 8–12 Sept. 1986, Springer Proc. Phys. Vol. 15 (Springer, Berlin, Heidelberg 1987) p. 265

5.13 W.M. Brandenburg, M.P. Bailey, P.D. Texeira: IEEE J. QE-**8**, 414 (1972)

5.14 C.J. Buczek, R.J. Freiburg, P.P. Chenausky, R.J. Wayne: Proc. IEEE **59**, 659 (1971)

5.15 M. Hishii: Techn. Dig. of Symp. on Gas Flow Lasers and Chemical Lasers Keio University, Yokohama (1982) p. 68

Chapter 6

6.1 R. Dumanchin, J. Rocca-Serra: Compte R Acad. Sci. **269**, 916 (1969)

6.2 A.J. Beaulieu: Appl. Phys. Lett. **16**, 504 (1970)

6.3 J.J. Lowke, A.V. Phelps, B.W. Irwin: J. Appl. Phys. **44**, 4664 (1973)

6.4 T. Holstein: Phys. Rev. **70**, 367 (1946)

6.5 E.W. McDaniel: *Collision Phenomena in Ionized Gases* (Wiley, New York 1964) p. 23

6.6 L.S. Frost, A. Phelps: Phys. Rev. **127**, 1621 (1962); Phys. Rev. **136**, 1538 (1964)

6.7 R.D. Hake, Jr., A.V. Phelps: Phys. Rev. **158**, 70 (1967)

6.8 A. Andrick, D. Danner, H. Ehrhardt: Phys. Lett. **29A**, 346 (1969)

6.9 A. Stamatovic, G.J. Schulz: Phys. Rev. **188**, 213 (1969)

6.10 D. Spence, J.L. Mauer, G.J. Schulz: J. Chem. Phys. **57**, 5516 (1972)

6.11 A.G. Engelhardt, A.V. Phelps, C.G. Risk: Phys. Rev. **135**, A1566 (1964)

6.12 P.T. Smith: Phys. Rev. **36**, 1293 (1930)

6.13   G.J. Schulz, R.E. Fox: Phys. Rev. **106**, 1179 (1957)

6.14   G.J. Schulz, J.W. Philbrick: Phys. Rev. Lett. **13**, 477 (1964)

6.15   J.D. Jobe, R.M. St. John: Phys. Rev. **164**, 117 (1967)

6.16   G.J. Schulz: Phys. Rev. **128**, 178 (1962)

6.17   J.H. Parker, Jr., J.J. Lowke: Phys. Rev. **181**, 290 (1969)

6.18   J.N. Bardsley, M.A. Biondi: *Advances in Atomic and Molecular Physics*, ed. by D.R. Bates (Academic, New York 1970) Vol. 6, p. 1

6.19   G.J. Ernst, A.G. Boer: Opt. Commun. **34**, 235 (1980)

6.20   W.J. Witteman: Philips Res. Repts. **21**, 73 (1966)

6.21   K.R. Manes, H.J. Seguin: J. Appl. Phys. **43**, 5073 (1972)

6.22   G.J. Schulz: Phys. Rev. **135**, A988 (1964)

6.23   R.L. Taylor, S. Bitterman: Rev. Mod. Phys. **41**, 26 (1969)

6.24   R. Dumanchin, J.C. Farcy, M. Michon, J. Rocca-Serra: Proc. VI Intl. Quantum Elec. Conf., Kyoto, Japan (1970)

6.25   R. Dumanchin, M. Michon, J.C. Farcy, G. Boudinet, J. Rocca-Serra: IEEE J. QE-8, 163 (1972)

6.26   A.K. Laflamme: Rev. Sci. Instrum. **41**, 1578 (1970)

6.27   H.M. Lamberton, P.R. Pearson: Electron. Lett. **7**, 141 (1971)

6.28   W. Rogowski: Arch. Electrotech. **12**, 1 (1923) or J.D. Cobine: *Gaseous Conductors – Theory and Engineering Applications* (McGraw-Hill, New York 1941)

6.29   M.C. Richardson, A.J. Alcock, K. Leopold, P. Burtyn: IEEE J. QE-9, 236 (1973)

6.30   T.Y. Chang: Rev. Sci., Instrum. **44**, 405 (1973)

6.31   W.R. Smythe: *Static and Dynamic Electricity* (McGraw-Hill, New York 1950)

6.32   G.J. Ernst: Opt. Commun. **49**, 275 (1984)

6.33   G.J. Ernst: Rev. Sci. Instrum. **48**, 1281 (1977).

6.34   G.J. Ernst, A.G. Boer: Opt. Commun. **27**, 105 (1978)

6.35   G.J. Ernst, A.G. Boer: Opt. Commun. **34**, 221 (1980)

6.36   G.J. Ernst, A.G. Boer: Opt. Commun. **44**, 125 (1982)

6.37   D.J. Brink, V. Hasson: J. Appl. Phys. **49**, 2250 (1978)

6.38   V. Hasson, H.M. von Bergmann: Rev. Sci. Instrum. **50**, (1979)

6.39   V. Hasson, H.M. von Bergmann: J. Phys. E9, 73 (1976)

6.40   R.E. Beverly: Light Emission from High-Current Surface-Spark Discharges, Chapter VI, *Prog. Opt. 16*, (North Holland, Amsterdam 1978)

6.41   W.R. Kaminski: Corona preionization technique for carbon dioxide TEA lasers, Report No. 82R-980701-02, United Technologies Research Center, West Palm Beach, FL (1982)

6.42   R.V. Babcock, I. Liberman, W.D. Partlow: IEEE J. QE-**12**, (1976)

6.43   C.A. Fenstermacher, M.J. Nutter, J.P. Rink, K. Boyer: Bull. Am. Phys. Soc. **16**, 42 (1971)

6.44   C.A. Fenstermacher, M.J. Nutter, W.T. Leland, K. Boyer: Appl. Phys. Lett. **20**, 56 (1972)

6.45 J.D. Daugherty, E.R. Pugh, D.H. Douglas-Hamilton: Bull. Am. Phys. Soc. **17**, 399 (1972)

6.46 N.G. Basov, E.M. Belenov, V.A. Danilychev, O.M. Kerimov, I.B. Vovsh, A.F. Suchkov: JETP Lett. **14**, 285 (1971)

6.47 F.A. van Goor: private communication, University of Twente, Enschede, The Netherlands

6.48 S. Singer, C.J. Elliott, J. Figueira, L. Liberman, J.V. Parker, G.T. Schappert: High Power, Short-pulse CO_2 Laser Systems for Inertial-Confinement Fusion, in *Developments in High-Power Lasers and their Applications*, ed by C. Pellegrini (North Holland, Amsterdam 1981) p. 190

6.49 J.D. Daugherty: *Electron Beam Ionized Lasers:* in *Principles of Laser Plasmas*, ed. by G. Bekefi (Wiley, New York 1976)

6.50 C. Cason, G.J. Dezenberg, R.J. Huff: Appl. Phys. Lett. **23**, 110 (1973)

Chapter 7

7.1 J.C. Slater: Rev. Mod. Phys. **30**, 1 (1958)

7.2 D.A. Pinnow: IEEE J. QE-**6**, 223 (1970)

7.3 A.A. Oliner (ed.): *Acoustic Surface Waves*, Topics Appl. Phys. Vol. 24 (Springer, Berlin, Heidelberg 1978)

7.4 Y.R. Shen (ed.): *Nonlinear Infrared Generation*, Topics Appl. Phys. Vol. 16 (Springer, Berlin, Heidelberg 1977)

7.5 D.J. Kuizenga, A.E. Siegman: IEEE J. QE-**6**, 694 (1970)

7.6 W.J. Witteman, A.H.M. Olbertz: IEEE J. QE-**13**, 381 (1977)

7.7 T.J. Nelson: IEEE J. QE-**8**, 29 (1972)

7.8 O.R. Wood, R.L. Abrams, T.J. Bridges: Appl. Phys. Lett. **17**, 376 (1970), also
R.L. Abrams, O.R. Wood: Appl. Phys. Lett. **19**, 518 (1971)

7.9 F.A. van Goor: Opt. Commun. **45**, 404 (1983)

7.10 P. Bernard, P.A. Belanger: Opt. Lett. **4**, 196 (1979)

7.11 A. Girard: Opt. Commun. **11**, 346 (1974)

7.12 J.A. Weiss, J.M. Schur: Appl. Phys. Lett. **22**, 453 (1973)

7.13 F.A. van Goor, R.J.M. Bonnie, W.J. Witteman: IEEE J. QE-**21**, (1985)

7.14 P.A. Belanger, J. Boivin: Can. J. Phys. **54**, 720 (1976)

7.15 A.J. Alcock, K. Leopold, M.C. Richardson: Appl. Phys. Lett. **23**, 562 (1973)

7.16 F.A. van Goor: Opt. Commun. **41**, 205 (1982)

7.17 R.J.M. Bonnie, F.A. van Goor: Opt. Commun. **57**, 64 (1986)

Chapter 8

8.1 A. Yariv: *Quantum Electronics* (Wiley, New York 1967)

8.2 B.K. Vainshtein: *Modern Crystallography I*, Springer Ser. Solid-State Sci., Vol. 15 (Springer, Berlin, Heidelberg 1981)

8.3   D.J. Kuizenga, A.E. Siegman: IEEE J. QE-**6**, 694 (1970)

8.4   W.J. Witteman, A.H.M. Olbertz: Opt. and Quantum Electron. **12**, 259 (1980)

8.5   T.Y. Chang: Rev. Scient. Instr. **44**, 405 (1973)

Chapter 9

9.1   V.S. Letokhov: Sov. Phys. JETP **28**, 562 (1969)

9.2   J.A. Fleck: Appl. Phys. Lett. **12**, 178, (1968); J. Appl. Phys. **39**, 3318 (1968)

9.3   B. Hausherr, E. Mathieu, H. Weber: IEEE J. QE-**9**, 445 (1973)

9.4   W.H. Glenn: IEEE J. QE-**11**, 8 (1975)

9.5   H.A. Haus: J. Appl. Phys. **46**, 3049 (1975)

9.6   O.R. Wood, S.E. Schwartz: Appl. Phys. Lett. **12**, 263 (1968)

9.7   J.H. McCory: Appl. Phys. Lett. **15**, 353 (1969)

9.8   A.F. Gibson, M.F. Kimmett, C.A. Rosito: Appl. Phys. Lett. **18**, 546 (1971)

9.9   A.F. Gibson, M.F. Kimmett, B. Norris: Appl. Phys. Lett. **24**, 306 (1974)

9.10   B.J. Feldman, F.J. Figueira: Appl. Phys. Lett. **25**, 301 (1974)

9.11   F. Keilmann: IEEE J. QE-**12**, 592 (1976)

9.12   C.R. Phipps. S.J. Thomas, J. Ladish, S.J. Czuchlewski, F.J. Figueira: IEEE J. QE-**13**, (1977)

9.13   M. Sargent III: Opt. Commun. **20**, 298 (1977)

9.14   R.S. Taylor, B.K. Garside, E.A. Ballik: IEEE J. QE-**14**, 532 (1978)

9.15   A.J. Alcock, A.C. Walker: Appl. Phys. Lett. **25**, 299 (1974)

9.16   R.L. Fork, B.I. Greene, C.V. Shank: Appl. Phys. Lett. **38**, 671 (1981)

9.17   A.E. Siegman: Opt. Lett. **6**, 334 (1981)

Chapter 10

10.1   L.M. Frantz, J.S. Nodvik: J. Appl. Phys. **34**, 2346 (1963)

10.2   G.J. Schappert: Appl. Phys. Lett. **23**, 319 (1973)

10.3   R.R. Jacobs, K.J. Pettipiece, S.J. Thomas: Appl. Phys. Lett. **24**, 375 (1974)

10.4   R.K. Preston, R.T. Pack: J. Chem. Phys. **69**, 2823 (1978)

10.5   F.A. Hopf, C.K. Rhodes: Phys. Rev. A**8**, 912 (1973)

10.6   R.A. Rooth: Dissertation, University of Twente, Enschede, The Netherlands (1986)

10.7   B.J. Feldman: Opt. Commun. **14**, 13 (1975)

10.8   M. Piltch: Opt. Commun. **7**, 397 (1973)

10.9   A.H.M. Olbertz: Opt. and Quantum Electron. **9**, 536 (1977)

10.10 R. Giles, A.A. Offenberger: Appl. Phys. Lett. **40**, 944 (1982)

10.11 R.L. Sheffield, S. Nazemi, A. Javan: Appl. Phys. Lett. **29**, 588 (1976)

10.12 R.A. Rooth, F.A. van Goor, W.J. Witteman: IEEE J. QE-**19**, 1610 (1983)

10.13 D.L. Smith, D.T. Davis: IEEE J. QE-**10**, 138 (1974)

10.14 J.F. Figueira, A.V. Nowak: Appl. Opt. **19**, 420 (1980)

10.15 S. Singer, C.J. Elliott, J. Figueira, L. Liberman, J.V. Parker, G.T. Schappert: High Power, Short-pulse CO_2 Laser Systems for Inertial-Confinement Fusion, in *Developments in High-Power Lasers and their Applications*, ed. by C. Pellegrini (North Holland, Amsterdam 1981) p. 190

10.16 R.A. Rooth, W.J. Witteman: J. Appl. Phys. **58**, 1120 (1985)

10.17 C. Dang, J. Reid, B.K. Garside: Appl. Phys. B**31**, 163 (1983); also R.A. Rooth, J.A. van der Pol, E.H. Haselhoff, W.J. Witteman: IEEE J. QE-**23**, Aug. (1987)

10.18 N.E. Aver'Yanov, Yu. A. Baloshin: Sov. Phys. Techn. Phys. **25**, 1123 (1980)

10.19 R.J. Harach: IEEE J. QE-**11**, 349 (1975)

Subject Index

ABCD-law 94, 97, 102
Absolute frequencies 22-52
Absorption 57, 61, 70, 76
Acousto-optic modulation 196-203
AM mode locking 195, 282
   modulation 196
Angular momentum 13, 15
Anharmonic vibrational constants 11
Anode 82, 83
Applications 2
Attachment 155, 157, 161

Bessel distribution 84
Boltzmann's equation 141-155
Bragg angle 197, 198
Brewster angle 197, 202, 286
Broadening 59-61
   collisional 59, 60
   Doppler 59, 60
   homogeneous 63, 108
   pressure 60

Cathode 82, 83, 105, 130, 186, 187, 190
CdTe 233, 235, 249, 285, 286, 290
Chang profile 170
CO 65, 66, 104
   vibrational excitation 66, 67
CO_2
   dissociation 104, 105
   excitation cross sections 151, 152
   Hamiltonian 11, 12
   molecule 9
   vibrational excitation 64-66
   vibrations 8, 9
   wave equation 10, 11
Cooling 3, 108, 127
   convection 127-129
Coriolis force 15
Current − voltage characteristic 82, 120, 161

Degeneracy
   rotational 12, 16, 53
   vibrational 13, 15
Diffusion
   ambipolar 84, 86, 140

coefficient 155
   electron 84
   ion 84
Dipole moment 20, 53
Discharge
   e-beam sustained 185-194
   glow 81, 139
   self-sustained 139, 161, 168, 169, 182, 183
Doppler shift 59, 63
Drift current 83
   velocity 86, 132, 140, 155, 156, 189

Efficiency 2, 3, 108, 157
Einstein A coefficient 53, 56
Einstein B coefficient 56, 57
Electrode 105, 168-177
Electron
   current 83, 84
   distribution 141, 157
Energy extraction 272
   multiband 279, 280
Ernst profile 174

Fast flow 3, 127
   axial 128
   transverse 128-133
Fermi
   relaxation 294-296
   resonance 16, 18, 19, 68, 294
FM mode locking 232
FM mode modulation 196, 232
Frequency stabilization 119
Fresnel number 5, 6

Gain 20, 21, 58, 62, 63, 70, 97, 134, 135
   Gaussian beam 94-97
   profile 92-95, 98
   saturation 62, 63, 76, 96, 103, 134, 135
   saturation measurements 101-103
   small signal 63, 76, 96, 269, 291, 292
Gas composition 105-108
Gas dynamic laser 127
Germanium 80, 109, 137, 264, 288
   outcoupling plate 80, 99, 114
Gladstone-Dale constant 89

H$_2$ 62,105
H$_2$O 65,75,76,81,109
Hamiltonian 11,18
Heat dissipation 3,88
Helium 74,81,105,109
    excitation cross section 153
Hermite polynomial 13

Infrared bands 15,21
Ion current 83,84
Ionization
    coefficient 155,157
    energy 86
Isotopes 16,20–52

Laser
    fusion 2
    power 160
Legendre polynomial 12
Lifetime
    lower state 69,75
    sealed-off operation 106
    upper state 63
Line
    competition 109,112,283–285
    shape 58–61,63
Littrow arrangement 113

Marx generator 169,170,183,184
Medical application 2
Mobility 84
Mode
    competition 108–110
    Gaussian 5,92,97,117,129
    multi 5,89
    single 5,89,108
Mode locking
    AM 195–231
    FM 238–250
    numerical results 229–231
    passive 251–266
    transient evolution 224–228
Modulator 196,235
Moment of inertia 11
Multi-atmosphere 139
Multi-line 281–285,290–293
Multiple pass 293–295

NaCl window 80,283
Nitrogen 64,65,75,81,105,109
    excitation cross sections 152
    vibrational excitation 65,158,159
Normal
    coordinates 10,11
    modes 8,9

O$_2$ 105
OH 107

Opto-galvanic effect 119,121
Opto-voltaic
    effect 119
    input 123,124,126
    signal 125
Outcoupling 6,7,70,112–117,119
Output
    energy 184,193
    power 106,108,110

Passive mode locking 251–266
    CO$_2$ system 262
    criteria 261,262
Parasitic radiation 288,289
Positive column 82,83,86
Power extraction 76–80
Preionization
    double discharge 169
    corona 178–181
Prepulse 288
Pressure broadening 60,61,63
Profiles
    minimum width 174–178
    Rogowski 169
    uniform field 170–174
Propagation constant 93,94
Pulse
    amplification 267–272
    energy extraction 272
    Gaussian 204,210,236,241,273,276–278
    propagation 267–270
    selection 285–288
    spectral width 207,217,238,240,246
    time width 207,216,238,240,246,258
Pyrex 105

Q-switching 76
Quantum number
    magnetic 12,53
    rotational 12,54

Raman bands 15,21
Recombination 83,140,187
Refractive index 89,91,92,98,100
Regular bands 19,20,111,112
Relaxation 62,66,75
    intramode 295,296
    rate constant 75,76
Resistance 82,120
Resonator
    stable 4
    unstable 4–7
Retropulse 288
Riccati equation 87
Rogowski profile 170

Rotational
    constant   12, 14, 15, 23–52
    level   16
    line distribution   16, 17
    relaxation   68, 270–272, 276–279

Saturable absorption   251–257, 288
Saturation   62
    energy   269, 275
    intensity   63
Sealed-off systems   104–107
Sequence bands   22, 111, 112
Similarity rules   85–88
Spontaneous emission   53–56, 164
Stabilization
    current   120, 122, 131
    frequency   118–126
    mode locking   219–223
    voltage   121
Stimulated emission   56, 57, 62, 63, 95
    cross section   57, 63

Temperature
    model   64, 68, 69
    profile   88, 89, 91
    translational   64, 68–74, 162–168
    vibrational   68–74
Thermal effects   87
Transit time   134–136

Transition
    P-branch   19, 22, 55, 58
    R-branch   19, 22, 55, 58
    rotational   19, 64, 117, 283
    selection   112, 113, 117
Transport coefficients   155, 157
Triodes   82
Tunable laser   119

UV-preionization   139, 168–170, 178, 184

Vibrational
    degrees of freedom   64
    excitation   64–66, 140
    levels   2, 15, 16
    rotational energy   14
    temperature   68–74
Vibrations
    antisymmetric   16, 19
    symmetric   16

Waste heat   3
Wave function   11, 14
    rotational   12
    vibrational   12–14
Welding   2

Xenon   74, 75, 81, 109

ZeSe   80, 137